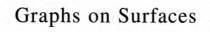

Graphs on Surfaces

Johns Hopkins Studies in the Mathematical Sciences

in association with the Department of Mathematical Sciences

The Johns Hopkins University

# Graphs on Surfaces

Bojan Mohar

and

Carsten Thomassen

The Johns Hopkins University Press

Baltimore and London

The publisher thanks the authors for supplying the camera-ready copy from which this book was prepared.

The Johns Hopkins University Press
2715 North Charles Street
Baltimore, Maryland 21218-4363
www.press.jhu.edu

Library of Congress Cataloging-in-Publication Data

Mohar, Bojan, 1956–
    Graphs on surfaces / Bojan Mohar, Carsten Thomassen.
        p.    cm. — (Johns Hopkins studies in the mathematical sciences)
    Includes bibliographical references and index.
    ISBN 0-8018-6689-8 (acid-free paper)
        1. Graph theory. 2. Surfaces.    I. Thomassen, Carsten, 1948–    II. Title.
III. Series.
QA166.M64   2001
511′.5—dc21                                                                        00-048799

A catalog record for this book is available from the British Library.

To the memory
of our fathers

# Contents

# Preface

Graphs on surfaces form a natural link between discrete and continuous mathematics. They enable us to understand both graphs and surfaces better. It would be difficult to prove the celebrated classification theorem for (compact) surfaces without the use of graphs. Map color problems are usually formulated and solved as problems concerning graphs. These and several other combinatorial properties of surfaces are surveyed in this work.

There are numerous examples showing the importance of planar graphs (or, equivalently, graphs on the sphere) in discrete mathematics. One beautiful example is Steinitz' theorem that the graphs (or 1-skeletons) of the 3-dimensional polytopes are precisely the 3-connected planar graphs. Planar graphs and their duals form an important link between graphs and matroids.

Also graphs on higher surfaces are useful. Consider, for example, the following seemingly simple question of characterizing those graphs that do not contain two disjoint cycles of odd length. This problem was solved by Lovász using Seymour's deep characterization of regular matroids. The interesting graphs in this family are those which can be drawn in the projective plane such that all faces are bounded by even cycles. The fact that no such graph has two disjoint odd cycles could be proved by a tedious combinatorial argument. The natural proof, however, is merely a reference to the fact that the projective plane has no two disjoint noncontractible cycles, an immediate consequence of the classification theorem.

Graphs on surfaces play a central role in the deep Robertson-Seymour theory on graph minors. One of the main results of Robertson and Seymour is the following: Any infinite collection of graphs contains two graphs, say $G$ and $H$, such that $H$ is a minor of $G$, that is, $H$ can be obtained from $G$ by contracting and deleting edges and deleting isolated vertices. An important consequence (and source of motivation) of this impressive result is an extension of Kuratowski's theorem to higher surfaces: For every fixed surface $S$, there is a finite collection $\mathbf{Forb}_0(S)$ of graphs that cannot be drawn on $S$ such that any graph $G$ can be drawn on $S$ if and only if $G$ has no graph in $\mathbf{Forb}_0(S)$ as a minor.

The literature on graphs on surfaces is immense and has already been treated in some monographs. Ringel [Ri74] presented the complete proof of the Heawood conjecture. White [Wh73] emphasized the algebraic aspects, and so did Gross and Tucker [GT87] who also discussed voltage and current graphs. Tutte [Tu84] and Bonnington and Little [BL95] treated graphs on surfaces from a purely combinatorial point of view. Our work has little overlap with these books and includes also more recent results in topological graph theory.

Graphs on surfaces can be described and treated combinatorially using the so-called Heffter-Edmonds-Ringel rotation principle. But, to prove this rigorously requires a substantial amount of work. This is the subject of the first three chapters of the present work. Chapter 1 introduces basic graph theoretic concepts. Chapters 2 and 3 contain basic topological results: The Jordan Curve Theorem, the Jordan-Schoenfliess Theorem, and the classification theorem for compact 2-dimensional surfaces, following [Th92a].

Chapter 2 also treats planar graphs and includes a fundamental result of Koebe [Ko36] and its applications to circle packing representations and extensions of Steinitz' theorem. In Chapter 3 we complete the discussion needed for the purely combinatorial description of graphs on surfaces.

In the last five chapters, graphs on surfaces are formally treated purely combinatorially. Although one may ignore the surface completely, it is nevertheless useful to think of and refer to the surface. Chapter 4 builds on this combinatorial approach and treats basic topological concepts like contractibility of cycles, the genus and the maximum genus of graphs. Chapter 5 is about the edge-width and the face-width of an embedded graph. Graphs of large width are of particular interest in that they share many properties with planar graphs.

Chapter 6 addresses the problem of extending an embedding of a subgraph to an embedding of the whole graph in the same surface. The results about embedding extensions are applied in Section 6.5 where we prove that the list of minimal forbidden subgraphs for the projective plane is finite.

Chapter 7 is devoted to the excluded minor theorem of Robertson and Seymour. A short proof of Thomassen [Th97c] with only little reference to surfaces is given. That proof depends on two other results of Robertson and Seymour, namely the excluded grid theorem [RS86b] and the result that there are only finitely many excluded minors of bounded tree-width, for each fixed surface. A simple proof of the former was obtained recently by Diestel, Gorbunov, Jensen and Thomassen [DJGT99] and it is presented in Section 7.1. A simple proof of the latter has recently been obtained by Mohar [Mo01], and it is presented in Section 7.2.

Coloring problems have always played an important role in the theory of graphs on surfaces. Most of them are mentioned in the monograph of Jensen and Toft [**JT95**]. Some of the most recent results are surveyed in Chapter 8.

We shall maintain an Internet page where we shall attempt to report on any errors and necessary revisions of the text in the book. There we shall also discuss any major progress related to the subject of the book, solutions, partial solutions and generalizations of problems stated in the book, and add links to related sites. A link to the home page of the book will be posted on the following address:

http://www.fmf.uni-lj.si/~mohar

We wish to thank Dan Archdeacon and Bruce Richter for their careful reading and many suggestions on parts of the manuscript. Thanks are also due to Mike Albertson, Adrien Douady, Martin Juvan, John Maharry, Neil Robertson, Riste Škrekovski, and Robin Thomas for valuable comments and to Gašper Fijavž for drawings of the configurations in Appendix B. The figures in Appendix A were produced by the authors using the graph drawing package Pajek, written by Vladimir Batagelj and Andrej Mrvar. Support from the Danish Natural Science Research Council and from the University of Ljubljana is gratefully acknowledged.

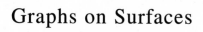

Graphs on Surfaces

# CHAPTER 1

# Introduction

## 1.1. Basic definitions

A *graph* $G$ is a pair of sets, $V(G)$ and $E(G)$, where $V(G)$ is nonempty and $E(G)$ is a set of 2-element subsets of $V(G)$. Unless stated otherwise, we assume that $V(G)$ is finite. The elements of the set $V(G)$ are called *vertices* of the graph $G$, the elements of $E(G)$ are the *edges* of $G$. For an edge $e = \{u, v\} \in E(G)$, vertices $u$ and $v$ are called *endvertices* or *ends* of $e$. The number of vertices, $n = |V(G)|$, is the *order* of the graph $G$. Vertices $u$ and $v$ of $G$ are *adjacent* in $G$ if there is an edge $e$ of $G$ having $u$ and $v$ as its ends. The edge $e$ is also said to *join* or *connect* its ends. If $v$ is an endvertex of the edge $e$, then $v$ and $e$ are said to be *incident*. Two edges with a common end are called *adjacent*. If vertices $u$ and $v$ are adjacent in $G$, then $u$ is a *neighbor* of $v$ (and *vice versa*, $v$ is a neighbor of $u$). Instead of writing the edges of a graph as unordered pairs, e.g. $e = \{u, v\}$, we shortly write $e = uv = vu$.

A generalization of the notion of a graph is a *multigraph* where we allow distinct edges to have the same pair of endvertices. Such edges are said to be *parallel* and form *multiple edges*. In multigraphs we also allow *loops* — edges with only one end. (Some authors use the term *graph* in this generalized sense, and then multigraphs without multiple edges and loops are distinguished as *simple graphs*.) Formally, a *multigraph* is an ordered triple $G = (V(G), E(G), \partial)$, where $V(G)$ is a (finite) nonempty set of vertices, $E(G)$ is a (finite) set of edges, and $\partial$ is a function that assigns to each edge $e \in E(G)$ a pair of vertices (the *ends* of $e$).

Definitions that will be introduced for graphs usually make sense also for multigraphs. Most of the results that we shall derive for graphs carry over to multigraphs in a natural and obvious way. Because of slightly easier presentation and since the more general framework of multigraphs does not yield any stronger results, we decided to limit our presentation to graphs whenever possible.

The number of edges of $G$ having a vertex $v \in V(G)$ as an end (with each loop being counted twice) is the *degree* of $v$ in the (multi)graph $G$, and it is denoted by $\deg_G(v)$ or simply $\deg(v)$. A vertex of degree 0 is an *isolated vertex*. If all vertices of $G$ have the same degree $k$, the graph is *$k$-regular*; 3-regular graphs are also called *cubic*.

Let $H$ and $G$ be graphs. We say that $H$ is a *subgraph* of $G$ and write $H \subseteq G$ if $V(H) \subseteq V(G)$ and $E(H) \subseteq E(G)$. Two special classes of subgraphs are of particular interest. If $V(H) = V(G)$, then $H$ is a *spanning subgraph*. On the other hand, if $V(H)$ is an arbitrary subset of $V(G)$ but $E(H)$ consists of all those edges of $G$ which have both ends in $V(H)$, then $H$ is an *induced subgraph*. If $U \subseteq V(G)$, then we denote by $G(U)$ ($G$ restricted to $U$) the induced subgraph of $G$ with vertex set $U$.

An *isomorphism* of graphs $G$ and $H$ is a 1-1 mapping $\psi$ of $V(G)$ onto $V(H)$ such that adjacent pairs of vertices of $G$ are mapped to adjacent vertices in $H$, and nonadjacent pairs of vertices have nonadjacent images. Graphs $G$ and $H$ are *isomorphic* if there is an isomorphism between $G$ and $H$. We often write $G = H$ to denote that $G$ and $H$ are isomorphic.

The *path* $P_n$ on $n$ vertices is the graph with vertices $\{v_1, v_2, \ldots, v_n\}$ and $n - 1$ edges $v_i v_{i+1}$, $1 \le i < n$. The vertices $v_1$ and $v_n$ are the *endvertices* (or just *ends*) of the path $P_n$. If we add to $P_n$ the edge $v_1 v_n$, we obtain the *cycle* $C_n$ of length $n$. We also call it an *$n$-cycle*. Cycles can have any length $n \ge 3$. In case of multigraphs we also have 1-cycles and 2-cycles; $C_1$ is a loop, and $C_2$ is the multigraph of order 2 with two parallel edges.

A subgraph $P$ of a graph $G$ isomorphic to a path $P_n$ is said to be a *path* in $G$ *joining* (or *connecting*) the ends of $P$. A graph is *connected* if any two of its vertices are connected by a path in $G$. A *connected component* of a graph $G$ is a maximal connected subgraph of $G$.

A subgraph $C$ of $G$ isomorphic to $C_n$ is said to be an *$n$-cycle* in $G$. An edge $e$ of $G$ joining two nonconsecutive vertices of $C$ is called a *chord* of $C$. If $C$ has no chords, we say that it is *chordless* or that it is an *induced cycle* in $G$.

A sequence $W = v_0 e_1 v_1 \ldots e_k v_k$ ($k \ge 0$) of vertices and edges of $G$ such that $e_i$ is an edge joining the vertices $v_{i-1}$ and $v_i$ ($1 \le i \le k$) is said to be a *walk* in $G$. If $v_k = v_0$, the walk $W$ is *closed*. When there is no confusion, we may omit the edges and write $W = v_0 v_1 \ldots v_k$ to denote the walk through the vertices $v_0, v_1, \ldots, v_k$. Analogous notation is used to denote paths and cycles in $G$. For example, if $C$ is a $k$-cycle with (consecutive) vertices $v_0, v_1, \ldots, v_k$, where $v_0 = v_k$, we write $C = v_0 v_1 \ldots v_k$.

The *complete graph* $K_n$ is the graph of order $n$ in which any two vertices are adjacent. Complete subgraphs of $G$ are called *cliques* in $G$. The *complete bipartite graph* $K_{n,m}$ is the graph with vertices $\{v_1, \ldots, v_n\} \cup \{u_1, \ldots, u_m\}$ and all possible pairs $v_i u_j$ as its edges, $1 \le i \le n$, $1 \le j \le m$.

For $n \ge 3$, the *wheel* of $n$ spokes is the graph $W_n$ obtained from the cycle $C_n$ by adding a new vertex (called the *center of the wheel*) and joining it to all vertices of $C_n$. For example, $W_3 = K_4$.

In Figure 1.1 some of the graphs introduced above are represented. Vertices may be considered as points in the plane (and shown as small

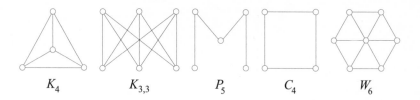

$K_4$        $K_{3,3}$        $P_5$        $C_4$        $W_6$

FIGURE 1.1. Some graphs

circles) and edges are simple (polygonal) curves joining the points corresponding to their ends. Curves representing the edges are allowed to cross each other but their interiors do not contain any vertices of the graph. Such a representation of a graph $G$ is called a *drawing* of $G$.

We now define some operations on graphs. Let $G_1, \ldots, G_k$ be subgraphs of the same graph $G$. The *union* $G_1 \cup \cdots \cup G_k$ is the graph $H \subseteq G$ with $V(H) = V(G_1) \cup \cdots \cup V(G_k)$ and $E(H) = E(G_1) \cup \cdots \cup E(G_k)$. Similarly we define the *intersection* $G_1 \cap \cdots \cap G_k$.

Let $G$ and $H$ be graphs. The *Cartesian product* of $G$ and $H$ is the graph $G \,\square\, H$ with vertex set $V(G \,\square\, H) = V(G) \times V(H)$, in which two vertices $(u, v)$ and $(u', v')$ are adjacent if either $u = u'$ and $vv' \in E(H)$, or $uu' \in E(G)$ and $v = v'$. Cartesian products of two paths, $P_k \,\square\, P_l$, are called *grid graphs*. The grid graph $P_6 \,\square\, P_4$ is drawn in Figure 1.2.

For $X \subseteq V(G)$ we denote by $G - X$ the subgraph of $G$ obtained by deleting from $G$ the vertices in $X$ and all edges incident with them. If $v$ is a vertex of $G$, we simply write $G - v$ for $G - \{v\}$, and we call it a *vertex-deleted subgraph* of $G$. Similar notation is used for edge-deleted subgraphs: If $A \subseteq E(G)$, then $G - A$ is the subgraph of $G$ obtained by removing from $G$ the edges in the set $A$. Again, $G - e$ ($e \in E(G)$) is used for $G - \{e\}$. If $u, v \in V(G)$ are nonadjacent vertices of $G$, then we denote by $G + uv$ the graph obtained from $G$ by adding the edge joining $u$ and $v$.

Let $e = uv$ be an edge of $G$, and let $G/e$ be the (multi)graph obtained from $G$ by removing the edge $e$ and identifying its ends $u, v$ into a new vertex. We call this operation an *edge contraction*. If $A$ is a set of edges of $G$, we denote by $G/A$ the (multi)graph obtained by successively contracting all edges in $A$. It is easy to see that the order in which we contract edges in $A$ is not important. In general, $G/A$ is a multigraph even if $G$ is a graph. For example, in $G/e$ we get multiple edges if $e$ belongs to a 3-cycle in $G$. Figure 1.2 represents a graph $G$ and a subset $A$ of its edges (shown as thick lines). The multigraph on the right is $G/A$.

A graph $H$ is a *minor* of $G$ if $H$ is isomorphic to a graph obtained from a subgraph $G'$ of $G$ by contracting a set of edges $A \subseteq E(G')$. If $A, B \subseteq E(G)$ are disjoint edge-sets, then $(G/A) - B = (G - B)/A$. Hence, every minor can be obtained from $G$ by successively contracting edges and

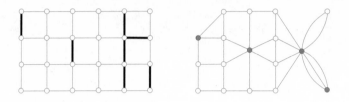

FIGURE 1.2. Contraction of edges

deleting edges and isolated vertices, and the order in which these operations are performed is unimportant. Minors of a graph $G$ are therefore
determined (up to isolated vertices) by specifying a set $A \subseteq E(G)$ of edges
which will be contracted and a set $B \subseteq E(G)\backslash A$ of edges to be removed.

The inverse operation of edge contraction is called *vertex splitting*.
This is the replacement of a vertex $v \in V(G)$ by adjacent vertices $v'$ and
$v''$ and the replacement of each edge $e = vu$ incident to $v$ either by the
edge $v'u$ or by $v''u$. See Figure 1.3 for an example.

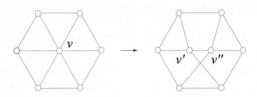

FIGURE 1.3. A vertex splitting

Let $G$ be a graph. A graph $H$ is a *subdivision* of $G$ if $H = G$ or
$H$ can be obtained from $G$ by inserting vertices of degree two on some
edges. Equivalently, some edges of $G$ are replaced by paths such that each
of these paths has only its endpoints in common with $G$. Two graphs
$G, H$ are *homeomorphic* if there is a third graph $K$ which is isomorphic
to a subdivision of $G$ and isomorphic to a subdivision of $H$. In Figure 1.4
there is an example of homeomorphic graphs such that neither of them is
a subdivision of the other.

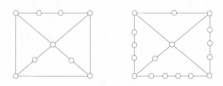

FIGURE 1.4. Homeomorphic graphs

Let $H$ be a subgraph of $G$. An $H$-*bridge* in $G$ (also called an $H$-*component* in $G$) is a subgraph of $G$ which is either an edge not in $H$ but with both ends in $H$ (and its ends also belong to the bridge), or a connected component of $G - V(H)$ together with all edges (and their endvertices in $H$) which have one end in this component and the other end in $H$. Let $B$ be an $H$-bridge. Vertices of $B \cap H$ are *vertices of attachment* of $B$ (shortly *attachments*), and each edge of $B$ incident with a vertex of attachment is a *foot* of $B$. The edge sets of the $H$-bridges partition $E(G)\backslash E(H)$. Bridges of graphs were introduced and studied by Tutte [**Tu66a**]. They are used in numerous arguments on graphs; see, e.g., the monograph of Voss [**Vo91**].

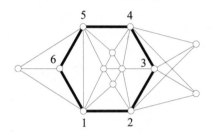

FIGURE 1.5. A graph $G$

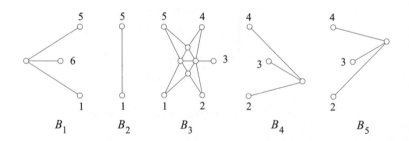

FIGURE 1.6. The bridges of a cycle in $G$

Let $C$ be a cycle in a graph $G$. Two $C$-bridges $B_1$ and $B_2$ *overlap* if at least one of the following conditions is satisfied:

(i) $B_1$ and $B_2$ have three vertices of attachment in common.

(ii) $C$ contains distinct vertices $a, b, c, d$ (in this cyclic order) such that $a$ and $c$ are vertices of attachment of $B_1$ and $b, d$ are vertices of attachment of $B_2$.

In case (ii), we also say that $B_1$ and $B_2$ *skew-overlap*. For example, if $C$ is the thick cycle in the graph of Figure 1.5 and its bridges are denoted as in Figure 1.6, the following bridges overlap: $B_3$ and $B_4$, $B_3$ and $B_5$, and $B_4$ and $B_5$. The first two pairs also skew-overlap, while $B_4$ and $B_5$ do not.

## 1.2. Trees and bipartite graphs

A *tree* is a connected graph without cycles. Figure 1.7 shows three trees.

Let $T$ be a tree. Vertices of degree 1 in $T$ are called *endvertices* of $T$. If $P$ is a longest path in $T$, and $T$ is of order at least 2, then the two ends of $P$ have degree one in $T$. This shows that every tree except $K_1$ has at least two endvertices. Many statements on trees follow by induction by deleting one or more endvertices. Another important observation is that every connected graph contains a *spanning tree*, i.e., a spanning subgraph which is a tree. See Figure 1.8 for an example of a spanning tree in a graph. The edges in the spanning tree are thick.

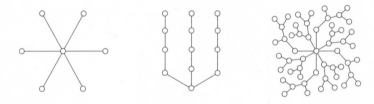

FIGURE 1.7. Trees

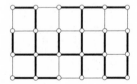

FIGURE 1.8. A spanning tree of a graph

PROPOSITION 1.2.1. *Let $T$ be a graph of order $n$. Then the following statements are equivalent:*

   (i) *$T$ is a tree.*
  (ii) *$T$ is connected and has $n-1$ edges.*
 (iii) *$T$ contains no cycles and has $n-1$ edges.*
  (iv) *$T$ is connected but every edge-deletion results in a disconnected graph.*
   (v) *$T$ contains no cycles but every edge-addition results in a graph with a cycle.*
  (vi) *Any two vertices in $T$ are connected by exactly one path.*

SKETCH OF THE PROOF. It is easy to prove (for example by deleting an endvertex and using induction) that (i) implies each of (ii)–(vi). To prove the converse, it is convenient to use the fact that every connected graph has a spanning tree. For example, to see that (ii) implies (i), take a spanning tree $T'$ of $T$. Since the tree $T'$ has $n - 1$ edges, it contains all edges of $T$ and thus $T = T'$ is a tree. The other implications are similar.                                                                                            □

The following result relates trees and minors.

PROPOSITION 1.2.2. *A graph $H$ with vertices $v_1, \ldots, v_k$ is a minor of $G$ if and only if $G$ contains pairwise disjoint trees $T_1, \ldots, T_k$ such that for $1 \le i < j \le k$ there is an edge between a vertex of $T_i$ and a vertex of $T_j$ whenever $v_i$ and $v_j$ are adjacent in $H$.*

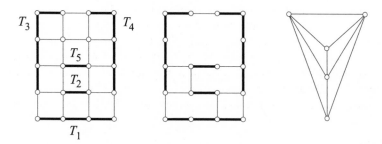

FIGURE 1.9. A grid with $K_5$ minus an edge as a minor

In Figure 1.9, a graph $G$ with subtrees $T_1, \ldots, T_5$ (represented by thick lines) is exhibited to show that the graph $K_5$ minus an edge is a minor of $G$.

A graph $G$ is *bipartite* if its vertex set can be partitioned into sets $V_1, V_2$ such that each edge of $G$ has one end in $V_1$ and the other end in $V_2$. In this case, $V_1$ and $V_2$ are said to form a *bipartition* of $G$. It is easy to see that every tree is bipartite. More generally, König proved:

THEOREM 1.2.3 (König [**Kö36**]). *A graph is bipartite if and only if it contains no cycle of odd length.*

PROOF. Clearly, a bipartite graph has no cycle of odd length. Suppose conversely that $G$ is a graph with no cycle of odd length. Without loss of generality we may assume that $G$ is connected. Let $T$ be a spanning tree of $G$. Then $T$ is bipartite and has a bipartition $V(T) = V(G) = V_1 \cup V_2$. If this is not a bipartition of $G$, then $G$ has an edge $e = xy$ with both ends in $V_1$ (say). The path $P$ in $T$ from $x$ to $y$ has even length and hence the cycle $P + e$ in $G$ has odd length.                                                                    □

A set of edges $M \subseteq E(G)$ is called a *matching* in $G$ if no two edges in $M$ have an endvertex in common; $M$ is a *perfect matching* if every vertex of $G$ is an end of an edge in $M$. A set of vertices $U \subseteq V(G)$ is said to be a *vertex cover* in $G$ if every edge of $G$ has at least one end in $U$. Matchings and vertex covers in bipartite graphs are closely related.

**The König-Egerváry Theorem.** *If $G$ is a bipartite graph, then the maximum number of edges in a matching in $G$ is equal to the minimum number of vertices in a vertex cover in $G$.*

For a proof of this theorem we refer to Bondy and Murty [**BM81**]. It also follows easily from Menger's Theorem below.

## 1.3. Blocks

If $G$ is a graph we define a relation $\sim$ on the edge set $E(G)$ as follows: $e_1 \sim e_2$ if $e_1 = e_2$ or $G$ has a cycle containing $e_1$ and $e_2$. It is easy to see that $\sim$ is an equivalence relation. An equivalence class together with all the endvertices of the edges in the class is a subgraph of $G$ and it is said to be a *block* of $G$. It is convenient to treat isolated vertices as blocks of the graph as well, so that any graph is the union of its blocks. An edge $e$ which is equivalent to no other edge is a *cutedge* of $G$. A vertex which belongs to more than one block is a *cutvertex*. The following statements are easily verified.

PROPOSITION 1.3.1. *Let $G$ be a graph.*

(a) *An edge $e$ of $G$ is a cutedge if and only if the ends of $e$ belong to distinct components of $G - e$.*

(b) *A vertex $v$ of $G$ is a cutvertex if and only if $G - v$ has more components than $G$.*

(c) *If $u, v$ are vertices in a block $B$ with at least two edges, then $B$ has a cycle containing $u$ and $v$.*

(d) *Two blocks of $G$ have at most one vertex in common and such a common vertex is a cutvertex of $G$.*

(e) *If $B_1, B_2, \ldots, B_k$ $(k \geq 3)$ are distinct blocks in $G$ such that $B_i \cap B_{i+1} \neq \emptyset$ for $i = 1, 2, \ldots, k - 1$, and $B_1 \cap B_2$, $B_2 \cap B_3$, $\ldots$, $B_{k-1} \cap B_k$ are distinct, then $B_1 \cap B_k = \emptyset$.*

PROOF. (a), (b), and (c) follow from the definition of blocks. To prove (d), assume that $B_1$ and $B_2$ are distinct blocks having distinct vertices $u, v$ in common. By (c), $B_i$ has a cycle $C_i$ containing $u$ and $v$ for $i = 1, 2$. It is then easy to find a cycle containing an edge of $C_1$ and an edge of $C_2$, a contradiction. Statement (e) is proved in a similar way. $\quad\square$

If $G$ is a graph, we may define a new graph $T(G)$ as follows. The vertex set of $T(G)$ consists of the cutvertices of $G$ together with the blocks of $G$. A block is joined to all the cutvertices it contains. Statements (d) and (e) of Proposition 1.3.1 imply that the new graph has no cycles. Clearly, $T(G)$ has the same number of components as $G$. So, if $G$ is connected, the new

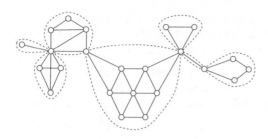

FIGURE 1.10. The blocks of a graph

graph is a tree. It is called the *block-cutvertex tree* of $G$. An endvertex of this tree is a block of $G$ (because every cutvertex is contained in at least two blocks) and it is called an *endblock* of $G$. Figure 1.10 shows a graph and its blocks. That graph has four endblocks.

## 1.4. Connectivity

If $G$ is a graph and $k$ a natural number, then $G$ is *$k$-connected* if $G$ is of order at least $k + 1$ and, for any set $S$ of at most $k - 1$ vertices, $G - S$ is connected. Thus, being 1-connected is the same as being connected (except for the graph $K_1$). The *connectivity* of a graph $G$ is the largest number $k$ such that $G$ is $k$-connected. The concept of connectivity is extended to multigraphs as follows. A multigraph is *2-connected* if it has no loops and the underlying graph is 2-connected. Note that 2-connected multigraphs have at least three vertices. For $k \geq 3$, a multigraph is *$k$-connected* if and only if it is a $k$-connected graph (and thus has neither loops nor multiple edges).

A set $S \subseteq V(G)$ of vertices of a connected graph $G$ is *separating* if $G - S$ is disconnected. If $u, v$ are vertices in distinct components of $G - S$, then we say that $S$ *separates* $u$ and $v$.

By Proposition 1.3.1 we have:

PROPOSITION 1.4.1. *Let $G$ be a connected graph of order at least three. Then the following statements are equivalent.*

(a) *$G$ is 2-connected.*
(b) *Any two vertices of $G$ are on a common cycle.*
(c) *Any two edges of $G$ are on a common cycle.*
(d) *$G$ has no cutvertices.*
(e) *For every vertex $v$ of $G$, $G - v$ is connected.*
(f) *$G$ has only one block.*

Clearly, a cycle of length at least three is 2-connected. If $H$ is a 2-connected graph and $P$ is a path such that $P \cap H$ consists of the two ends of $P$, then clearly $H \cup P$ is 2-connected. So we can construct 2-connected graphs by starting with a cycle of length at least three and successively adding a path having only the ends in common with the current graph. We get all 2-connected graphs this way.

PROPOSITION 1.4.2. *If $G$ is a 2-connected graph, then it can be obtained from a cycle of length at least three by successively adding a path having only its ends in common with the current graph.*

PROOF. Let $H$ be a cycle in $G$. If $H = G$ there is nothing to prove. So assume that $H \neq G$. Since $G$ is connected, it contains an edge $uv \in E(G) \backslash E(H)$ such that $u \in V(H)$. Since $G$ is 2-connected, $G - u$ is connected. Let $Q$ be a shortest path in $G - u$ from $v$ to some vertex in $H$. The path $P$ consisting of the edge $uv$ and $Q$ is a path which has precisely its ends in common with $H$. Add this path to the current graph. Repeating this procedure we obtain the entire graph $G$.                     □

The decomposition of a 2-connected graph in a cycle and a sequence of paths is called an *ear decomposition* of the graph.

Proposition 1.4.2 has an analogue for 3-connected graphs. Tutte proved the following result.

THEOREM 1.4.3 (Tutte [**Tu61a**, **Tu66a**]). *Every 3-connected graph can be obtained from a wheel by a sequence of vertex splittings and edge-additions so that all intermediate graphs are 3-connected.*

A relatively simple proof of Theorem 1.4.3 was found by Halin [**Ha69**].

Thomassen [**Th80a**] showed that the following "generalized contraction" operation is sufficient to reduce every 3-connected graph to $K_4$. If $G$ is a graph and $e \in E(G)$, let $G/\!/e$ be the graph obtained from the edge-contracted graph $G/e$ by replacing all multiple edges by single edges joining the same pairs of vertices. Then $W_n/\!/e = W_{n-1}$ (if $n > 3$) for every edge $e$ not incident with the center.

LEMMA 1.4.4. *Every 3-connected graph $G$ of order at least five contains an edge $e$ such that the graph $G/\!/e$ is 3-connected.*

Lemma 1.4.4 shows that every 3-connected graph can be obtained from the smallest 3-connected graph $K_4$ by a sequence of "generalized vertex splittings" where "generalized" means that after splitting a vertex $v$ some of the edges incident to $v$ in $G$ result in two new edges, and so the new edge obtained by a splitting $v$ can be in one or more 3-cycles. This result may sound stronger than Theorem 1.4.3. However, the converse holds: Lemma 1.4.4 follows easily from Theorem 1.4.3 while it takes extra work to derive Theorem 1.4.3 from Lemma 1.4.4. But, Lemma 1.4.4 has a very short proof (as shown below) and interesting applications, for example the simplest known proof of Kuratowski's Theorem, as shown later.

PROOF OF LEMMA 1.4.4 (Thomassen [**Th80a**]). Suppose (reductio ad absurdum) that for every edge $e = xy$, the graph $G/\!/e$ has connectivity

2. Since it has at least four vertices, it has a separating set consisting of two vertices. Then one of these vertices must be the vertex obtained by identifying $x$ and $y$ and hence $G$ has a separating set $\{x, y, z\}$. We choose $e = xy$ and $z$ in such a way that the largest component $H$ of $G - \{x, y, z\}$ is as large as possible. Let $H'$ be another component and let $u$ be a vertex of $H'$ adjacent to $z$. Now $G$ has a separating set of the form $\{z, u, v\}$, since $G /\!/ zu$ also has connectivity 2. Consider the induced subgraph of $G$ on the vertices $(V(H) \cup \{x, y\}) \backslash \{v\}$. It is easily seen that this subgraph is connected, and since it does not contain any of the vertices $z, u, v$, it is contained in a component $H''$ of $G - \{z, u, v\}$. But $|V(H'')| > |V(H)|$ and we have a contradiction to the maximality property of $H$.                $\square$

An edge $e$ in a graph $G$ is said to be *contractible* if the connectivity of $G /\!/ e$ is at least the connectivity of $G$. Lemma 1.4.4 stimulated further investigations. For instance, the distribution of contractible edges has been studied in several papers. See, e.g., [**Sa90, De90, ES91, EOSY95**] and their references.

We shall strengthen Lemma 1.4.4 slightly. For this purpose we make use of the following observation.

LEMMA 1.4.5. *If $G$ is a 2-connected graph, then $G$ has a cycle $C$ such that, for each edge $e$ of $C$, $G/e$ has no cutvertices.*

PROOF. The proof is by induction on $|E(G)|$. If $|E(G)| < 3$, there is nothing to prove, so assume that $|E(G)| \geq 3$. We may assume that $G$ contains an edge $e = xy$ such that $G/e$ is not 2-connected. Then $G - \{x, y\}$ is disconnected. So, $e$ is a chord in some cycle $C'$ of $G$, and hence $G - e$ is 2-connected. Now apply induction to $G - e$.                $\square$

The proof of Lemma 1.4.4 shows that the following holds: If $xy$ is a noncontractible edge in a 3-connected graph $G$ with at least five vertices, and $z$ is a vertex such that $G - \{x, y, z\}$ is disconnected, then for any two components $H_1$, $H_2$ of $G - \{x, y, z\}$, $G$ has a contractible edge $e_i$, where either $e_i$ is in $E(H_i)$ or joins $H_i$ to $\{x, y, z\}$ for $i = 1, 2$. In particular, $G$ has at least two contractible edges. We can go a little further:

PROPOSITION 1.4.6. *If $G$ is a 3-connected graph with at least five vertices and $e_0 = x_0 y_0$ is an edge of $G$, then $G$ has a contractible edge with no end in common with $e_0$.*

PROOF. Let $e = xy$ be an edge of $G' = G - \{x_0, y_0\}$. Suppose that $G /\!/ e$ is not 3-connected. Let $z$ be a vertex such that $G - \{x, y, z\}$ is disconnected. Using the remark preceding Proposition 1.4.6, we may assume that $e_0$ is incident with $z$ (since otherwise one of $e_1, e_2$ has no end in common with $e_0$). So we may assume the following: If $e = xy$ is any edge of $G'$, then for either $z = x_0$ or $z = y_0$, $G - \{x, y, z\}$ is disconnected.

Now, let $H$ be an endblock of $G'$. If $G'$ is 2-connected, let $q$ be an arbitrary vertex of $G'$. Otherwise, let $q$ be the cutvertex of $G'$ that is contained in $H$. We select an edge $e_1 = x_1 y_1$ in $H$ as follows. If $H$ has only one edge, let $e_1$ be that edge, and we assume that $x_1 = q$ in this case. If $H$ is 2-connected, we let $C$ be a cycle satisfying the conclusion of Lemma 1.4.5 and we let $e_1$ be any edge of $C$ not incident with $q$. As assumed above, there is a vertex $z \in \{x_0, y_0\}$, say $z = x_0$, such that $G - \{x_1, y_1, z\}$ is disconnected. If $H$ has only one edge, then this implies that $G - \{x_0, x_1\}$ is disconnected, a contradiction. Similarly, if $H \neq G'$ and $H$ is 2-connected, then $H - \{x_1, y_1\}$ is connected and hence $y_0$ is joined only to $x_0, x_1, y_1$. Therefore $G - \{x_0, q\}$ is disconnected, a contradiction. Finally, if $H = G'$, then $G - \{x_1, y_1, x_0\}$ has precisely two components, one of which is $y_0$ (since $H - \{x_1, y_1\} = G - \{x_0, y_0, x_1, y_1\}$ is connected). So $y_0$ is joined to precisely $x_0, x_1, y_1$. But now we repeat this argument for another edge of $C$ and obtain a contradiction. $\qquad\square$

One of the most fundamental theorems on graphs is Menger's Theorem [**Me27**]:

**Menger's Theorem.** *Let $s$ and $t$ be distinct nonadjacent vertices in a graph $G$ and let $k$ be a natural number. Then the following assertions are equivalent:*

(i) *$G$ has $k$ paths from $s$ to $t$ having only $s$ and $t$ in common pair by pair.*

(ii) *For each subset $S \subseteq V(G)\backslash\{s, t\}$ with at most $k-1$ vertices, $G - S$ has a path from $s$ to $t$.*

There are many proofs of this theorem. A recent short proof with references to some of the others is given by McCuaig [**MC84**].

Menger's Theorem implies a characterization of $k$-connected graphs:

THEOREM 1.4.7. *Let $k$ be a natural number and $G$ be a graph with at least $k + 1$ vertices. Then the following statements are equivalent:*

(a) *$G$ is $k$-connected.*

(b) *If $x$ and $y$ are distinct vertices of $G$, then $G$ has $k$ paths from $x$ to $y$ such that no two of them have a vertex except $x$ and $y$ in common.*

(c) *If $A = \{a_1, \ldots, a_p\}$ and $B = \{b_1, \ldots, b_q\}$ are disjoint sets of vertices of $G$ and $s_1, \ldots, s_p$, $t_1, \ldots, t_q$ are natural numbers such that $s_1 + \cdots + s_p = t_1 + \cdots + t_q = k$, then $G$ has $k$ paths from $A$ to $B$ such that $s_i$ of them start at $a_i$ $(i = 1, \ldots, p)$, $t_j$ end in $b_j$ $(j = 1, \ldots, q)$, and no two of these paths have a vertex in $V(G)\backslash(A \cup B)$ in common.*

PROOF. Clearly, (c) implies (b), and (b) implies (a). Dirac [**Di60**] proved that (a) implies (c) by induction on $k$ as follows: If $G$ has an edge $e = a_i b_j$ where $1 \leq i \leq p$, $1 \leq j \leq q$, $2 \leq s_i$, $2 \leq t_j$, then we apply induction to $G - e$ replacing $s_i$ and $t_j$ with $s_i - 1$ and $t_j - 1$, respectively. So assume that $G$ has no such edge $e$. Then we replace each $a_i$ with $s_i$

pairwise distinct and mutually adjacent vertices, each having the same neighbors as $a_i$ ($i = 1, \ldots, p$). Likewise, $b_j$ is replaced by $t_j$ vertices ($j = 1, \ldots, q$). Then we add two new vertices $s, t$ and join $s$ to all vertices replacing $a_1, \ldots, a_p$, and we join $t$ to all vertices replacing $b_1, \ldots, b_q$. It is easy to verify that the new graph satisfies (ii) in Menger's Theorem. The equivalent statement (i) of that theorem gives $k$ paths from $s$ to $t$ which in turn give rise to $k$ paths in $G$ having the desired properties.    □

CHAPTER 2

# Planar graphs

A graph is *planar* if it can be drawn in the plane in such a way that no edges intersect, except of course at a common endvertex. Planar graphs corresponding to the regular polyhedra and other geometric figures have been investigated since the time of the ancient Greeks. Several mathematical works from the 18th and the 19th Century exhibit properties of graphs corresponding to convex 3-polyhedra. A classical example is Euler's polyhedral formula. A more recent cornerstone in geometry is Steintz' theorem that the 3-connected planar graphs are the 1-skeletons of convex polyhedra in the 3-space. Planar graphs also appear in some applied disciplines. An example is VLSI design where one would like to design a large electric network on a planar electric board so that the connections between the components of the network do not intersect (or intersect as little as possible). Some results on planar graphs were inspired by such practical problems. However, the main source of inspiration for planar graphs was the Four Color Conjecture (now Theorem; cf. Chapter 8) that the vertices of any planar graph can be colored with four colors in such a way that no two adjacent vertices have the same color.

Several books and surveys contain results on planar graphs. We refer, for example, to Tutte [**Tu84**]. Algorithmic aspects are treated in [**NC88, Wi85, Ya95**]. Kelmans' survey [**Ke93**] includes material that has been published only in Russian journals. Of course, the Four Color Problem, which is the subject of the monographs [**AH89, Ore67, SK77**], involves some theory on planar graphs. In this chapter we concentrate on a few fundamental results which are useful when going to higher surfaces. We also present rigorous proofs of the topological prerequisites for studying graphs in the plane and on surfaces beginning with the Jordan Curve Theorem, which says that a simple closed curve in the Euclidean plane partitions the plane into precisely two parts: the interior and the exterior part of the plane. Although this fundamental result seems intuitively obvious, it is fascinatingly difficult to prove. There are several proofs in the literature. A relatively elementary proof (involving only approximation with polygons) has been published by Tverberg [**Tv80**]. Our proof is taken from [**Th92a**]. Apart from very basic point set topology and a few simple properties of the plane, it uses only basic tools from graph theory.

The properties of the plane needed in our proof are that $\mathbb{R}^2$ is an arcwise connected Hausdorff space, that $K_{3,3}$ cannot be embedded in $\mathbb{R}^2$, and that no simple arc separates $\mathbb{R}^2$. These properties are indeed sufficient for every simple closed curve in a topological space $X$ to separate $X$ as shown in [**Th90a**]. These ideas are used in [**Th90d**] to obtain a characterization of the 2-sphere in terms of Jordan curve separation. They are presented in Section 2.10.

We also present the short proof by Thomassen [**Th92a**] of the Jordan–Schönflies Theorem which says that a homeomorphism of a simple closed curve in the plane onto a circle in the plane can be extended to a homeomorphism of the entire plane. Again, this result seems intuitively clear but a rigorous proof involves a number of technical details. While the Jordan Curve Theorem is also valid for spheres in $\mathbb{R}^3$, the Jordan–Schönflies Theorem does not extend to $\mathbb{R}^3$ (as shown by the so-called Alexander Horned Sphere, see, e.g., Moise [**Mo77**]).

Sections 2.3–2.7 give basic combinatorial and geometric results about planar graphs: We present the short proof in [**Th80a**] of Kuratowski's theorem which characterizes planar graphs in terms of forbidden subgraphs. Several other characterizations of planar graphs are presented. We establish many properties of planar graphs and discuss various kinds of representations of planar graphs. In Section 2.8 we give a proof of the Koebe-Andreev-Thurston Circle Packing Theorem in a very strong primal dual version due to Brightwell and Scheinerman [**BS93**] (Theorem 2.8.8). This result has important applications. In particular, a short proof of Steinitz' theorem follows. In Section 2.9 we indicate how Rodin and Sullivan [**RS87**] used it to give a simpler proof of the Riemann Mapping Theorem.

## 2.1. Planar graphs and the Jordan Curve Theorem

Let $X$ be a topological space. A *curve* (or an *arc*) in $X$ is the image of a continuous function $f : [0, 1] \to X$. The curve is *simple* if $f$ is 1-1. The arc $A = f([0, 1])$ is said to *join* (or *connect*) its endpoints $f(0)$ and $f(1)$. If $0 \le a < b \le 1$, then $f([a, b])$ is the *segment* of $A$ from $f(a)$ to $f(b)$. A *simple closed curve* is defined analogously except that now we have $f(0) = f(1)$. Also, a *segment* of a simple closed curve is either $f([a, b])$, or $f([0, a] \cup [b, 1])$ for some $0 \le a < b < 1$. Therefore there are two possible segments from $f(a)$ to $f(b)$ on the simple closed curve.

A topological space $X$ is *arcwise connected* if any two elements of $X$ are connected by a simple arc in $X$. The existence of a simple arc between two points of $X$ determines an equivalence relation whose equivalence classes are called the *arcwise connected components*, or the *regions* of $X$. A set $C \subseteq X$ *separates* $X$ if $X \backslash C$ is not arcwise connected. A *face* of $C \subseteq X$ is an arcwise connected component of $X \backslash C$.

If $X$ is a metric space, $z \in X$, and $\varepsilon > 0$, then $D(z, \varepsilon)$ denotes the set of points in $X$ of distance less than $\varepsilon$ from $z$, and $C(z, \varepsilon)$ denotes the set of points of distance $\varepsilon$ from $z$.

We shall repeatedly make use of the following simple fact: If $F_1, F_2$ are disjoint closed sets in $X$, and $A$ is an arc from an element of $F_1$ to an element of $F_2$, then $A$ contains a segment $A'$ from an element of $F_1$ to an element of $F_2$ such that $A'$ has no intermediate point in common with $F_1 \cup F_2$. Furthermore, the union of a finite number of arcs is compact and hence closed if $X$ is a Hausdorff space.

A graph $G$ is *embedded* in a topological space $X$ if the vertices of $G$ are distinct elements of $X$ and every edge of $G$ is a simple arc connecting in $X$ the two vertices which it joins in $G$, such that its interior is disjoint from other edges and vertices. An *embedding* of a graph $G$ in the topological space $X$ is an isomorphism of $G$ with a graph $G'$ embedded in $X$. In this case, $G'$ is said to be a *representation* of $G$ in $X$. If there is an embedding of $G$ into $X$, we say that $G$ *can be embedded* into $X$.

It is easy to see that every graph has an embedding into the 3-space. There are many ways of embedding it with all edges being straight line segments. For example, one can just let the vertices be distinct points on the curve $\{(\sin t, \cos t, t) \mid 0 \le t \le \pi/2\}$.

Our main concern in this chapter are graphs which can be embedded in the (Euclidean) *plane* $\mathbb{R}^2$. Such graphs are called *planar*. A graph $G$ that is embedded in the plane is a *plane* graph. The point set of a plane graph $G$ is compact. Therefore exactly one face of $G$ (i.e., a region of $\mathbb{R}^2 \backslash G$) is unbounded. It is called the *unbounded face* of $G$ (also the *exterior* or the *outer face* of $G$).

Simple arcs in the plane may be quite complicated sets. But, every plane graph can be redrawn such that the edges are not too complicated. An arc in $\mathbb{R}^2$ is a *polygonal arc* if it is the union of a finite number of straight line segments.

LEMMA 2.1.1. *If $G$ is a planar graph, then $G$ has a representation in the plane such that all edges are simple polygonal arcs.*

In the proof of Lemma 2.1.1 we will make use of the following result.

LEMMA 2.1.2. *If $\Omega$ is an open arcwise connected set in the plane, then any two points in $\Omega$ are joined by a simple polygonal arc in $\Omega$.*

PROOF. Let $x, y \in \Omega$ and let $A$ be a simple arc joining $x$ and $y$ in $\Omega$. For each point $z \in \Omega$ there is an $\varepsilon > 0$ such that the open disc $D_z = D(z, \varepsilon)$ is contained in $\Omega$. A continuous image of a compact set is compact, so $A$ is compact in $\Omega$. Therefore its open cover $\{D_z \mid z \in A\}$ has a finite subcover $\{D_{z_1}, D_{z_2}, \dots, D_{z_k}\}$. Now it is easy to construct a simple polygonal arc in the union $D_{z_1} \cup \cdots \cup D_{z_k}$ connecting $x$ and $y$. $\square$

PROOF OF LEMMA 2.1.1. Let $\Gamma$ be a plane graph isomorphic to $G$.
Let $p$ be some vertex of $\Gamma$, and let $D_p$ be a closed disc with $p$ as the
center such that $D_p$ intersects only those edges that are incident with $p$.
Furthermore, assume that $D_p \cap D_q = \emptyset$ for every pair of distinct vertices
$p, q$ of $\Gamma$. For each edge $pq$ of $\Gamma$ let $C_{pq}$ be a segment of the edge $pq$ such
that $C_{pq}$ joins $D_p$ with $D_q$ and has only its ends in common with $D_p \cup D_q$.
We can now redraw $G$ so that we use all arcs $C_{pq}$ and such that the parts
of the edges in the discs $D_p$ are straight line segments. Using Lemma
2.1.2 it is now easy to replace each of the arcs $C_{pq}$ by a simple polygonal
arc.                                                                      $\square$

Next we prove a special case of the Jordan Curve Theorem.

LEMMA 2.1.3. *If $C$ is a simple closed polygonal arc in the plane, then
$\mathbb{R}^2 \backslash C$ consists of precisely two arcwise connected components each of which
has $C$ as its boundary.*

PROOF. Let $P_1, P_2, \ldots, P_n$ be the straight line segments of $C$. Assume without loss of generality that none of $P_1, P_2, \ldots, P_n$ is horizontal.
For each $z \in \mathbb{R}^2 \backslash C$ denote by $\pi(z)$ the number of segments $P_i$ $(1 \leq i \leq n)$
such that the horizontal right half-line in $\mathbb{R}^2$ starting at $z$ contains a point
of $P_i$ but does not contain the endpoint of $P_i$ that has the largest second
coordinate. We let $\overline{\pi}(z)$ be $\pi(z)$ reduced modulo 2. If $A$ is a polygonal
arc in $\mathbb{R}^2 \backslash C$, then it is easy to see that $\overline{\pi}(z)$ is constant on $A$. By Lemma
2.1.2, $\overline{\pi}$ is constant on arcwise connected components of $\mathbb{R}^2 \backslash C$. It is obvious that points close to $C$ must have different value of $\overline{\pi}$ on each side. It
follows that $\mathbb{R}^2 \backslash C$ has at least two arcwise connected components.
Select a disc $D$ such that $D \cap C$ is a straight line segment. If $a, b, c$ are
points in $\mathbb{R}^2 \backslash C$, then we easily find three polygonal arcs in $\mathbb{R}^2 \backslash C$ starting
at these points and terminating in $D$. (We first come close to $C$ and then
follow $C$ close enough until we reach $D$.) Two of the three arcs can be
combined to get an arc in $\mathbb{R}^2 \backslash C$ joining two of the points $a, b, c$. Therefore
$\mathbb{R}^2 \backslash C$ has at most two arcwise connected components. This argument
also shows that every point on $C$ belongs to the boundary of each face of
$\mathbb{R}^2 \backslash C$.                                                      $\square$

COROLLARY 2.1.4. *Let $C$ be a simple closed polygonal arc in the plane,
and let $P$ be a simple polygonal arc joining distinct points $p, q \in C$ such
that $P \cap C = \{p, q\}$. Let $S_1, S_2$ be the two segments of $C$ from $p$ to $q$.
Then $C \cup P$ has precisely three faces whose boundaries are $C$, $P \cup S_1$, and
$P \cup S_2$, respectively.*

PROOF. By Lemma 2.1.3, each of $C$, $P \cup S_1$, $P \cup S_2$ has exactly two
faces. Each face of $C \cup P$ is contained in the intersection of a face of $P \cup S_1$
and a face of $P \cup S_2$ since $(P \cup S_1) \cup (P \cup S_2) = C \cup P$. We may assume that
$P \backslash \{p, q\}$ is contained in the bounded face of $C$. For $i = 1, 2$, let $X_i, Y_i$ be

the bounded and the unbounded face, respectively, of $P \cup S_i$. Then $Y_1 \cap Y_2$ contains the unbounded face $Y$ of $C$. Therefore $Y_1 \cap Y_2 = Y$, and this is a face of $C \cup P$. Since $Y_2 \supseteq Y$, it is clear that $S_1 \backslash \{p, q\} \subseteq Y_2$ and therefore $X_1 \subseteq Y_2$. Therefore $X_1 \cap Y_2 = X_1$ is a face of $C \cup P$. Similarly we see that $X_2 \cap Y_1 = X_2$ is a face of $C \cup P$. The obtained faces are bounded by $C$, $S_1 \cup P$, and $S_2 \cup P$, respectively. Since $X_2 \subseteq Y_1$, $X_1$ and $X_2$ are disjoint. Hence our list of faces of $C \cup P$ is complete. □

Corollary 2.1.4 implies a special case of *Euler's formula*.

PROPOSITION 2.1.5. *If $\Gamma$ is a 2-connected plane graph where all edges are polygonal arcs, then $\Gamma$ has exactly $|E(\Gamma)| - |V(\Gamma)| + 2$ faces. Each face has a cycle of $\Gamma$ as its boundary.*

PROOF. We use induction on the number of cycles in $\Gamma$. If $\Gamma$ has only one cycle $C$ then we are done by Lemma 2.1.3. Otherwise, Proposition 1.4.2 guarantees that $\Gamma$ has a 2-connected subgraph $\Gamma'$ such that $\Gamma$ is obtained from $\Gamma'$ by adding a path $P$ between two vertices of $\Gamma'$ and no intermediate vertex of $P$ belongs to $\Gamma'$. Since $\Gamma'$ has fewer cycles than $\Gamma$, we apply the induction hypothesis to $\Gamma'$ and complete the proof by applying Corollary 2.1.4 to the cycle $C$ of $\Gamma'$ bounding the face of $\Gamma'$ which contains $P$. □

The cycles of a 2-connected plane graph $G$ bounding the faces of $G$ are said to be *facial*.

We define a *triangulation* or a *quadrangulation* as a connected plane graph with polygonal edges such that each face boundary is a 3-cycle or, respectively, a 4-cycle.

PROPOSITION 2.1.6. *Let $\Gamma$ be a plane graph with polygonal edges and with $n \geq 4$ vertices. Denote by $q$ the number of edges of $\Gamma$. Then:*

(a) *$q \leq 3n - 6$ with equality holding if and only if $\Gamma$ is a triangulation.*
(b) *If $\Gamma$ has no 3-cycles, then $q \leq 2n - 4$ with equality if and only if $\Gamma$ is a quadrangulation.*

PROOF. Denote by $f$ the number of faces of $\Gamma$. Assume first that $\Gamma$ is 2-connected. By Proposition 2.1.5 we have

$$n - q + f = 2 \,. \tag{2.1}$$

Proposition 2.1.5 implies that every edge is in precisely two facial cycles. Therefore

$$3f \leq 2q \tag{2.2}$$

with equality if and only if $\Gamma$ is a triangulation. Combining (2.1) and (2.2) we get (a). Similarly we get (b).

If $\Gamma$ is not 2-connected, then (a) or (b) (or a slightly weaker inequality in case of less than 4 vertices) applied to each block of $\Gamma$ shows that there is strict inequality in (a) or (b). □

Lemma 2.1.1 and Proposition 2.1.6 applied to the graphs $K_5$ and $K_{3,3}$, respectively, show that these graphs have too many edges to be planar.

COROLLARY 2.1.7. *The complete graph $K_5$ and the complete bipartite graph $K_{3,3}$ are nonplanar.*

Now we are ready to prove the Jordan Curve Theorem.

If $C$ is a closed curve in the plane, the unbounded face of $C$ is called the *exterior* of $C$ and is denoted by $\text{ext}(C)$. The union of all other faces is the *interior* of $C$ and is denoted by $\text{int}(C)$. Furthermore, we write

$$\overline{\text{int}}(C) = C \cup \text{int}(C), \quad \text{and} \quad \overline{\text{ext}}(C) = C \cup \text{ext}(C).$$

PROPOSITION 2.1.8. *If $C$ is a simple closed curve in the plane $\mathbb{R}^2$, then $\mathbb{R}^2 \backslash C$ is not arcwise connected.*

PROOF. $C$ is the image of a continuous 1-1 mapping of a compact set. Therefore $C$ contains a point $p_1$ with the minimal $x$-coordinate among all points on $C \subseteq \mathbb{R}^2$, and it contains a point $p_2$ with the maximal $x$-coordinate, and $p_1 \neq p_2$. Clearly, there is a polygonal arc $P$ in $\text{ext}(C) \cup \{p_1, p_2\}$ from $p_1$ to $p_2$ as indicated in Figure 2.1. $P$ can be chosen such that $Q$ is below the large horizontal segment of $P$. Let $L$ be a vertical straight line separating $p_1$ and $p_2$. Denote by $P_1$ and $P_2$ the two segments on $C$ from $p_1$ to $p_2$. Then $P_1 \cap L$ and $P_2 \cap L$ are nonempty, compact, and disjoint. Therefore $L$ contains a segment $Q$ joining $P_1$ with $P_2$ and having only its ends in common with $C$ (see Figure 2.1). By the choice of $P$, $Q \cap P = \emptyset$.

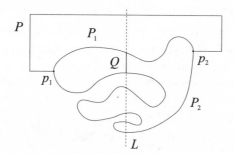

FIGURE 2.1. A simple closed curve separates the plane.

Suppose now that $\mathbb{R}^2 \backslash C$ is arcwise connected. Then there is a simple arc $R$ in $\mathbb{R}^2 \backslash C$ connecting a point of $Q \backslash C$ and a point of $P \backslash \{p_1, p_2\}$ such that only the endpoints of $R$ belong to $Q \cup P$. Then $C \cup Q \cup P \cup R$ is a plane graph isomorphic to $K_{3,3}$ which is impossible by Corollary 2.1.7. □

In Proposition 2.1.12 below we shall prove even more: $\text{int}(C)$ consists of a single face.

LEMMA 2.1.9. *Let* $\Gamma_1, \Gamma_2$ *be plane graphs such that each edge is a simple polygonal arc. Then the union of the point sets of* $\Gamma_1$ *and* $\Gamma_2$ *is the point set of a plane graph* $\Gamma_3$, *and* $\Gamma_3$ *is uniquely determined up to homeomorphism. If both* $\Gamma_1$ *and* $\Gamma_2$ *are 2-connected and have at least two points in common, then also* $\Gamma_3$ *is 2-connected.*

PROOF. For $i = 1, 2$, let $\Gamma_i'$ denote the plane graph which is a subdivision of $\Gamma_i$ such that each edge of $\Gamma_i'$ is a straight line segment. Let $\Gamma_i''$ be the subdivision of $\Gamma_i'$ such that a point $p$ on an edge $e$ of $\Gamma_i'$ is a vertex of $\Gamma_i''$ if either $p$ is a vertex of $\Gamma_1'$ or $\Gamma_2'$, or $p$ is on an edge of $\Gamma_{3-i}'$ that crosses $e$. Then the usual union of graphs $\Gamma_1''$ and $\Gamma_2''$ is a graph that can play the role of $\Gamma_3$. (Note that $\Gamma_3$ has no loops or multiple edges since its edges are straight line segments.)

It is obvious that $\Gamma_3$ is uniquely determined up to a homeomorphism. The last statement about 2-connectivity is left as an easy exercise.    □

Lemma 2.1.9 does not hold for general plane graphs since two arcs can intersect infinitely often.

We shall use the notation $\Gamma_3 = \Gamma_1 \sqcup \Gamma_2$ to denote the graph $\Gamma_3$ arising from $\Gamma_1$ and $\Gamma_2$ as described in the proof of Lemma 2.1.9. Note that the "union" of plane graphs defined above is associative.

LEMMA 2.1.10 (Thomassen [**Th92a**]). *Let* $\Gamma_1, \Gamma_2, \ldots, \Gamma_k$ $(k \geq 2)$ *be 2-connected polygonal arc embedded plane graphs such that, for* $i = 2, 3, \ldots, k - 1$, *the graph* $\Gamma_i$ *has at least two points in common with each of* $\Gamma_{i-1}$ *and* $\Gamma_{i+1}$ *and no point in common with any other* $\Gamma_j$, $|j - i| \geq 2$. *Then any point which is in the outer face of each of* $\Gamma_1 \sqcup \Gamma_2$, $\Gamma_2 \sqcup \Gamma_3$, $\ldots$, $\Gamma_{k-1} \sqcup \Gamma_k$ *is also in the outer face of* $\Gamma_1 \sqcup \Gamma_2 \sqcup \cdots \sqcup \Gamma_k$.

PROOF. Suppose $p$ is a point in a bounded face of $\Gamma = \Gamma_1 \sqcup \cdots \sqcup \Gamma_k$. By Lemma 2.1.9, $\Gamma$ is 2-connected, and by Proposition 2.1.5, there is a cycle $C$ in $\Gamma$ such that $p \in \text{int}(C)$. Choose $C$ such that $C$ is in $\Gamma_i \sqcup \Gamma_{i+1} \sqcup \cdots \sqcup \Gamma_j$ and such that $j - i$ is minimum. We will show that $j - i \leq 1$. So assume that $j - i \geq 2$. Among all cycles in $\Gamma_i \sqcup \cdots \sqcup \Gamma_j$ having $p$ in the interior we assume that $C$ is chosen in such a way that the number of edges in $C$ and not in $\Gamma_{j-1}$ is minimum. Since $C$ intersects both $\Gamma_j \backslash \Gamma_{j-1}$ and $\Gamma_{j-2} \backslash \Gamma_{j-1}$, $C$ has at least two disjoint maximal segments $P_1$ and $P_2$ in $\Gamma_{j-1}$. Since $\Gamma_{j-1}$ is connected, it contains a path from $P_1$ to $P_2$. Let $P_3$ be a shortest path in $\Gamma_{j-1}$ from $P_1$ to $C - V(P_1)$. Then $C \cup P_3$ has three cycles two of which have $p$ in the interior. The one that contains $P_3$ has fewer edges not in $\Gamma_{j-1}$ than $C$. This contradicts the minimality of $C$.    □

THEOREM 2.1.11. *If* $P$ *is a simple arc in the plane, then* $\mathbb{R}^2 \backslash P$ *is arcwise connected.*

PROOF (from [**Th92a**]). Let $p, q$ be two points of $\mathbb{R}^2 \backslash P$, and let $d$ be a positive number such that each of $p, q$ has distance $> 3d$ from $P$.

We shall join $p, q$ by a simple polygonal arc in $\mathbb{R}^2 \backslash P$. Since $P$ is the image of a continuous (and hence uniformly continuous) map, we can partition $P$ into segments $P_1, P_2, \ldots, P_k$ such that $P_i$ joins $p_i$ and $p_{i+1}$ for $i = 1, 2, \ldots, k$ and such that each point on $P_i$ has distance less than $d$ from $p_i$ $(i = 1, 2, \ldots, k)$. Let $d'$ be a positive number smaller than the minimal distance between $P_i$ and $P_j$, $1 \leq i$, $i + 2 \leq j \leq k$, and $(i, j) \neq (1, k)$. Note that $d' \leq d$. For each $i = 1, 2, \ldots, k$, we partition $P_i$ into segments $P_{i,1}, P_{i,2}, \ldots, P_{i,k_i}$ such that $P_{i,j}$ joins a point $p_{i,j}$ with $p_{i,j+1}$ for $j = 1, 2, \ldots, k_i - 1$ and such that each point on $P_{i,j}$ has distance less than $d'/4$ to $p_{i,j}$. Let $\Gamma_i$ be the graph which is the union of the boundaries of the squares that consist of horizontal and vertical line segments of length $d'/2$ and have a point $p_{i,j}$ $(1 \leq j \leq k_i)$ as the center. Then the graphs $\Gamma_1, \Gamma_2, \ldots, \Gamma_k$ satisfy the assumptions of Lemma 2.1.10. Hence both of $p$ and $q$ are in the outer face of $\Gamma_1 \sqcup \ldots \sqcup \Gamma_k$ (because they are outside the disc of radius $3d$ and with center $p_{i+1}$, while $\Gamma_i \sqcup \Gamma_{i+1}$ is inside that disc, $i = 1, 2, \ldots, k - 1$), and $P$ does not intersect that face. Therefore $p$ and $q$ can be joined by a simple polygonal arc disjoint from $P$. $\qquad\square$

If $C$ is a closed subset of the plane, and $\Omega$ is a region in $\mathbb{R}^2 \backslash C$, then a point $p$ in $C$ is *accessible* from $\Omega$ if for some (and hence each) point $q$ in $\Omega$, there is a polygonal arc from $q$ to $p$ having only $p$ in common with $C$. If $C$ is a simple closed curve, then $p$ need not be accessible from $\Omega$. However, if $P$ is any segment of $C$ containing $p$, then Theorem 2.1.11 implies that $(\mathbb{R}^2 \backslash C) \cup P$ contains a simple polygonal arc $P'$ from $q$ to a region of $\mathbb{R}^2 \backslash C$ distinct from $\Omega$. Then $P'$ intersects $C$ in a point on $P$. Since $P$ can be chosen to be arbitrarily small, we conclude that the points on $C$ accessible from $\Omega$ are dense on $C$. We also get:

PROPOSITION 2.1.12. *If $C$ is a simple closed curve in the plane, then $\mathbb{R}^2 \backslash C$ has at most two regions.*

PROOF. Assume (*reductio ad absurdum*) that $q_1, q_2, q_3$ are points in distinct regions $\Omega_1, \Omega_2, \Omega_3$ of $\mathbb{R}^2 \backslash C$. Let $p_1, p_2, p_3$ be distinct points on $C$ and let $D_i$ be a disc around $p_i$ $(i = 1, 2, 3)$ such that $D_1, D_2, D_3$ are pairwise disjoint and contain none of $q_1, q_2, q_3$. By the remark following Theorem 2.1.11, in $\Omega_i$ there is a simple polygonal arc $P_{i,j}$ from $q_i$ to $D_j$ for $i, j = 1, 2, 3$. We may assume that $P_{i,j} \cap P_{i,j'} = \{q_i\}$ for $j \neq j'$, and $P_{i,j} \cap P_{i',j'} = \emptyset$ when $i \neq i'$. We can now extend (by adding three segments of $C$) the union of the curves $P_{i,j}$ $(i, j = 1, 2, 3)$ to a plane graph with vertex set $\{q_1, q_2, q_3, p_1, p_2, p_3\}$ isomorphic to $K_{3,3}$. This contradicts Corollary 2.1.7. $\qquad\square$

Propositions 2.1.8 and 2.1.12 constitute what is usually called the *Jordan Curve Theorem*.

**The Jordan Curve Theorem.** *Any simple closed curve $C$ in the plane divides the plane into exactly two arcwise connected components. Both of these regions have $C$ as the boundary.*

The Jordan Curve Theorem is named after Camille Jordan. Apparently, the first correct proof was given by Veblen in 1905 [**Ve05**]. This result is a special case of the Jordan-Schönflies Theorem which we prove in the next section.

Discrete versions of the Jordan Curve Theorem have been considered by Little [**Li88**], Stahl [**St83**], and Vince [**Vi89**].

## 2.2. The Jordan-Schönflies Theorem

Now that we have proved the Jordan Curve Theorem we can extend some of the previous results. For example, Corollary 2.1.4 easily extends to the case where $C$ is any simple closed curve and $P$ is a simple arc in $\overline{\text{int}}(C)$ such that only its ends lie on $C$. Also Lemma 2.1.9 remains valid if $\Gamma_1$ and $\Gamma_2$ are plane graphs consisting of a simple closed curve $C$ (which is the outer cycle in both $\Gamma_1$ and $\Gamma_2$) and polygonal curves in $\overline{\text{int}}(C)$. (Lemma 2.1.9 would not be valid if $\Gamma_1$ and $\Gamma_2$ had distinct outer cycles, or if the interior edges are not polygonal arcs.)

If $C$ and $C'$ are simple closed curves, and $\Gamma$ and $\Gamma'$ are 2-connected plane graphs whose exterior faces are bounded by $C$ and $C'$, respectively, then $\Gamma$ and $\Gamma'$ are said to be *plane-isomorphic* if there is an isomorphism $\gamma$ of $\Gamma$ to $\Gamma'$ which maps $C$ onto $C'$ such that a cycle in $\Gamma$ bounds a face of $\Gamma$ if and only if the image of the cycle is a face boundary in $\Gamma'$. The isomorphism $\gamma$ is said to be a *plane-isomorphism* of $\Gamma$ and $\Gamma'$.

**The Jordan-Schönflies Theorem.** *If $f$ is a homeomorphism of a simple closed curve $C$ in the plane onto a closed curve $C'$ in the plane, then $f$ can be extended to a homeomorphism of the entire plane.*

PROOF (from [**Th92a**]). Without loss of generality we may assume that $C'$ is a convex polygon. We shall first extend $f$ to a homeomorphism of $\overline{\text{int}}(C)$ to $\overline{\text{int}}(C')$. Let $B$ denote a countable dense set in $\text{int}(C)$ (for example the points with rational coordinates). As mentioned before Proposition 2.1.12, the points on $C$ accessible from $\text{int}(C)$ are dense on $C$. Therefore, there exists a countable set $A \subseteq C$ which is dense in $C$ consisting of points accessible from $\text{int}(C)$. Let $p_1, p_2, \ldots$ be a sequence of points in $A \cup B$ such that each point in $A \cup B$ occurs infinitely often in this sequence. Let $\Gamma_0$ denote any 2-connected graph consisting of $C$ and some simple polygonal curves in $\overline{\text{int}}(C)$. Let $\Gamma_0'$ be a graph consisting of $C'$ and simple polygonal curves in $\overline{\text{int}}(C')$ such that $\Gamma_0$ and $\Gamma_0'$ are plane-isomorphic (with isomorphism $g_0$) such that $g_0$ and $f$ coincide on $C$. We now extend $f$ to $C \cup V(\Gamma_0)$ such that $g_0$ and $f$ coincide on $V(\Gamma_0)$. We shall define a sequence of 2-connected graphs $\Gamma_0, \Gamma_1, \Gamma_2, \ldots$ and $\Gamma_0', \Gamma_1', \Gamma_2', \ldots$

such that, for each $n \geq 1$, $\Gamma_n$ is an extension of a subdivision of $\Gamma_{n-1}$, $\Gamma'_n$ is an extension of a subdivision of $\Gamma'_{n-1}$, there is a plane-isomorphism $g_n$ of $\Gamma_n$ onto $\Gamma'_n$ which coincides with $g_{n-1}$ on $V(\Gamma_{n-1})$, and $\Gamma_n$ (respectively $\Gamma'_n$) consists of $C$ (respectively $C'$) and simple polygonal curves in $\overline{\mathrm{int}}(C)$ (respectively $\overline{\mathrm{int}}(C')$). Also, we shall assume that $\Gamma'_n \backslash C'$ is connected for each $n$. We then extend $f$ such that $f$ and $g_n$ coincide on $V(\Gamma_n)$.

Suppose that we have already defined $\Gamma_0, \ldots, \Gamma_{n-1}, \Gamma'_0, \ldots, \Gamma'_{n-1}$, and $g_0, \ldots, g_{n-1}$. We shall define $\Gamma_n$, $\Gamma'_n$ and $g_n$ as follows. We consider the point $p_n$. If $p_n \in A$, then we let $P$ be a simple polygonal curve from $p_n$ to a point $q_n$ of $\Gamma_{n-1} \backslash C$ such that $\Gamma_{n-1} \cap P = \{p_n, q_n\}$. We let $\Gamma_n$ denote the graph $\Gamma_{n-1} \cup P$. The arc $P$ is drawn in a face of $\Gamma_{n-1}$. By Proposition 2.1.5, this face is bounded by a cycle $S$, say. We add to $\Gamma'_{n-1}$ a simple polygonal curve $P'$ in the face bounded by $g_{n-1}(S)$ such that $P'$ joins $f(p_n)$ with $g_{n-1}(q_n)$ (if $q_n$ is a vertex of $\Gamma_{n-1}$) or a point on $g_{n-1}(a)$ (if $a$ is an edge of $\Gamma_{n-1}$ containing the point $q_n$). Then we put $\Gamma'_n = \Gamma'_{n-1} \cup P'$ and we define the plane-isomorphism $g_n$ from $\Gamma_n$ to $\Gamma'_n$ in the obvious way. We extend $f$ to $C \cup V(\Gamma_n)$ such that $f(q_n) = g_n(q_n)$.

If $p_n \in B$, we consider the largest square with vertical and horizontal sides, which has $p_n$ as the center and which is in $\overline{\mathrm{int}}(C)$. Inside this square (whose sides we are not going to add to $\Gamma_{n-1}$ as they may contain infinitely many points of $C$) we draw a new square with vertical and horizontal sides each of which has distance $< 1/n$ from the sides of the first square. Inside the new square we draw vertical and horizontal lines such that $p_n$ is on both a vertical line and a horizontal line and such that all regions in the square have diameter $< 1/n$. We let $H_n$ be the union of $\Gamma_{n-1}$ and the new horizontal and vertical straight line segments possibly together with an additional polygonal curve in $\mathrm{int}(C)$ in order to make $H_n$ 2-connected and $H_n \backslash C$ connected. By Proposition 1.4.2, $H_n$ can be obtained from $\Gamma_{n-1}$ by successively adding paths in faces. We add the corresponding paths to $\Gamma'_{n-1}$ and obtain a graph $H'_n$ which is plane-isomorphic to $H_n$. Then we add vertical and horizontal line segments in $\overline{\mathrm{int}}(C')$ to $H'_n$ such that the resulting graph has no (bounded) region of diameter $\geq 1/2n$. If necessary, we displace some of the lines a little such that they intersect $C'$ only in $f(A)$ and such that all bounded regions have diameter $< 1/n$ and such that each of the new lines has only finite intersection with $H'_n$. This extends $H'_n$ into a graph that we denote by $\Gamma'_n$. We add to $H_n$ polygonal curves such that we obtain a graph $\Gamma_n$ plane-isomorphic to $\Gamma'_n$. Then we extend $f$ such that it is defined on $C \cup V(\Gamma_n)$ and coincides with the plane-isomorphism $g_n$ on $V(\Gamma_n)$.

When we extend $H'_n$ into $\Gamma'_n$ and $H_n$ into $\Gamma_n$, we are adding many edges and it is perhaps difficult to visualize what is going on. However, Proposition 1.4.2 tells us that we can look at the extension of $H'_n$ into $\Gamma'_n$ as the result of a sequence of path additions (each of which is a straight line segment in a face). We then just perform successively the corresponding additions in $H_n$. Note that we have plenty of freedom for that since the current mapping $f$ is only defined on the current vertex set. The images of the points on the current edges have not been specified yet. In this way we extend $f$ to a 1-1 map defined on $F = C \cup V(\Gamma_0) \cup V(\Gamma_1) \cup \cdots$ whose image is the set $C' \cup V(\Gamma'_0) \cup V(\Gamma'_1) \cup \cdots$. These sets are dense in $\overline{\text{int}}(C)$ and $\overline{\text{int}}(C')$, respectively.

If $p$ is a point in $\text{int}(C)$ on which $f$ is not yet defined, then we consider a sequence $q_1, q_2, \ldots$ converging to $p$ and consisting of points from $V(\Gamma_0) \cup V(\Gamma_1) \cup \cdots$. We shall show that $f(q_1), f(q_2), \ldots$ converges and we let $f(p)$ be the limit. Let $d$ be the distance from $p$ to $C$, and let $p_n$ be a point of $B$ at distance $< d/3$ from $p$. Then $p$ is inside the largest square in $\overline{\text{int}}(C)$ having $p_n$ as the center (and also inside what we called the new square if $n$ is sufficiently large). By the construction of $\Gamma_n$ and $\Gamma'_n$ it follows that $\Gamma_n$ has a cycle $S$ such that $p \in \text{int}(S)$ and such that both $S$ and $g_n(S)$ are in discs of radius $< 1/n$. Since $f$ maps $F \cap \text{int}(S)$ into $\text{int}(g_n(S))$ and $F \cap \overline{\text{ext}}(S)$ into $\overline{\text{ext}}(g_n(S))$, it follows in particular, that the sequence $f(q_m), f(q_{m+1}), \ldots$ is in $\text{int}(g_n(S))$ for some $m$. Since $n$ can be chosen arbitrarily large, $f(q_1), f(q_2), \ldots$ is a Cauchy sequence and hence convergent. It follows that $f$ is well-defined. Moreover, using the above notation, $f$ maps $\text{int}(S)$ into $\text{int}(g_n(S))$. Hence $f$ is continuous in $\text{int}(C)$. Since $V(\Gamma'_0) \cup V(\Gamma'_1) \cup \cdots$ is dense in $\text{int}(C')$, the same argument shows that $f$ maps $\text{int}(C)$ onto $\text{int}(C')$ and that $f$ is 1-1 and that $f^{-1}$ is continuous on $\text{int}(C')$.

It only remains to be shown that $f$ is continuous on $C$. (Then also $f^{-1}$ is continuous since $\overline{\text{int}}(C)$ is compact.) In order to prove this, it is sufficient to consider a sequence $q_1, q_2, \ldots$ of points in $\text{int}(C)$ converging to $q$ on $C$ and then show that $f(q_1), f(q_2), \ldots$ converges to $f(q)$. Suppose therefore that this is not the case. Since $\overline{\text{int}}(C')$ is compact, we may assume (by considering an appropriate subsequence, if necessary) that $\lim_{n \to \infty} f(q_n) = q' \neq f(q)$. Since $f^{-1}$ is continuous on $\text{int}(C')$, $q'$ is on $C'$. Since $A$ is dense in $C$, $f(A)$ is dense in $C'$ and hence each of the two curves on $C'$ from $q'$ to $f(q)$ contain a point $f(q_1)$ and $f(q_2)$, respectively, in $f(A)$. For some $n$, $\Gamma_n$ has a path $P$ from $q_1$ to $q_2$ having only $q_1$ and $q_2$ in common with $C$. As we have noted at the beginning of this section, $P$ separates $\text{int}(C)$ in two regions. These two regions are mapped on the two distinct regions of $\text{int}(C') \backslash g_n(P)$. Hence we cannot have $\lim_{n \to \infty} f(q_n) = q'$. This contradiction shows that $f$ has the appropriate extension to $\text{int}(C)$.

By similar arguments, $f$ can be extended to $\text{ext}(C)$: Without loss of generality we may assume that $\text{int}(C)$ contains the origin and that both $C$ and $C'$ are in the interior of the quadrangle $T$ with corners $(\pm 1, \pm 1)$. Let $L_1, L_2$ be the line segments (on lines through the origin) from $(1, 1)$ and $(-1, -1)$, respectively, to $C$. Let $p_i$ be the end of $L_i$ on $C$, $i = 1, 2$. Let $L'_1, L'_2$ be simple polygonal arcs from $f(p_1)$ to $(1, 1)$ and from $f(p_2)$

to $(-1,-1)$, respectively, such that $L_1' \cap L_2' = \emptyset$ and these arcs have only their ends in common with $C \cup T$. It is easy to see that we can extend $f$ to a homeomorphism $C \cup L_1 \cup L_2 \cup T \to C' \cup L_1' \cup L_2' \cup T$ so that $f$ is the identity on $T$. Now we can use the method of the first part of the proof to extend $f$ to a homeomorphism of $\overline{\text{int}}(T)$ (onto itself). Since $f$ is the identity on $T$, it can be extended to the entire plane such that it is the identity on $\text{ext}(T)$. This determines a required homeomorphism. $\square$

If $F$ is a closed set in the plane, then we say that a point $p$ in $F$ is *curve-accessible* if, for each point $q \notin F$, there is a simple arc from $q$ to $p$ having only $p$ in common with $F$. The Jordan–Schönflies Theorem implies that every point on a simple closed curve is curve-accessible. Hence, the following is an extension of Proposition 2.1.12.

THEOREM 2.2.1 (Thomassen [**Th92a**]). *If $F$ is a closed set in the plane with at least three curve-accessible points, then $\mathbb{R}^2 \backslash F$ has at most two regions.*

PROOF. If $p_1$, $p_2$, $p_3$ are curve-accessible points in $F$ and $q_1$, $q_2$, $q_3$ belong to distinct regions of $\mathbb{R}^2 \backslash F$, then we get, as in the proof of Proposition 2.1.12, a plane graph isomorphic to $K_{3,3}$ with vertices $p_1$, $p_2$, $p_3$, $q_1$, $q_2$, $q_3$, a contradiction to Corollary 2.1.7. $\square$

In Theorem 2.2.1, "three" cannot be replaced by "two". To see this, we let $F$ be a collection of three or more internally disjoint simple arcs between two fixed points.

Some other consequences of the Jordan–Schönflies Theorem are presented below. First we generalize Corollary 2.1.4.

PROPOSITION 2.2.2. *Let $P_1, P_2, P_3$ be simple arcs with ends $p, q$ such that $P_i \cap P_j = \{p, q\}$ for $1 \le i < j \le 3$. Then $P_1 \cup P_2 \cup P_3$ has precisely three faces with boundaries $P_1 \cup P_2$, $P_1 \cup P_3$ and $P_2 \cup P_3$, respectively. If the outer face is bounded by $P_1 \cup P_2$, and $P_1', P_2', P_3'$ are simple arcs joining $p', q'$ such that $P_i' \cap P_j' = \{p', q'\}$ for $1 \le i < j \le 3$, and such that $P_3' \subseteq \overline{\text{int}}(P_1' \cup P_2')$, then any homeomorphism $f$ of $P_1 \cup P_2 \cup P_3$ onto $P_1' \cup P_2' \cup P_3'$ such that $f(P_i) = P_i'$ $(i = 1, 2, 3)$ can be extended to a homeomorphism of $\mathbb{R}^2$ onto itself.*

PROOF. If $P_1 \subseteq \overline{\text{ext}}(P_2 \cup P_3)$, $P_2 \subseteq \overline{\text{ext}}(P_1 \cup P_3)$, and $P_3 \subseteq \overline{\text{ext}}(P_1 \cup P_2)$, then it is easy to extend $P_1 \cup P_2 \cup P_3$ to a $K_{3,3}$ in the plane, a contradiction. So we may assume that $P_3 \subseteq \overline{\text{int}}(P_1 \cup P_2)$. The first part of the proposition follows easily. To prove the last part, it is sufficient to consider the case where $P_1', P_2', P_3'$ are polygonal arcs. This case is done by using the Jordan-Schönflies Theorem to $\overline{\text{int}}(P_1 \cup P_3)$, $\overline{\text{int}}(P_2 \cup P_3)$, and $\overline{\text{ext}}(P_1 \cup P_2)$, respectively. $\square$

Now we generalize Proposition 2.1.5.

THEOREM 2.2.3. *If $G$ is a 2-connected plane graph, then $G$ has $|E(G)| - |V(G)| + 2$ faces. Each face is bounded by a cycle of $G$. If $G'$ is a plane graph isomorphic to $G$, such that each facial cycle in $G$ corresponds to a facial cycle in $G'$, and such that the cycle bounding the outer face in $G$ corresponds to the boundary of the outer face of $G'$, then any homeomorphism of $G$ onto $G'$ (extending the isomorphism of $G$ onto $G'$) can be extended to a homeomorphism of the entire plane.*

PROOF. The proof is similar to that of Proposition 2.1.5 except that we use Proposition 2.2.2 instead of Corollary 2.1.4. □

The above formula relating $|V(G)|$, $|E(G)|$, and the number of faces of plane graphs is known as *Euler's (polyhedral) formula* and holds for arbitrary connected plane graphs. The proof is left to the reader.

PROPOSITION 2.2.4. *Let $P$ be a simple arc in $\mathbb{R}^2$ from $p$ to $q$ and let $\varepsilon$ be a positive real number. Then there exists a simple closed curve $C$ such that $P \subseteq \text{int}(C)$, every point in $\overline{\text{int}}(C)$ has distance less than $\varepsilon$ from $P$, and $\text{int}(C) \backslash P$ is arcwise connected.*

PROOF. $C$ is found by the same method as in the proof of Lemma 2.1.2. To prove the last statement, it suffices to consider the case where $C$ is a circle, by the Jordan-Schönflies Theorem. For that we consider a circle $C'$, $C' \subseteq \text{int}(C)$, such that $P \subseteq \text{int}(C')$, and we use the fact that $\mathbb{R}^2 \backslash P$ is arcwise connected (Theorem 2.1.11). □

Recall that for $r \in \mathbb{R}^2$ and $\varepsilon$ a positive number, $D(r, \varepsilon)$ denotes the disc centered at $r$ and with radius $\varepsilon$ (with respect to the Euclidean distance in $\mathbb{R}^2$).

LEMMA 2.2.5. *Let $P$ be a simple arc from $p$ to $q$. Let $r \in P \backslash \{p, q\}$ and let $\varepsilon$ be any positive real number such that $\text{dist}(r, p) > \varepsilon$, $\text{dist}(r, q) > \varepsilon$. Then there exists a homeomorphism $f$ of $\mathbb{R}^2$ onto itself such that $\mathbb{R}^2 \backslash D(r, \varepsilon)$ is fixed under $f$ and such that $f$ takes some segment of $P$ containing $r$ onto a straight line segment.*

PROOF. Let $P_1, P_2$ be the two minimal segments of $P$ starting at $r$ and intersecting the boundary $C(r, \varepsilon)$ of $D(r, \varepsilon)$. Then apply Proposition 2.2.2 to $P_1 \cup P_2 \cup C(r, \varepsilon)$. □

Now we can prove a result, similar to the Jordan-Schönflies Theorem.

THEOREM 2.2.6. *Let $f : A \to B$ be a homeomorphism between simple arcs $A$ and $B$ in the plane. Then $f$ can be extended to a homeomorphism of the entire plane.*

PROOF. It suffices to consider the case $A = [0, 1]$ (viewed as a subset in $\mathbb{R}^2$). Let $a := f(0)$ and $b := f(1)$ be the endpoints of $B$. It suffices to prove that $a$ and $b$ are curve-accessible on $B$ since we may then embed $B$

into a simple closed curve $C$ in $\mathbb{R}^2$ and use the Jordan-Schönflies Theorem to complete the proof.

Let $t_n$ be a decreasing sequence in $]0,1]$ such that $t_n \to 0$ as $n \to \infty$. Using Lemma 2.2.5 successively, we may assume that $B$ has a straight line segment $L_n$ containing $f(t_n)$. Let $d_n$ be the distance between the two segments of $B \backslash L_n$. Let $\varepsilon_1 > \varepsilon_2 > \cdots > 0$ such that $\varepsilon_n < d_n/10$. Without loss of generality assume that $f(t_n)$ is the midpoint of $L_n$. Let $s_1 \in D(f(t_1), \varepsilon_1) \backslash B$. By Proposition 2.2.4, there exists, for each $n \geq 2$, a simple polygonal arc $P_n$ in $\mathbb{R}^2 \backslash B$ from $s_1$ to $D(f(t_n), \varepsilon_n) \backslash B$ such that each point on $P_n$ is either in $D(f(t_1), \varepsilon_1)$ or has distance less than $\varepsilon_n$ from $P$.

Consider any natural number $m$. Then $D(f(t_m), \varepsilon_m) \backslash L_m$ has two components (halfdiscs). One of them is called the *good halfdisc* of $D(f(t_m), \varepsilon_m)$ if, for infinitely many integers $n \geq m$, the arc $P_n$ (when traversed from $D(f(t_m), \varepsilon_n)$) first meets $D(f(t_m), \varepsilon_m)$ in this halfdisc. Infinitely many of these arcs $P_n$ contain an arc from the good side of $D(f(t_{m+1}), \varepsilon_{m+1})$ to the good side of $D(f(t_m), \varepsilon_m)$. Let $Q_m$ denote such a segment. By a local modification in $D(f(t_{m+1}), \varepsilon_{m+1})$ we may assume that $Q_m \cup Q_{m+1}$ is a simple arc. Then $Q_1 \cup Q_2 \cup \cdots \cup \{a\}$ is a simple arc having only $a$ in common with $B$.

We prove similarly that $b$ is curve-accessible. $\qquad\square$

Theorem 2.2.6 in particular implies that every point on a simple arc is curve-accessible.

## 2.3. The Theorem of Kuratowski

The best-known result about planar graphs and one of the most beautiful theorems in graph theory is the theorem of Kuratowski [**Ku30**] which characterizes planar graphs in terms of forbidden subgraphs. This result was independently discovered at the same time by Frink and Smith [**FS30**], but it appeared only in the form of an abstract since their proof was similar to Kuratowski's, which had just appeared at that time. It seems that Pontrjagin independently came to the same result (see Burstein [**Bu78**]). The first relatively simple proof of Kuratowski's theorem was published in 1954 by Dirac and Schuster [**DSch54**]. Other short proofs are surveyed in [**Th81**]. A recent short proof was obtained by Makarychev [**Ma97**]. The proof given here is due to Thomassen [**Th80a**].

**Kuratowski's Theorem.** *A graph is planar if and only if it does not contain a subdivision of $K_5$ or a subdivision of $K_{3,3}$ as a subgraph.*

The forbidden subgraphs $K_5$ and $K_{3,3}$ are called the *Kuratowski graphs*. We already proved the "easy part" of Kuratowski's theorem; namely that $K_{3,3}$ and $K_5$ are nonplanar (Corollary 2.1.7). Also their subdivisions cannot be represented in the plane. The converse will follow from Lemmas 2.3.1 and 2.3.3 below. From these lemmas we shall derive some other results about planar graphs. For this purpose we introduce two additional concepts. A plane graph $G$ is *straight line embedded* if each edge is a straight line segment. If each bounded face of a straight line embedded graph $G$ is convex, and the unbounded face is a complement of a convex set, then the embedding of $G$ is said to be *convex*.

LEMMA 2.3.1. *Let $G$ be a 3-connected graph containing no subdivision of $K_{3,3}$ or $K_5$ as a subgraph. Then $G$ has a convex embedding in the plane.*

PROOF. We prove the lemma by induction on $n = |V(G)|$. If $n = 4$ or 5, the statement is easily verified so we proceed to the induction step and assume that $n \geq 6$.

By Lemma 1.4.4, $G$ contains an edge $e = xy$ such that the graph $G' = G/\!/e$ ($G/e$ with parallel edges removed) is 3-connected. Let $z$ denote the vertex of $G'$ obtained by identifying $x$ and $y$. If $G'$ contains a subdivision of $K_5$ or $K_{3,3}$, it is easy to find such a subdivision in $G$ as well, so we can assume by the induction hypothesis that $G'$ is a convex embedded plane graph. The graph $G' - z$ is 2-connected. By Proposition 2.1.5, the face of $G' - z$ containing the point $z$ is bounded by a cycle $C$ of $G'$. Clearly, $C$ is also a cycle in $G$. Let $x_1, \ldots, x_k$ be the neighbors of $x$, occurring on $C$ in that cyclic order, and let $P_i$ be the segment of $C$ joining $x_i$ and $x_{i+1}$ and not containing any $x_j$, $j \neq i, i+1$, where indices are taken modulo $k$. If all neighbors of $y$ (other than $x$) are contained in one of the paths $P_i$, it is easy to get a convex embedding of $G$ from the convex representation of $G'$: represent $x$ by $z$ and $y$ by a point close enough to $x$. (The case when $z$ is adjacent to the unbounded face of $G'$ has to be considered separately.) On the other hand, if this is not the case, then either $y$ is joined to three or more vertices among $x_1, \ldots, x_k$ in which case $C$ together with $x$ and $y$ determines a subdivision of $K_5$, or else $y$ is joined to vertices $u, v$ on $C$ such that there are indices $i, j$ ($1 \leq i < j \leq k$) such that $x_i, x_j$ are distinct from $u, v$ and $u, x_i, v, x_j$ appear on $C$ in that order. In the latter case, $C$ together with $x$ and $y$ determines a subdivision of $K_{3,3}$ in $G$. This completes the proof.                                    □

An immediate corollary of this result is a theorem of Stein [**St51**] and Tutte [**Tu60**].

THEOREM 2.3.2 (Stein [**St51**], Tutte [**Tu60**]). *Every 3-connected planar graph has a convex embedding in the plane.*

Lemma 2.3.1 proves Kuratowski's theorem for 3-connected graphs. The general case follows from Lemma 2.3.3 below.

LEMMA 2.3.3. *Let $G$ be a graph of order $n \geq 4$ containing no subdivision of $K_5$ or $K_{3,3}$ such that the addition of any edge joining two nonadjacent vertices of $G$ creates such a subdivision. Then $G$ is 3-connected.*

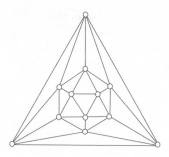

FIGURE 2.2. The icosahedron

PROOF. We use induction on $n = |V(G)|$. The statement is easily verified for $n = 4, 5$ so we assume $n \geq 6$ and proceed to the induction step.

It is an easy exercise to prove that $G$ is 2-connected and that $x$ and $y$ are adjacent if $G - \{x, y\}$ is disconnected. Now suppose that $G$ is not 3-connected. Then it has adjacent vertices $x, y$ such that $G - \{x, y\}$ is disconnected. Then $G = G_1 \cup G_2$ where $G_1$ and $G_2$ have precisely $x$, $y$, and the edge $xy$ in common, and $|V(G_i)| \geq 3$, $i = 1, 2$. It is easy to see that the addition of an edge to $G_i$ ($i = 1$ or 2) creates a subdivision of $K_5$ or $K_{3,3}$ in $G_i$. Hence each $G_i$ is either isomorphic to $K_3$ or it is 3-connected by the induction hypothesis. By Lemma 2.3.1, $G_i$ has a convex embedding in the plane. Let $z_i$ be a vertex distinct from $x, y$ but on the boundary of the same face of $G_i$ in a convex embedding of $G_i$. By assumption, the graph $G + z_1 z_2$ contains a subdivision $K$ of $K_5$ or $K_{3,3}$. If the vertices of $K$ of degree at least 3 are all in $G_1$ (or all in $G_2$), we easily get a subdivision of $K_5$ or $K_{3,3}$ in $G_1$ (or in $G_2$), a contradiction. Also, since $K_5$ and $K_{3,3}$ are 3-connected, one of $V(G_1) \backslash V(G_2)$ or $V(G_2) \backslash V(G_1)$, say $V(G_2) \backslash V(G_1)$, contains precisely one vertex of $K$ of degree more than two. Moreover, the subdivision we are considering is $K_{3,3}$ because $K_5$ is 4-connected. Now, we can find a subdivision of $K_{3,3}$ in the graph obtained from $G_1$ by adding a new vertex and joining it to $x$, $y$, and $z_1$. But since all these neighbors are on the boundary of the same face of $G_1$, this graph is planar, a contradiction.                                    □

A graph $G$ as in Lemma 2.3.3 is said to be *maximal planar*. It is easy to see that planar representations of maximal planar graphs are triangulations of the plane (if $|V(G)| \geq 3$). Conversely, a triangulation with $n$ vertices has $3n - 6$ edges and is therefore maximal planar by Proposition 2.1.6. It follows that the graph of any triangulation of the plane with at least 4 vertices is 3-connected. An example of a planar triangulation is represented in Figure 2.2. That graph is the graph of the icosahedron.

Lemmas 2.3.1 and 2.3.3 also imply the following result due to Wagner [**Wa36**], usually called *Fáry's theorem*.

THEOREM 2.3.4 (Wagner [**Wa36**], Fáry [**Fa48**]). *Every planar graph has a straight line embedding in the plane.*

A simple proof of this result was also found by Bryant [**Br89**].

Extensions of Theorems 2.3.2 and 2.3.4 will be discussed in Section 2.8. The reader is also referred to [**Th84a**].

## 2.4. Characterizations of planar graphs

In this section we present other characterizations of planar graphs. Most of them are rather simple consequences of Kuratowski's theorem. On the other hand, some of them easily imply Kuratowski's theorem.

THEOREM 2.4.1 (Wagner [**Wa37a**]). *A graph $G$ is planar if and only if neither $K_{3,3}$ nor $K_5$ is a minor of $G$.*

PROOF. Assume first that $G$ is not planar. By Kuratowski's theorem, $G$ contains a subdivision of one of the two Kuratowski graphs. Contracting some edges of this subdivision, we get $K_{3,3}$ or $K_5$ as a minor of $G$.

Conversely, let $G$ be planar. It is easily seen that if $H$ is a planar graph and $e \in E(H)$, then $H - e$ and $H/e$ are both planar. Therefore every minor of $G$ is planar. By Corollary 2.1.7, $K_{3,3}$ and $K_5$ are not minors of $G$.  □

Tutte [**Tu75**] gave a condition for a planar graph to have a representation such that two prescribed vertices are incident to a common face, and used this result to derive a new proof of Kuratowski's theorem. Conversely, Tutte's result can be derived from Kuratowski's theorem as shown in Theorem 2.4.3 below.

LEMMA 2.4.2. *Let $C$ be a cycle of a plane graph $G$ and let $B_1, B_2$ be overlapping $C$-bridges of $G$. Then $B_1$ and $B_2$ are embedded in distinct faces of $C$.*

PROOF. If $B_1$ and $B_2$ are both in, say, $\text{ext}(C)$, then we obtain a planar embedding of $K_{3,3}$ or $K_5$ by adding to $C \cup B_1 \cup B_2$ a vertex in $\text{int}(C)$ and joining it to all vertices of $C$. This contradiction to Corollary 2.1.7 proves the lemma.  □

THEOREM 2.4.3 (Tutte [**Tu75**]). *Let $x, y$ be vertices of a planar graph $G$. Then $G$ has an embedding in the plane such that $x$ and $y$ are incident to a common face unless $G$ contains a cycle $C$ separating $x$ and $y$ such that the $C$-bridges of $G$ containing $x$ and $y$, respectively, are overlapping.*

PROOF. If $G$ has no embedding with $x$ and $y$ on a common face, then $G + xy$ is nonplanar and therefore contains a subdivision $H$ of $K_{3,3}$ or $K_5$. Since $G$ is planar, $xy \in E(H)$ and $H - xy$ is planar. It is easily seen that

$H - xy$ contains a cycle $C$ such that $x$ and $y$ belong to overlapping $C$-bridges in $H - xy$, say $B_1, B_2$. By Lemma 2.4.2, $B_1$ and $B_2$ are embedded in different faces of $C$. Therefore $B_1$ and $B_2$ are contained in distinct $C$-bridges in $G$, and these two $C$-bridges overlap since $B_1$ and $B_2$ overlap.  □

Let $G$ be a graph and $C$ a cycle of $G$. The *overlap graph* $O(G, C)$ of $C$ in $G$ has $C$-bridges as its vertices, and two of them are adjacent if they overlap. Recall that, if $C$ is a cycle in a graph $G$, we say that two $C$-bridges $B_1, B_2$ *skew-overlap* if $C$ contains four vertices $a, b, c, d$ in this cyclic order such that $a$ and $c$ are vertices of attachment of $B_1$, and $b$, $d$ are vertices of attachment of $B_2$ on $C$. The spanning subgraph of the overlap graph $O(G, C)$ containing only those edges which correspond to skew-overlapping $C$-bridges is called the *skew-overlap graph* of $C$ in $G$.

If $C$ is a cycle of a plane graph $G$, then by the Jordan Curve Theorem $\mathbb{R}^2 \backslash C$ consists of two components. By Lemma 2.4.2, any two overlapping $C$-bridges lie in distinct faces of $C$. Consequently, the overlap graph of $C$-bridges is bipartite. Tutte [**Tu58**] proved that also the converse statement holds, namely that every nonplanar graph contains a cycle whose overlap graph is nonbipartite. Using Kuratowski's theorem, this last statement was refined by Thomassen [**Th80a**] and further refined by Williamson [**Wi93**] as follows.

THEOREM 2.4.4 (Tutte [**Tu58**], Thomassen [**Th80a**], Williamson [**Wi93**]). *For a graph $G$ the following assertions are equivalent:*

   (a) *$G$ is nonplanar.*
   (b) *$G$ contains a cycle whose overlap graph is nonbipartite.*
   (c) *$G$ contains a cycle whose skew-overlap graph is nonbipartite.*
   (d) *$G$ contains a cycle whose skew-overlap graph contains a cycle of length 3.*

PROOF. Trivially, (d) implies (c), and (c) implies (b). The Jordan Curve Theorem combined with Lemma 2.4.2 shows that (b) implies (a). Therefore, it is sufficient to prove that (a) implies (d). We do this by induction on $m$, the number of edges needed to be removed from $G$ to obtain a subdivision of $K_5$ or $K_{3,3}$. By the theorem of Kuratowski this number is well-defined. If $G$ is a subdivision of $K_5$ or $K_{3,3}$ the statement is easy to verify so we proceed to the induction step. Now, $G$ contains an edge $e$ such that $G - e$ is nonplanar but having a smaller $m$. By the induction hypothesis, $G - e$ has a cycle $C$ whose skew-overlap graph $H$ in $G - e$ contains a cycle $T$ of length 3. Let $H'$ be the skew-overlap graph of $C$ in $G$. Then either $H \subseteq H'$ in which case we are finished, or else $H'$ is obtained from $H$ by identifying two vertices. We may assume that the latter possibility holds and that the two vertices which are identified both belong to $T$. So $e$ joins two vertices $x$ and $y$ belonging to distinct skew-overlapping $C$-bridges, say $R_x$ and $R_y$, respectively, in $G - e$. Then

$C$ contains four vertices $z_1, z_2, z_3, z_4$ (in this order) such that $z_1, z_3$ are in $R_x$ and $z_2, z_4$ are in $R_y$. Let $P_1$ denote the segment of $C$ from $z_2$ to $z_1$ containing $z_3$ and $z_4$. Denote by $P_2$ the other segment of $C$ connecting $z_1$ and $z_2$. We may assume that no intermediate vertex of $P_2$ belongs to $R_x$ or $R_y$. Now, let $C'$ be any cycle containing $P_1$ and $e$. Such a cycle exists since $z_1$ is in $R_x$ and $z_2$ is in $R_y$. Let $e_3$ (respectively $e_4$) be an edge of $R_x$ (respectively $R_y$) containing the vertex $z_3$ (respectively $z_4$). Then the $C'$-bridges containing $P_2$, $e_3$, and $e_4$, respectively, are distinct and pairwise skew-overlapping. This completes the proof. □

Another planarity criterion is due to MacLane [**ML37**]. This characterization uses the consequence of Theorem 2.2.3 that the cycle space of a 2-connected planar graph is generated by the facial cycles. For simplicity we present this result only for 2-connected graphs.

Let $G$ be a graph. A spanning subgraph $D$ of $G$ is an *Eulerian subgraph* of $G$ if every vertex of $D$ has even degree. For example, the *trivial subgraph* (the spanning subgraph without edges) and any cycle (together with some isolated vertices) are Eulerian subgraphs of $G$. If $D_1$ and $D_2$ are Eulerian subgraphs of $G$, then we define the Eulerian subgraph $D_1 + D_2$ of $G$ as the spanning subgraph of $G$ with edge set $E(D_1) + E(D_2)$ (where $+$ denotes the symmetric difference of sets). This operation on Eulerian subgraphs of $G$ is called the *sum*. It is easy to see that the set of all Eulerian subgraphs of $G$ forms a vector space over the field $GF(2)$ with respect to the sum. It is called the *cycle space* of $G$ and it is denoted by $Z(G)$. It is also easy to see that every Eulerian subgraph of $G$ can be expressed as a sum of cycles of $G$. In particular, the cycles (viewed as Eulerian subgraphs) generate $Z(G)$.

Let $G$ be a connected graph and $T$ a spanning tree of $G$. For each edge $e \in E(G) \backslash E(T)$ we let $C(e, T)$ be the unique cycle in $T + e$. (That cycle is called the *fundamental cycle* of $e$ with respect to the spanning tree $T$.) Since $C(e, T)$ is the only fundamental cycle with respect to $T$ which contains the edge $e$, the cycles $C(f, T)$, $f \in E(G) \backslash E(T)$, are linearly independent in $Z(G)$. On the other hand, it is easy to see that every Eulerian subgraph $D$ is the sum of cycles $C(e, T)$, $e \in E(D) \backslash E(T)$. Therefore, the cycles $C(e, T)$, $e \in E(G) \backslash E(T)$, form a basis of the cycle space of $G$. In particular, the dimension of $Z(G)$ is equal to $|E(G)| - |V(G)| + 1$. Euler's formula implies that for a plane graph $G$, this number equals the number of faces of $G$ decreased by one. Also, by Theorem 2.2.3, the facial cycles generate the cycle space of $G$ since every cycle $C$ in $G$ is the sum of all facial cycles in $\overline{\text{int}}(C)$.

We say that a collection **B** of cycles of a graph $G$ is a *2-basis* of $G$ if **B** is a basis of the cycle space $Z(G)$ and every edge of $G$ is contained in at most two cycles of **B**.

THEOREM 2.4.5 (MacLane [**ML37**]). *A 2-connected graph G has a 2-basis if and only if it is planar. In this case, any 2-basis of G consists of all facial cycles, except one, of some planar embedding of G.*

PROOF. First we show that $K_{3,3}$ and $K_5$ do not admit a 2-basis. The dimension of $Z(K_{3,3})$ is four. If four cycles of $K_{3,3}$ constitute a 2-basis, then at least seven edges of $K_{3,3}$ would have to appear twice in these cycles since each cycle of $K_{3,3}$ contains at least 4 edges. But then the sum of these four cycles contains at most two edges. The only Eulerian graph with at most two edges is the subgraph with no edges, so our four original cycles are not linearly independent, a contradiction. A similar proof shows that $K_5$ does not have a 2-basis.

Since $K_{3,3}$ and $K_5$ do not have a 2-basis, their subdivisions cannot have 2-basis as well. To show that no nonplanar graph has a 2-basis it suffices to prove, by Kuratowski's theorem, that if $G$ has a 2-basis, then each edge-deleted subgraph of $G$ has one. Assume now that $G$ has a 2-basis **B**, and let $e \in E(G)$. If $e$ is contained in none, or in exactly one cycle $C \in \mathbf{B}$, then **B**, or $\mathbf{B} \backslash \{C\}$ (respectively) is a 2-basis of $G - e$. If $e$ is in two cycles from the 2-basis, say in cycles $C, C' \in \mathbf{B}$, let $C_1, C_2, \ldots, C_r$ be edge-disjoint cycles of $G$ such that $C + C' = C_1 + C_2 + \cdots + C_r$. Then $(\mathbf{B} \backslash \{C, C'\}) \cup \{C_1, C_2, \ldots, C_r\}$ contains a 2-basis of $G - e$. This completes the proof of the necessity of MacLane's condition.

Conversely, let $G$ be a 2-connected plane graph. Let $C_1, \ldots, C_r$ be the facial cycles of $G$ and let $C_r$ be the outer cycle. By Euler's formula, $r = |E(G)| - |V(G)| + 2 = 1 + \dim Z(G)$. We have previously observed that $C_1, C_2, \ldots, C_{r-1}$ generate $Z(G)$. Hence they form a basis of $Z(G)$. Since $C_1 + C_2 + \cdots + C_r = 0$, the cycles $C_1, C_2, \ldots, C_r$ minus any one of them form a 2-basis of $G$.

Now, let **B** be an arbitrary 2-basis of a graph $G$. We can easily transform **B** into a 2-basis $\mathbf{B}'$ of the graph $G'$ which is obtained from $G$ by subdividing each edge by inserting a new vertex on the edge. We construct a new graph $G''$ as follows. For each cycle $C \in \mathbf{B}'$ we add a new vertex $v_C$ to $G'$ and join it to all vertices on $C$. Let $\mathbf{B}''$ be the set of cycles of $G''$ consisting of all 3-cycles containing a vertex $v_C$ and an edge on $C$, where $C \in \mathbf{B}'$. It is easy to see that $\mathbf{B}''$ is a 2-basis of $G''$, and hence $G''$ is planar.

Choose a planar representation of $G''$. For each $C \in \mathbf{B}'$, the graph $G'' - v_C$ is 2-connected. This implies that $C$ is a facial cycle in the planar representation of $G'' - v_C$ induced by the planar representation of $G''$. It follows that **B** is a subset of the set of facial cycles of the planar representation of $G$ obtained from that of $G''$. Finally, **B** contains all but one of the facial cycles because the cardinality of **B** is equal to $\dim Z(G)$.  □

COROLLARY 2.4.6. *A 2-connected plane graph is bipartite if and only if every facial cycle is even.*

PROOF. Clearly, if some facial cycle is odd, then $G$ is not bipartite. Conversely, if $G$ is nonbipartite, it contains an odd cycle $C$. By Theorem 2.4.5 (the easy part), $C$ is the sum of facial cycles and hence one of the facial cycles must be odd.                                                          □

It should be noted that Corollary 2.4.6 also has an easy direct proof by using the Jordan Curve Theorem.

Fournier [**Fo74**] formulated another planarity criterion which was obtained from a characterization of graphic matroids.

THEOREM 2.4.7. *A graph is nonplanar if and only if it contains three cycles $C_1, C_2, C_3$ such that they all have an edge $e$ in common and no edge of one of the cycles is a chord in one of the other two cycles, and whenever we contract the edge set of one of the cycles, then those edges of the other two cycles which are not contracted belong to the same block of the resulting graph.*

PROOF. By Kuratowski's theorem, a nonplanar graph contains a subdivision of $K_{3,3}$ or $K_5$ and in that subdivision it is easy to find cycles satisfying the conditions of the theorem. On the other hand, it is easy to see that a plane graph cannot have three such cycles.                          □

In the 3-connected case there is a more precise version of Theorem 2.4.7, namely Corollary 2.5.4.

An elegant characterization of planar graphs in terms of the dimension of partially ordered sets was discovered by Schnyder [**Sch89**].

Let $P = (X, <)$ be a *partial order*, i.e., $<$ is a transitive, asymmetric and irreflexive binary relation on $X$. A partial order is a *linear order* if every pair of distinct elements of $X$ is comparable. Dushnik and Miller [**DuMi41**] defined the *dimension* $\dim P$ of a partial order $P$ as the minimum number of linear orders on $X$ whose intersection is the relation $<$. If $G$ is a graph, let $X_G = V(G) \cup E(G)$. The incidence relation of $G$ defines a partial order $<_G$ on $X_G$, i.e. $a <_G b$ if and only if $a \in V(G)$, $b \in E(G)$ and $a$ is an endvertex of $b$. We now define the *order-dimension* of $G$ as the dimension of the partial order $(X_G, <_G)$.

THEOREM 2.4.8 (Schnyder [**Sch89**]). *A graph is planar if and only if its order-dimension is at most three. Moreover, for each 3-dimensional representation $\{<_1, <_2, <_3\}$ of $<_G$ there is a straight line embedding of $G$ in the plane, $v \mapsto (v_1, v_2)$ (where $v \in V(G)$ and $(v_1, v_2) \in \mathbb{R}^2$), such that for all $u, v \in V(G)$ we have $u <_i v$ if and only if $u_i < v_i$ $(i = 1, 2)$.*

PROOF. We include only a proof of the "if" part. (For the "only if" part we refer to [**Sch89**] or the monograph of Trotter [**Tr92**, Ch. 6].) Now suppose that the order-dimension of $G$ is at most three, i.e., $<_G = \;<_1 \cap <_2 \cap <_3$ where $<_1, <_2, <_3$ are linear orders. For $v \in V(G)$ and $i = 1, 2$, let $v_i = 2^t$ where $t$ is the ordinal of $v$ with respect to $<_i$, and let $f(v) := (v_1, v_2) \in \mathbb{R}^2$. It suffices to show that $f$ can be extended to a straight line embedding of $G$, i.e., given any two disjoint edges $uv$, $u'v'$ of $G$, the line segments $\overline{f(u)f(v)}$ and $\overline{f(u')f(v')}$ do not intersect. (For, if an edge $\overline{f(v)f(u)}$ is then contained in $\overline{f(v)f(w)}$, then either $u$ has degree 1, or else $u$ has degree 2 and is joined to $v$ and $w$. Then we move $f(u)$ a little.) Assume that $u$ is the maximum of $\{u, v, u', v'\}$ with respect to $<_1$. Since $<_G = <_1 \cap <_2 \cap <_3$ and $<_1$, $<_2$, $<_3$ are linear orders on $X_G$, we have $v >_i u'v'$, $u' >_j uv$, $v' >_k uv$ for some choice of $i, j, k \in \{1, 2, 3\}$. Since $uv >_j u$ and $uv >_k u$, we have $j \neq 1$, $k \neq 1$ by $<_1$–maximality of $u$. Assuming $\overline{f(u)f(v)} \cap \overline{f(u')f(v')} \neq \emptyset$, a similar argument shows that $i \neq 1$ (since otherwise either $v' >_1 v$ or $u' >_1 v$). Clearly, $i \neq j, k$. Hence we can assume that $i = 2$, $j = k = 3$ or $i = 3$ and $j = k = 2$. The latter case is easy to dispose of, so we assume the former. Then $v >_2 u'v'$. This implies $v_2 > u_2'$ and $v_2 > v_2'$. Since $\overline{f(u)f(v)} \cap \overline{f(u')f(v')} \neq \emptyset$ we have $v >_2 u$. Summarizing, $u_1 > v_1$, $v_2 > u_2$, $v_2 > u_2'$, $v_2 > v_2'$, $u_1 > u_1'$, $u_1 > v_1'$. Since the two edges intersect, one of $f(u')$, $f(v')$ must lie in the triangle $f(u)$, $f(v)$, $(u_1, v_2)$. Say $f(u')$ does. Then a geometric argument shows that $\frac{u_1}{2} < u_1' < u_1$ or $\frac{v_2}{2} < u_2' < v_2$ which contradicts the definition of $f$.  $\square$

Several other characterizations of planarity are known. Whitney's planarity theorem is presented in Theorem 2.6.5. Using this, Jaeger [**Ja79**] obtained a new planarity criterion. Whitney's theorem also plays an important role in the verification of Rosenstiehl's planarity criterion [**Ro76**] (with a detailed proof in [**RR78**]). De Fraysseix and Rosenstiehl [**FR82, FR85**] derived a characterization of planar graphs using depth-first-search trees. Similar characterizations have been obtained by Archdeacon, Bonnington, and Little [**ABL93**] (cf. also Keir and Richter [**KR96**]), Holton and Little [**HL77**], and Archdeacon and Širáň [**AS98**].

Recently, Colin de Verdière [**CV90, CV87**] (with an English translation of [**CV90**] in [**CV93**]) obtained another interesting characterization of planar graphs.

Let $G$ be a connected graph with vertex set $\{v_1, \ldots, v_n\}$, and let $\mathbf{M}(G)$ be the set of symmetric real $n \times n$ matrices $M = (m_{ij})_{i,j=1}^n$ with the following properties:

(i) If $i \neq j$ and $v_i$ is adjacent to $v_j$ in $G$, then $m_{ij} < 0$.

(ii) If $i \neq j$ and $v_i, v_j$ are not adjacent in $G$, then $m_{ij} = 0$.

(iii) $M$ has exactly one negative eigenvalue (and it has multiplicity 1).

(iv) There is no nonzero symmetric $n \times n$ matrix $X = (x_{ij})$ such that $MX = 0$ and such that $x_{ij} = 0$ whenever $i = j$ or $v_iv_j \in E(G)$.[1]

The graph invariant $\mu(G)$ introduced by Colin de Verdière [**CV90**] is defined as the largest corank of any matrix in $\mathbf{M}(G)$, where the *corank* of a matrix is the dimension of its kernel. As proved in [**CV90**], if $H$ is a

---

[1]Condition (iv) is called the *Strong Arnold Hypothesis* in [**CV90**]. The above formulation is taken from [**HLS99**].

minor of $G$, then $\mu(H) \leq \mu(G)$. It is an interesting exercise to show that $\mu(K_5) = 4$ and $\mu(K_{3,3}) = 4$. Therefore every nonplanar graph $G$ satisfies $\mu(G) \geq 4$, by the theorem of Kuratowski.

THEOREM 2.4.9 (Colin de Verdière [**CV90**]). *A connected graph $G$ is planar if and only if $\mu(G) \leq 3$.*

The proof of Theorem 2.4.9 in [**CV90**] is based on a difficult result from differential geometry. A simpler, combinatorial proof was found by van der Holst [**Ho95**]. Van der Holst, Lovász and Schrijver [**HLS99**] gave a survey on the parameter $\mu$. A related minor-monotone graph parameter $\lambda(G)$ is presented in Schrijver [**Sch97**].

## 2.5. 3-connected planar graphs

3-connected planar graphs have received special attention partly because of their relationship with convex polyhedra (cf. Steinitz' Theorem in Section 2.8). Whitney [**Wh33b**] proved that a 3-connected planar graph $G$ has only one planar representation in the sense that the facial cycles are uniquely determined (see also Theorem 2.5.1). This implies, with the aid of the Jordan–Schönflies Theorem, that, for any two representations $\Gamma_1, \Gamma_2$ of $G$ on the sphere (with isomorphisms $\phi_i : G \to \Gamma_i$, $i = 1, 2$), there exists a homeomorphism $\phi$ of the sphere onto itself taking $\Gamma_1$ onto $\Gamma_2$ such that the restriction of $\phi$ to $V(\Gamma_1)$ is $\phi_2\phi_1^{-1}$ (i.e., the "identity"). In this section we discuss this and similar results.

Recall that a cycle $C$ in a graph $G$ is an *induced cycle* if it has no chords. The cycle $C$ is *nonseparating* if $G - V(C)$ is connected. It is clear that $C$ is induced and nonseparating if and only if there is at most one $C$-bridge in $G$. Tutte [**Tu63**], showed that such cycles play an important role in 3-connected planar graphs.

THEOREM 2.5.1 (Tutte [**Tu63**]). *Let $G$ be a 3-connected planar graph. A cycle $C$ of $G$ is a facial cycle of some planar representation of $G$ if and only if it is induced and nonseparating. In this case, $C$ is a facial cycle in every planar representation of $G$.*

PROOF. By the Jordan Curve Theorem, every induced nonseparating cycle of a plane graph must be facial. Conversely, let $C$ be a facial cycle of a plane 3-connected graph, and assume that it is not an induced nonseparating cycle. Then there are at least two $C$-bridges $H_1$ and $H_2$. Since they both lie in ext($C$), they do not overlap by Lemma 2.4.2. Hence some pair of consecutive vertices of attachment of $H_1$ on $C$ separates $G$, which is a contradiction to the 3-connectedness of $G$. $\qquad\square$

Theorem 2.5.1 implies the aforementioned result of Whitney [**Wh33b**] and hence, by the Jordan-Schönflies Theorem, any isomorphism (not required to be a plane-isomorphism) of 3-connected graphs embedded in the plane such that the outer cycle of the first graph is mapped to the outer cycle of the image, can be extended to a homeomorphism $\mathbb{R}^2 \to \mathbb{R}^2$, and is therefore also a plane-isomorphism.

Induced nonseparating cycles are important also for nonplanar graphs. For example, Tutte [**Tu63**] proved that the induced nonseparating cycles generate the cycle space of an arbitrary 3-connected graph. In fact, Tutte proved the following stronger result.

THEOREM 2.5.2 (Tutte [**Tu63**]). *Let $G$ be a 3-connected graph. Then every edge $e$ of $G$ is contained in two induced nonseparating cycles having only $e$ and its ends in common. Moreover, the induced nonseparating cycles of $G$ generate the cycle space $Z(G)$ of $G$.*

Following [**Th80a**], we shall derive Theorem 2.5.2 from Proposition 1.4.6 combined with Lemma 2.5.3 below.

LEMMA 2.5.3. *Let $G$ be a 3-connected graph with at least 5 vertices. Let $e = xy \in E(G)$ be an edge of $G$, and let $G' = G /\!/ e$. Let $C'$ be an induced nonseparating cycle of $G'$. If the vertex $z \in V(G')$ corresponding to the contracted edge $e$ is not in $C'$, then $C'$ is also an induced nonseparating cycle in $G$. Otherwise, there is an induced nonseparating cycle $C$ of $G$ containing the edges of $C'$ not incident with $z$, and the other edges of $C$ are either adjacent to $e$, or equal to $e$.*

PROOF. If $z \notin V(C')$, then $C'$ is clearly an induced nonseparating cycle of $G$. Assume now that $z \in V(C')$ and denote by $a, b$ the two neighbors of $z$ on $C'$. If both $x$ and $y$ are adjacent to each of $a, b$, then either $x$ or $y$ has a neighbor in $G' - C'$. (Otherwise $\{a, b\}$ would be a separating set since $G$ has at least 5 vertices. But this is a contradiction to the 3-connectedness of $G$.) Suppose this vertex is $y$. Then $G$ contains a cycle using $C' - z$ and the edges $ax, xb$. This cycle is induced and nonseparating. Another possibility is that $x$ but not $y$ is joined in $G$ to both $a$ and $b$. Then $G$ contains a cycle $C$ with vertex set $(V(C')\backslash\{z\})\cup\{x\}$. This cycle is clearly induced. Since $C'$ is induced and nonseparating and $G$ is 3-connected, $C$ is also nonseparating. If none of $x$ and $y$ is adjacent in $G$ to both $a$ and $b$, then $C'$ gives rise to a unique cycle in $G$ containing the edge $e$. This cycle is induced and nonseparating.                                    □

PROOF OF THEOREM 2.5.2. We will use induction on $n = |V(G)|$. If $n \leq 4$ then $n = 4$ and $G = K_4$. The assertion of the theorem is easily verified in this case. Assume now that $n \geq 5$. By Proposition 1.4.6, $G$ contains an edge $e'$ having no ends in common with $e$ such that $G' = G /\!/ e'$ is 3-connected. Clearly $E(G') \subseteq E(G)\backslash\{e'\}$. By the induction hypothesis, the edge $e$ belongs to two induced nonseparating cycles $C_1', C_2'$ in $G'$ having only $e$ and its endvertices in common. By Lemma 2.5.3 there are induced nonseparating cycles $C_1, C_2$ in $G$ which differ from $C_1', C_2'$ only at the vertex $z$ from the lemma. Since $e$ and $e'$ are nonadjacent edges of $G$, $z$

belongs to at most one of $C_1', C_2'$. Therefore $C_1, C_2$ satisfy the conclusion of Theorem 2.5.2.

To prove that the induced nonseparating cycles in $G$ generate $Z(G)$, let $C$ be any cycle in $G$. We shall show that $G$ contains induced nonseparating cycles $C_1, \ldots, C_m$ such that $E(C) = E(C_1) + \cdots + E(C_m)$. As the induced cycles generate $Z(G)$, we may assume that $C$ is induced. By the induction hypothesis, $G'$ (as defined above) has a collection of induced nonseparating cycles $C_1', \ldots, C_m'$ such that $E(C_1') + \cdots + E(C_m') = E(C)$ or $E(C/\!/e')$. Let $C_i$ be the induced nonseparating cycle of $G$ such that $E(C_i) = E(C_i')$ or $E(C_i) = E(C_i') \cup \{e'\}$ for $i = 1, \ldots, m$. Then $E(C)$ is the sum of $E(C_1) + \cdots + E(C_m)$ and triangles containing $e'$. To see this, we observe that $E(C) + E(C_1) + \cdots + E(C_m)$ is Eulerian and contains only edges incident with the ends of $e'$. Hence, $E(C) + E(C_1) + \cdots + E(C_m)$ is is the sum of triangles containing $e'$. Note that since $G/\!/e'$ is 3-connected, all triangles containing $e'$ are nonseparating. This completes the proof. □

We obtain from Theorems 2.5.1, 2.5.2, and 2.4.5 the following characterization of planar graphs, due to Tutte.

COROLLARY 2.5.4 (Tutte [**Tu63**]). *A 3-connected graph is planar if and only if every edge is contained in precisely two induced nonseparating cycles.*

A special case of Kuratowski's theorem for cubic graphs was first discovered by Menger (see [**Kö36**]). In this case, only $K_{3,3}$ is needed. Hall [**Ha43**] and Wagner [**Wa37b**] showed that a 3-connected graph distinct from $K_5$ is planar if and only if it does not contain a $K_{3,3}$-subdivision. This follows immediately from the following observation:

LEMMA 2.5.5. *Let $G$ be a 3-connected graph of order six or more. If $G$ contains a subdivision of $K_5$, then it also contains a subdivision of $K_{3,3}$.*

PROOF. Let $K$ be a $K_5$-subdivision in $G$. Since $G$ is 3-connected and distinct from $K_5$, there is a $K$-bridge $B$ in $G$ whose vertices of attachment are not all in just one path of $K$ corresponding to an edge of $K_5$. It is easy to see that $K \cup B$ contains a subdivision of $K_{3,3}$. □

Kelmans [**Ke84a, Ke84b**] and Thomassen [**Th84b**] independently proved a stronger result.

THEOREM 2.5.6 (Kelmans [**Ke84a, Ke84b**], Thomassen [**Th84b**]). *Every 3-connected nonplanar graph $G$ distinct from $K_5$ contains a cycle with three chords which together form a subgraph of $G$ homeomorphic to $K_{3,3}$.*

Note that this result easily implies Theorem 2.4.4.

Kelmans [**Ke97**] proved that a 3-connected nonplanar graph without 3-cycles in which every separating set of 3 vertices separates a single vertex from the rest of the graph contains a subdivision of $K_{3,3}$ in which five edges forming a spanning tree of $K_{3,3}$ are not subdivided.

If true, the following conjecture in [**Th81**] is a planarity criterion for a special class of graphs that involves only $K_5$.

CONJECTURE 2.5.7. *Let $G$ be a 4-connected graph with $n$ vertices and at least $3n - 6$ edges. Then $G$ is planar if and only if $G$ contains no subdivision of $K_5$.*

Let us recall that a planar graph on $n$ vertices contains at most $3n - 6$ edges. Having that many edges, it is a triangulation and thus 3-connected.

P. D. Seymour (private communication, 1983) made the following conjecture:

CONJECTURE 2.5.8. *A 5-connected graph is planar if and only if it does not contain a subdivision of $K_5$.*

Note that Conjecture 2.5.8 becomes false if we replace "5-connected" by "4-connected" because of $K_{4,4}$.

It can be shown that each of Conjectures 2.5.7 and 2.5.8 implies the following theorem which was conjectured by Dirac [**Di64**] and proved recently by Mader [**Ma99, Ma98**].

**Mader's Theorem.** *Any graph with $n$ vertices and at least $3n - 5$ edges contains a subdivision of $K_5$.*

While Conjecture 2.5.7, if true, characterizes the 4-connected maximal planar graphs, the 4-connected planar graphs which are not maximal planar are characterized by a linking property found by Jung [**Ju70**]. We say that a graph $G$ is *2-linked* if, for any four distinct vertices $x_1, x_2, y_1, y_2 \in V(G)$, $G$ contains disjoint paths $P_1, P_2$ connecting $x_1, y_1$ and $x_2, y_2$, respectively.

THEOREM 2.5.9 (Jung [**Ju70**]). *A 4-connected graph $G$ is 2-linked if and only if $G$ is nonplanar or maximal planar.*

More generally, Seymour [**Se80**] and Thomassen [**Th80b**] independently characterized the graphs which are not 2-linked. This characterization results in a polynomial algorithm for this problem. The algorithmic part was also done by Shiloach [**Sh80**].

## 2.6. Dual graphs

Parsons [**Pa71**] showed that Whitney's planarity criterion in terms of graph duality (cf. Theorem 2.6.5) follows from Kuratowski's theorem. (See also Thomassen [**Th80a**].) We shall introduce graph duality in a concise way and give a short direct proof of Whitney's planarity criterion. Although this is not a deep result, it is important in that it provides an important link between graph theory and matroid theory. Dual graphs are also useful for describing embeddings.

Let $G$ be a connected multigraph. A set $E \subseteq E(G)$ of edges of $G$ is *separating* if $G - E$ is disconnected. If $V(G) = A \cup B$ where $A$ and $B$ are nonempty disjoint sets, then the set $E$ of edges between $A$ and $B$ is a *cut*. A cut $E$ (or a separating edge set) is *minimal* if no proper subset of $E$ is a cut (respectively, a separating set). Clearly, any cut is separating. The converse is not true. But, it is an easy exercise to show, that if $E$ is a minimal separating edge set, then $G - E$ has precisely two components with vertex sets $A$ and $B$, say, and $E$ is the set of edges between $A$ and $B$. In particular, $E$ is a cut. It follows that:

PROPOSITION 2.6.1. *The minimal cuts in a connected multigraph $G$ are precisely the minimal separating edge sets in $G$.*

If $e_1, e_2$ are edges in a cycle $C$ in $G$, then we can extend $C - e_1$ to a spanning tree $T$ of $G$. If $A$ and $B$ are the vertex sets of the two components of $T - e_2$, then the edges of $G$ between $A$ and $B$ form a minimal cut containing $e_1$ and $e_2$. So, any two edges in the same block of $G$ are contained in a minimal cut. Conversely, if $e_1$ and $e_2$ are contained in a minimal cut $E$, and we let $Q$ denote the block containing $e_1$, then $E \cap E(Q)$ is a separating set. By minimality, $E \cap E(Q) = E$. In particular, $e_1$ and $e_2$ both belong to $Q$. So, we have:

PROPOSITION 2.6.2. *Two edges $e_1$ and $e_2$ in a connected multigraph $G$ belong to a minimal cut in $G$ if and only if $e_1$ and $e_2$ are in the same block of $G$.*

Now, let $G$ and $G^*$ be connected multigraphs and let $\varphi : E(G) \rightarrow E(G^*)$ be a bijection. We denote $\varphi(e)$, $e \in E(G)$, by $e^*$, and for $E \subseteq E(G)$ we write $E^*$ to denote the set $\varphi(E)$. We say that $G^*$ is a *combinatorial dual* of $G$ if, for each set $E \subseteq E(G)$, $E$ is the edge set of a cycle in $G$ if and only if $E^*$ is a minimal cut in $G^*$. Let $e$ be an edge of $G$ such that $G - e$ is connected (or has an isolated vertex). If $G^*$ is a combinatorial dual of $G$, then it is easy to see that $G^*/e^*$ is a combinatorial dual of $G - e$. More generally:

PROPOSITION 2.6.3. *If $G^*$ is a combinatorial dual of $G$ and $E \subseteq E(G)$ is a set of edges of $G$ such that $G - E$ has only one component containing edges, then $G^*/E^*$ is a combinatorial dual of $G - E$ (minus isolated vertices).*

Proposition 2.6.2 shows that the blocks of $G^*$ are combinatorial duals of the blocks of $G$. Therefore, we shall consider only 2-connected multigraphs in the sequel.

Let $G$ be a 2-connected plane multigraph. Then we define the *geometric dual* $H$ of $G$ as follows. The multigraph $H$ is a plane multigraph that has precisely one vertex in each face of $G$. If $e$ is an edge of $G$, then $H$ has an edge $e^*$ crossing $e$ and joining the two vertices of $H$ in the two faces of $G$ that contain $e$ on the boundary. Moreover, $e^*$ has no other points in common with $G$, and all edges of $H$ are obtained in this way. An example of a plane graph and its geometric dual (shown dotted) is exhibited in Figure 2.3.

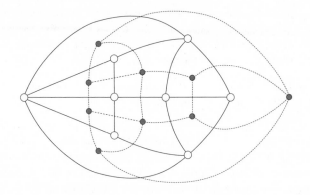

FIGURE 2.3. A plane graph and its geometric dual

We now have:

(A) *If $E \subseteq E(G)$ is the edge set of a cycle in $G$, then $E^*$ is a cut in*
    *$H$.*
(B) *If $E$ is the edge set of a forest in $G$, then $H - E^*$ is connected.*

Property (A) follows from the Jordan Curve Theorem, while (B) follows easily by induction on the number of edges of the forest (by removing an edge incident with an endvertex).

PROPOSITION 2.6.4. *Let $G$ be a 2-connected plane multigraph, and let $H$ be its geometric dual. Then $H$ is a combinatorial dual of $G$. Moreover, $G$ is a geometric dual graph (and hence a combinatorial dual) of $H$.*

PROOF. Since the minimal cuts of $G$ are the minimal separating sets of $G$, (A) and (B) imply that $H$ is a combinatorial dual of $G$. In particular, $H$ is 2-connected if it contains at least three vertices. (Otherwise, $G$ is a cycle and the claims are easy to verify.) To prove that $G$ is a geometric dual of $H$, it suffices to prove that, for each facial cycle $C^*$ in $H$, $G$ has only one vertex in the face $F$ of $H$ bounded by $C^*$. (Clearly, $G$ has no edge inside $F$.) But, if $G$ has two or more vertices in $F$, then some two vertices of $C^*$ can be joined by a simple arc inside $F$ having only its ends in common with $G \cup H$. But, this is impossible by the definition of $H$.  □

Whitney [**Wh33a**] proved that combinatorial duals are geometric duals. This gives rise to another characterization of planar graphs.

THEOREM 2.6.5 (Whitney [**Wh33a**]). *Let $G$ be a 2-connected multigraph. Then $G$ is planar if and only if it has a combinatorial dual. If $G^*$ is a combinatorial dual of $G$, then $G$ has an embedding in the plane such that $G^*$ is isomorphic to the geometric dual of $G$. In particular, also $G^*$ is planar, and $G$ is a combinatorial dual of $G^*$.*

PROOF. By Proposition 2.6.4, it suffices to prove the second part of the theorem. The proof will be done by induction on the number of edges of $G$. If $G$ is a cycle, then any two edges of $G^*$ are in a 2-cycle and hence $G^*$ has only two vertices. Clearly, $G$ and $G^*$ can be represented as a geometric dual pair.

If $G$ is not a cycle, then by Proposition 1.4.2, $G$ is the union of a 2-connected subgraph $G'$ and a path $P$ such that $G' \cap P$ consists of the two endvertices of $P$. By the induction hypothesis and Proposition 2.6.3, $H = G^*/E(P)^*$ is a combinatorial dual of $G'$. By the induction hypothesis, $G'$ and $H$ can be represented as a geometric dual pair, and $G'$ is also a combinatorial dual of $H$. If $e_1, e_2$ are two edges of $P$, then $e_1^*, e_2^*$ are two edges of $G^*$ which belong to a cycle $C^*$ of $G^*$. If $C^*$ has length at least 3, then it is easy to find a minimal cut in $G^*$ containing $e_1$ but not $e_2$. But this is impossible since any cycle in $G$ containing $e_1$ also contains $e_2$. Hence, all edges of $E(P)^*$ are parallel in $G^*$ and join two vertices $z_1, z_2$, say, in $G^*$. Let $z_0$ be the vertex in $H$ corresponding to $z_1, z_2$. The edges in $H$ incident with $z_0$ form a minimal cut in $H$. Let $C$ be the corresponding cycle in $G'$. As $E(C)^*$ separates $z_0$ from $H - z_0$ in $H$, $C$ is a simple closed curve separating $z_0$ from $H - z_0$. In particular, $C$ is facial in $G'$. Let $C_1, C_2$ be the two cycles in $C \cup P$ containing $P$ such that $E(C_i)^*$ is the minimal cut consisting of the edges incident with $z_i$, for $i = 1, 2$. Now we draw $P$ inside the face $F$ of $G'$ bounded by $C$ and represent $z_i$ inside $C_i$ for $i = 1, 2$. This way we obtain a representation of $G^*$ as a geometric dual of $G$.                                                  □

Tutte [**Tu63**] has shown that Theorem 2.6.5 easily follows from MacLane's planarity criterion (Theorem 2.4.5) as follows. Let $G^*$ be a combinatorial dual of $G$ and suppose that $G$ is not a cycle. (Otherwise, the theorem is easy to verify.) Then $G^*$ is 2-connected. Let $Z$ denote the collection of those cycles $C$ in $G$ such that $E(C)^*$ is the set of edges incident with a vertex. Every edge of $G$ is contained in precisely two cycles from $Z$. It is easily seen that $Z$ generates the cycle space of $G$ (because every cut in $G^*$ is a modulo 2 sum of cuts corresponding to some cycles in $Z$). So, the existence of a combinatorial dual $G^*$ of $G$ implies that $G$ has a 2-basis. By MacLane's criterion (Theorem 2.4.5), $G$ is planar. The cycles in the 2-basis are facial cycles of some planar representation of $G$, and this implies that $G^*$ is isomorphic to the geometric dual of this representation.

By Proposition 2.6.2, a combinatorial (and hence a geometric) dual of a block is a block. In particular, if $G$ is a 2-connected plane multigraph which is not a cycle, then its geometric dual is 2-connected. Duality also preserves 3-connectivity.

PROPOSITION 2.6.6. *Let $G$ be a 2-connected multigraph and let $G^*$ be its combinatorial dual. Then $G^*$ is 3-connected if and only if $G$ is 3-connected.*

PROOF. By Theorem 2.6.5, it suffices to show that $G$ is 3-connected whenever $G^*$ is 3-connected. Suppose that this is not the case. If $G$ has a vertex of degree 2, then $G^*$ has parallel edges, a contradiction. So, $G$ has minimum degree at least 3. Then we can write $G = G_1 \cup G_2$ where $G_1 \cap G_2$ consists of two vertices, $E(G_1) \cap E(G_2) = \emptyset$, and each of $G_1, G_2$ contains at least 3 vertices. By Theorem 2.6.5, $G$ is planar. Then $G$ has a facial cycle $C$ such that $C \cap G_i$ is a path $P_i$ for $i = 1, 2$. Clearly, $G/E(C)$ has two edges which are not in the same block. By Proposition 2.6.3 and Theorem 2.6.5, $G^* - E(C)^*$ has two edges which are not in the same block. As $E(C)^*$ is the set of edges incident with a vertex of $G^*$, $G^*$ is not 3-connected. □

Combining Theorem 2.6.5, Proposition 2.6.6 and Theorem 2.5.1 we get:

THEOREM 2.6.7. *If $G$ is a planar 3-connected graph, then $G$ has precisely one combinatorial dual $H$. The combinatorial dual is also a 3-connected graph and can be described as follows: The vertex set of $H$ corresponds to the collection of induced nonseparating cycles in $G$, such that two vertices of $H$ are adjacent if and only if the corresponding cycles in $G$ have an edge in common.*

If $G$ is a planar graph, then two plane representations $\Gamma_1$ and $\Gamma_2$ of $G$ are *equivalent* if there is a homeomorphism of $\mathbb{R}^2$ onto itself taking $\Gamma_1$ onto $\Gamma_2$. If $G$ is 2-connected, this is the same as saying that $\Gamma_1$ and $\Gamma_2$ are plane-isomorphic, i.e., the facial cycles in $\Gamma_1$ are the facial cycles in $\Gamma_2$, and the outer cycle in $\Gamma_1$ is the same as the outer cycle in $\Gamma_2$ (by the Jordan-Schönflies Theorem). We speak of *weak equivalence* if the facial cycles in the two embeddings are the same (with possibly different outer cycles). Now, Theorem 2.6.5 implies that there is a 1-1 correspondence between the weakly nonequivalent plane representations of $G$ and the combinatorial duals of $G$. We leave it to the reader to make this statement precise. (One has to work with edge-labeled dual multigraphs.)

Suppose that $G$ is a 2-connected plane multigraph and $C$ is a cycle of $G$ such that only two vertices, say $v$ and $w$, of $C$ have incident edges that are in ext($C$). Then we define a *flipping* of $G$ (with respect to $C$) as a re-embedding of $G$ such that the embedding in ext($C$) is unchanged and the embedding of $H := G \cap \overline{\text{int}}(C)$ is changed so that the new embedding of $H$ is equivalent with the original one but the clockwise orientations of all the facial cycles are reversed. Moreover, the outer face boundary of $H$ is the same as the outer face boundary of $H$ in the flipped graph. All edges in $H$ are redrawn and we say that they have been *flipped*. Note that the vertices $v$ and $w$ remain fixed. An example of a flipping is shown in Figure 2.4.

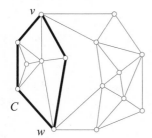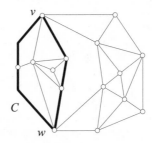

FIGURE 2.4. A flipping

THEOREM 2.6.8. *Let $G$ be a 2-connected plane multigraph and let $e$ be an edge on the outer facial cycle of $G$. Then any plane multigraph $G'$ isomorphic to $G$ is weakly equivalent to an embedding obtained from $G$ by a sequence of flippings such that the edge $e$ is never flipped.*

PROOF. The proof is by induction on the number of edges of $G$. If $G$ is a cycle, there is nothing to prove. Otherwise we may assume that the two ends $u, v$ of $e$ have degree at least three. Let $C_1$ and $C_2$ be the two facial cycles containing $e$ where $C_1$ is the outer cycle. Suppose first that $C_1$ and $C_2$ have only the edge $e$ in common. Let $P_i = C_i - e$, $i = 1, 2$. If $G - \{u, v\}$ has a path from $P_1$ to $P_2$, then we let $H \subseteq G$ be a largest possible subdivision of a 3-connected graph $M$ containing $C_1 \cup C_2$. Let $G'$ be another plane representation of $G$ and let $H'$ be the subgraph of $G'$ corresponding to $H$. By Theorem 2.5.1, $H$ and $H'$ are weakly equivalent. Assume without loss of generality that $H$ and $H'$ are equivalent. If $R$ is an $H$-bridge in $G$, then $M$ has an edge $a \neq e$ such that all vertices of attachment of $R$ are on the path $P_a$ in $H$ corresponding to $a$, by the maximality of $H$. Let $C_a$ be a cycle in $H - e$ containing $P_a$. We apply the induction hypothesis to $C_a$ and all $H$-bridges which have all vertices of attachment in $P_a$. The flippings are chosen such that $E(C_a) \backslash E(P_a)$ is not flipped. Doing this for every edge $a \in E(M) \backslash \{e\}$, we obtain a plane representation equivalent to $G'$.

A similar proof (with $H = C_1 \cup C_2$) works in the case when $G - \{u, v\}$ has no path from $P_1$ to $P_2$. We leave the details to the reader.

Suppose now that $C_1 \cap C_2$ contains a vertex $x$ which is not an end of $e$. Write $C_i = \{e\} \cup P_i \cup Q_i$ where $P_i$ joins $u$ and $x$ ($i = 1, 2$). Then $P_1$

and $Q_2$ are internally disjoint paths, and the subgraph $G_1$ of $G$ bounded by $\{e\} \cup P_1 \cup Q_2$ is 2-connected. The same holds for the subgraph $G_2$ of $G$ bounded by $P_2 \cup Q_1 \cup \{e\}$. Now we apply induction to $G_1$ and $G_1'$ (with $Q_2 \cup \{e\}$ not flipped) and to $G_2, G_2'$ (with $\{e\} \cup P_2$ not flipped). Note that, a flipping of $G_1$ keeping $Q_2 \cup \{e\}$ fixed is also a flipping of $G$. Also note that, after having performed the flippings in $G_1$, we can apply induction to $G_2$ because the bridges of $G_1$ in $G$ have all vertices of attachment in $Q_2$.                                                                                $\square$

The above proof also implies that the number of flippings needed to transform one plane representation of $G$ to any other one is less than the number of edges of $G$.

Let $G$ be a 2-connected graph and $x, y \in V(G)$. Let $H_0, H_1, \ldots, H_s$ be the $\{x, y\}$-bridges and assume that $s \geq 1$. Denote by $x_i$ and $y_i$ the copies of the vertices $x$ and $y$, respectively, in $H_i$ ($0 \leq i \leq s$). Denote by $G'$ the graph obtained from the disjoint union of $H_0, \ldots, H_s$ by identifying the vertices $y_0, x_1, x_2, \ldots, x_s$ into a single vertex, and $x_0, y_1, y_2, \ldots, y_s$ into another vertex. In other words, $G'$ is obtained from $G$ by "twisting" (or "switching") $H_0$ at the vertices $x$ and $y$. See Figure 2.5 for an example. We say that $G'$ is obtained from $G$ by a *2-switching* (also called *Whitney's 2-switching*).

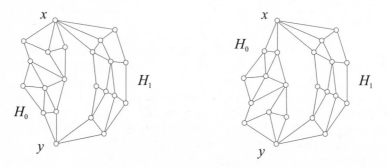

FIGURE 2.5. An example of Whitney's 2-switching

Two graphs $G$ and $G'$ are *cycle-isomorphic* if there is a bijection of their edge sets, $E(G) \to E(G')$, such that a set of edges $E \subseteq E(G)$ is the edge set of a cycle in $G$ if and only if its image in $E(G')$ is the edge set of a cycle in $G'$. If $G'$ is obtained from $G$ by a 2-switching, then it is easy to see that $G$ and $G'$ are cycle-isomorphic. Two graphs are *2-switching equivalent* if one can be obtained from the other by a sequence of 2-switchings. Such graphs are also cycle-isomorphic.

Let $G$ be a 2-connected plane multigraph. It is easy to see that any flipping on $G$ induces a 2-switching on the corresponding geometric duals, and *vice versa*. By Theorem 2.6.8, it follows that any two geometric duals

$H_1$, $H_2$ of $G$ are 2-switching equivalent such that all intermediate graphs are also geometric duals of $G$ with respect to different embeddings of $G$. Theorem 2.6.5 implies that different combinatorial duals of $G$ are cycle-isomorphic. Theorem 2.6.8 is therefore a special case of the following result, known as *Whitney's 2-switching theorem*.

THEOREM 2.6.9 (Whitney [**Wh33b**]). *Let $G$ be a 2-connected graph and $G'$ a graph without isolated vertices. Then $G'$ is cycle-isomorphic with $G$ if and only if $G'$ is 2-switching equivalent with $G$.*

A simple proof was found by Truemper [**Tr80**].

## 2.7. Planarity algorithms

In various practical situations it is important to decide if a given graph $G$ is planar, and if so, to find an embedding of $G$ in the plane. The proof of Kuratowski's theorem in Section 2.3 (and most other proofs as well) is constructive and can be turned into a polynomial time algorithm for testing planarity. However, faster algorithms are known. Hopcroft and Tarjan [**HT74**] developed a linear time planarity testing algorithm. Other efficient algorithms have been discovered by Booth and Lueker [**BL76**], de Fraysseix and Rosenstiehl [**FR82**], and Williamson [**Wi80, Wi84**]. Extensions of these algorithms also produce an embedding whenever the given graph is planar, or find a *Kuratowski subgraph* (a subgraph homeomorphic to $K_5$ or $K_{3,3}$) if the graph is nonplanar [**Wi80, Wi84**] (see also [**CNAO85, Ka90**]).

In this section we present the so-called *vertex addition algorithm* of Lempel, Even, and Cederbaum [**LEC67**]. Booth and Lueker [**BL76**] showed that one can implement it as a linear time algorithm by using an appropriate data structure called *PQ-trees*. We present only the mathematical ideas of the algorithm. For details concerning linear time implementation, the reader is referred to [**BL76**] or [**NC88**].

Let $G$ be a given graph whose planarity or nonplanarity has to be decided. Kuratowski's theorem implies that $G$ is planar if and only if each of its blocks is planar. The blocks can be found in linear time by a simple depth-first-search, see [**AHU74**]. Therefore we may assume henceforth that $G$ is 2-connected.

Let $G$ be a graph of order $n$. A numbering of its vertex set $V(G) = \{v_1, \ldots, v_n\}$ is an *st-ordering* if $v_1$ and $v_n$ are adjacent and for each $i = 2, \ldots, n-1$, $v_i$ has neighbors $v_j$ and $v_l$ such that $j < i < l$. The vertex $v_1$ is then called the *source*, and $v_n$ is the *sink* of the ordering.

LEMMA 2.7.1. *A graph with at least three vertices admits an st-ordering if and only if it is 2-connected. In that case any two adjacent vertices are the source and the sink of some st-ordering of the graph.*

PROOF. The "only if" part is obvious. Assume now that $G$ is 2-connected. We use induction on $|E(G)|$. Let $e_0 = uv$ be an arbitrary edge of $G$. If $G$ is a cycle, the result is obvious. Otherwise, Proposition 1.4.2 shows that $G = H \cup P$ where $H$ is 2-connected and $P$ is a path whose intersection with $H$ are precisely its ends. We may also assume that $e_0 \in E(H)$. By the induction hypothesis, $H$ has an st-ordering with $u$ and $v$ being the source and the sink, respectively. An st-ordering of $G$ is now obtained by inserting the intermediate vertices of $P$ between its ends in the ordering.                                                                □

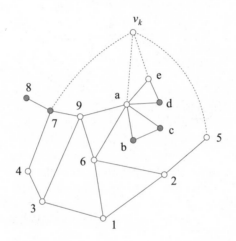

FIGURE 2.6. Adding a vertex

The vertex addition algorithm can be described as follows. Let $G$ be a 2-connected graph and let $v_1, \ldots, v_n$ be an st-ordering of its vertices. We start with the (trivial) embedding of $v_1$ and add successively $v_2, \ldots, v_n$. Let $G_k$ $(1 \leq k \leq n)$ be the subgraph of $G$ induced by $\{v_1, \ldots, v_k\}$ and let $B_k \subseteq V(G_k)$ be the vertices with a neighbor in $\{v_{k+1}, \ldots, v_n\}$. If $G$ is planar, then $G_k$ has an embedding in the plane such that all vertices in $B_k$ are on the outer face. Having such an embedding of $G_{k-1}$, we easily get an embedding of $G_k$ (see Figure 2.6 for an example). The only difficulty is to modify the embedding of $G_k$ such that all vertices of $B_k$ appear on the outer face. There is nothing to do if $v_k$ is adjacent only to one vertex of $G_{k-1}$. Otherwise, let $Q$ be the block of $G_k$ containing $v_k$. Denote by $U \subseteq V(Q)$ the set $B_k \cap V(Q)$ together with all those vertices $v \in V(Q)$ such that $v$ is connected by a path in $G_k - E(Q)$ to some vertex of $B_k$. If we find an embedding of $Q$ such that all vertices of $U$ are on the outer face boundary, we can use the embedding of $G_{k-1}$ and flip the blocks distinct from $Q$ that have a vertex in $B_k$ so that they are in the outer face of

$Q$ and such that all vertices of $B_k$ are on the outer face boundary. The reader may check the example of Figure 2.6 where $B_k$ consists of the black vertices $7, 8, b, c, d$ and $U = \{7, a, d\}$.

By the definition of an st-ordering, the subgraph of $G$ induced by $\{v_{k+1}, \ldots, v_n\}$ is connected. Therefore, if there is no embedding of $Q$ with $U$ on the outer face boundary, the graph $G$ is not planar.

## 2.8. Circle packing representations

Let $G$ be a plane graph. A *circle packing* (abbreviated CP) of $G$ is a set of circles $\{C_v \mid v \in V(G)\}$ in the plane such that:

(i) The interiors of the circles $C_v$, $v \in V(G)$, are pairwise disjoint open discs.

(ii) $C_u$ and $C_v$ $(u, v \in V(G))$ intersect if and only if $uv \in E(G)$.

(iii) By putting vertices $v \in V(G)$ in the centers of the corresponding circles $C_v$ and embed every edge $uv \in E(G)$ as a straight line segment joining $u$ and $v$ through $C_u \cap C_v$, we get a plane representation of $G$ which is equivalent to $G$ (where "equivalent" is defined after Theorem 2.6.7).

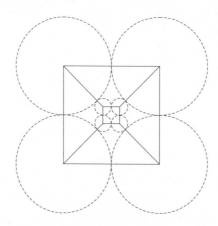

FIGURE 2.7. A CP representation of the 3-cube

Given a CP of $G$, the straight line representation of $G$ defined in (iii) is said to be a *circle packing representation* of $G$ (or just a *circle packing* of $G$). A circle packing representation of the graph of the 3-cube is shown in Figure 2.7.

Jackson and Ringel [**JR84, Ri85**] conjectured that every plane graph admits a circle packing representation. They used the term "coin representation" and the problem has also become known as *Ringel's coin problem*. However, as Sachs [**Sa94**] points out in his survey on this problem, it was solved already by Koebe [**Ko36**] who obtained the existence of circle packing representations as a corollary of a general theorem on conformal mapping of "contact domains". Section 2.9 shows that the relation between circle packings of graphs and conformal mappings is very strong.

In this section we present the result of Brightwell and Scheinerman [**BS93**] that every 3-connected planar graph and its dual have simultaneous straight line embeddings in the plane such that only dual pairs of edges intersect and every such pair is perpendicular. This proves an old conjecture of Tutte [**Tu63**]. This result is a corollary of the Primal-Dual Circle Packing Theorem 2.8.8. Another by-product of this result is Steinitz' Theorem [**St22**] which characterizes the graphs of the convex 3-dimensional polyhedra as the 3-connected planar graphs.

It is convenient to consider circle packings in the *extended plane* (the plane together with a point $\infty$ which we call *infinity*), and a circle packing may contain a special circle, denoted by $C_\omega$, which behaves differently. Instead of (i), we require that none of the circles intersects the exterior of $C_\omega$. We call $C_\omega$ a circle *centered at infinity*. To get the corresponding CP representation in (iii), each edge from a vertex $v$ to the vertex of $C_\omega$ is represented by the half-line from the center of $C_v$ through $C_v \cap C_\omega$ (towards infinity). See Figure 2.8 for an example of a CP representation with a circle centered at infinity.

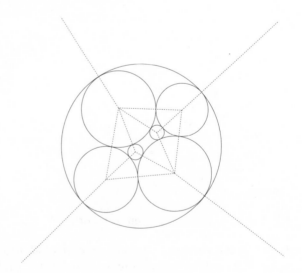

FIGURE 2.8. A CP with a circle centered at infinity

Let us view $\mathbb{R}^2$ as the complex plane $\mathbb{C}$ and the extended plane as $\mathbb{C}^* = \mathbb{C} \cup \{\infty\}$. Suppose that we have a transformation $w : \mathbb{C} \to \mathbb{C}$. It is well-known and easy to see that $w$ preserves circles (i.e., for every circle $C$ in the plane, $w(C)$ is a circle) if and only if $w$ can be expressed in the form:

$$w(z) = az + b \quad \text{or} \quad w(z) = a\bar{z} + b$$

where $a \neq 0$ and $b$ are complex numbers. In the extended plane $\mathbb{C}^*$ there are more general maps which preserve circles if we think of a line as a circle through $\infty$. Consider transformations $w : \mathbb{C}^* \to \mathbb{C}^*$ of the following form:

$$w(z) = \frac{az + b}{cz + d}, \qquad ad - bc \neq 0,$$

where $w(\infty) = a/c$ if $c \neq 0$ and $w(\infty) = \infty$ if $c = 0$. Also, $w(-d/c) = \infty$. These maps are called *fractional linear transformations* or *Möbius transformations*. It is well known (and easy to see) that every fractional linear transformation maps circles and lines to circles and lines in $\mathbb{C}^*$ (lines in $\mathbb{C}^*$ correspond to usual lines in the plane together with the point $\infty$). Every circle that does not contain the point $z = -d/c$ is mapped by $w$ onto a circle. Therefore every fractional linear transformation maps a CP onto another CP if $z = -d/c$ does not lie on any of the circles in the CP. If $z = -d/c$ is the center of a circle in a CP, then the transformed CP has that circle centered at infinity. An immediate corollary is the following:

LEMMA 2.8.1. *If a graph $G$ has a circle packing representation and $v$ is a vertex of $G$, then there is a CP representation of $G$ such that the circle corresponding to $v$ is centered at infinity.*

Let $G$ be a connected plane graph and $G^*$ its geometric dual. A *primal-dual CP* (abbreviated PDCP) of $G$ is a pair of simultaneous circle packings (in the extended plane) of $G$ and of $G^*$, respectively, such that for any dual pair of edges $e = uv \in E(G)$ and $e^* = u^*v^* \in E(G^*)$, the circles $C_u$ and $C_v$ corresponding to $e$ touch at the same point as the circles $C_{u^*}, C_{v^*}$ corresponding to $e^*$, and the line through the centers of $C_u$ and $C_v$ is perpendicular to the line through the centers of $C_{u^*}$ and $C_{v^*}$. We assume that the circle in a PDCP of $G$ and $G^*$ corresponding to the unbounded face of $G$ is centered at infinity.

Our aim is to show that every 3-connected plane graph $G$ admits a PDCP. This immediately yields simultaneous straight line representations of $G$ and its dual graph (with the vertex of $G^*$ corresponding to the unbounded face of $G$ at infinity) such that every pair of dual edges are perpendicular (cf. Theorem 2.8.10).

We need some auxiliary results. We assume in this section that $G$ is a 2-connected plane graph. We define the *vertex-face graph*[2] of $G$, $\Gamma = \Gamma(G)$, as the plane graph obtained as follows. The vertices of $\Gamma(G)$ correspond to vertices and faces of $G$, i.e., $V(\Gamma) = V(G) \cup V(G^*)$. The vertices of $G^*$ are obtained by selecting a point in each face of $G$. The edges in $\Gamma$ correspond to vertex-face incidence in $G$ and are embedded in the plane so that only their endpoints are in $G \cup G^*$. Then $\Gamma$ is bipartite and all

---

[2]The vertex-face graph is also known in the literature as the *radial graph*.

faces of $\Gamma$ are bounded by 4-cycles. The vertex-face graph of the graph of the 3-prism is represented in Figure 2.9. White vertices correspond to vertices of the 3-prism and black vertices to its faces.

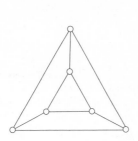 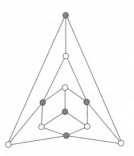

FIGURE 2.9. The 3-prism and its vertex-face graph

LEMMA 2.8.2. *Let $G$ be a 2-connected plane graph with at least 4 vertices and let $\Gamma$ be its vertex-face graph. Then the following assertions are equivalent:*

(a) *$G$ is 3-connected.*

(b) *Every 4-cycle in $\Gamma$ is facial.*

(c) *For every proper subset $S \subset V(\Gamma)$ that contains at least 5 vertices of $\Gamma$ we have[3]*

$$2|S| - |E(\Gamma(S))| \geq 5. \tag{2.3}$$

PROOF. If a 4-cycle in $\Gamma$ is nonfacial, then any two opposite vertices of the cycle separate $G$ or $G^*$. Hence (a) $\Rightarrow$ (b). If $G$ has a separating set $\{x, y\}$, then $\{x, y\}$ together with two faces whose boundaries intersect distinct components of $G - \{x, y\}$ form a nonfacial 4-cycle in $\Gamma$. Hence (b) $\Rightarrow$ (a).

By Proposition 2.1.6, $2|S| - |E(\Gamma(S))| \geq 4$ with equality if and only if $\Gamma(S)$ is a quadrangulation. So, if we have equality, one of the facial 4-cycles in $\Gamma(S)$ is nonfacial in $\Gamma$. Conversely, if $\Gamma$ has a nonfacial 4-cycle $C$, we obtain equality by letting $S$ be $V(C)$ together with the vertices in the interior or the exterior of $C$. This shows that (b) is equivalent to (c). □

In the sequel small Greek letters (e.g., $\nu$ or $\tau$) will be used for points in $\mathbb{C}^*$ which are vertices of $\Gamma$. We assume that the vertex of $\Gamma$ corresponding to the unbounded face of $G$ is the point at infinity and we denote it by $\omega$. The vertex-deleted subgraph $\Gamma - \omega$ will be denoted by $\Gamma'$. Given a PDCP $\{C_\nu \mid \nu \in V(\Gamma)\}$ of $G$, we will denote by $r_\nu$ the radius of $C_\nu$, $\nu \in V(\Gamma')$. It will be assumed that the circle $C_\omega$ of $\omega$ is centered at infinity and that $r_\omega$ is a negative number equal, in absolute value, to the radius of $C_\omega$.

---

[3]Recall that $\Gamma(S)$ is the subgraph of $\Gamma$ induced by $S$.

LEMMA 2.8.3. *Let* $r_\nu$, $\nu \in V(\Gamma)$, *be the radii of a PDCP of G. If* $\nu \in V(\Gamma')$ *and* $\nu\omega \notin E(\Gamma)$, *then*

$$\sum_{\substack{\tau \\ \nu\tau \in E(\Gamma)}} \operatorname{arctg} \frac{r_\tau}{r_\nu} = \pi. \tag{2.4}$$

*Let* $v_1, \ldots, v_k$ *be the vertices of* $\Gamma$ *such that* $v_i\omega \in E(\Gamma)$, $i = 1, \ldots, k$, *and let* $\alpha_i = 2\sum_\tau \operatorname{arctg}(r_\tau/r_{v_i})$ *where the sum is over all neighbors* $\tau$ *of* $v_i$ *in* $\Gamma'$. *Then*

$$0 < \alpha_i < \pi \quad (1 \le i \le k) \qquad and \qquad \sum_{i=1}^{k} \alpha_i = (k-2)\pi. \tag{2.5}$$

PROOF. We assume that $\nu$ is a vertex of $G$. (The proof is similar if $\nu$ is in $G^*$.) If $\nu$ is not a neighbor of $\omega$ (in $\Gamma$), then $C_\nu$ is the inscribed circle of the polygon in $G^*$ whose vertices are the neighbors (in $\Gamma$) of $\nu$. In Figure 2.10, one of these neighbors is shown. The angle $\alpha$ in Figure 2.10 equals $\operatorname{arctg}(r_\tau/r_\nu)$. This implies (2.4) since twice the sum of all such angles is equal to $2\pi$.

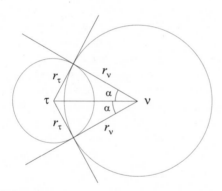

FIGURE 2.10. The angle of $\tau$ at $\nu$

If $v_i$ is adjacent to $\omega$, then $\alpha_i$ is the angle at $v_i$ of the outer facial cycle of $G$. Since $C_\omega$ is inscribed in that polygon, (2.5) holds. □

Suppose that we have simultaneous CP representations of $G$ and $G^* -$ $\omega$ such that for each edge $\nu\tau \in E(\Gamma')$, the circles $C_\nu$ and $C_\tau$ cross at the right angle. Then we say that we have a *weak PDCP*. We shall show that the existence of positive numbers $r_\nu$, $\nu \in V(\Gamma')$, satisfying (2.4) and (2.5) is sufficient for the existence of a weak PDCP. Suppose now that such "radii" $r_\nu$, $\nu \in V(\Gamma)$, exist. For each face $\Phi = \nu x \tau y$ of $\Gamma$, the radii uniquely determine the shape and the size of a quadrangle $Q(\Phi)$ corresponding to

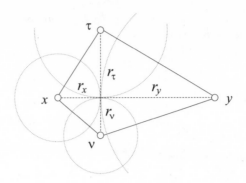

FIGURE 2.11. The quadrilateral $Q(\nu x \tau y)$

$\Phi$ in a possible PDCP representation of $G$ with given radii. Suppose that $\nu, x, y \neq \omega$. Then the angle of $Q(\Phi)$ at $\nu$ is equal to

$$\alpha(\Phi, \nu) = \operatorname{arctg} \frac{r_x}{r_\nu} + \operatorname{arctg} \frac{r_y}{r_\nu} \qquad (2.6)$$

and the length of the side $\nu x$ is equal to $\sqrt{r_\nu^2 + r_x^2}$. See Figure 2.11. If $\tau = \omega$, then $Q(\Phi)$ is a quadrangle with two bounded and two unbounded sides but the triangle $\nu x y$ is uniquely determined. By the angle condition (2.4), the quadrangles $Q(\Phi)$ fit together around each vertex $\nu \neq \omega$.

Next we prove that a locally plane representation of $\Gamma$ as obtained above by pasting together the quadrilaterals $Q(\Phi)$ determines a tiling of the entire plane. For this we need some auxiliary results on graphs drawn in the plane which may be of independent interest.

First we prove a special case of an elementary fact about covering spaces.

LEMMA 2.8.4. *Let $f : \mathbb{R}^2 \to \mathbb{R}^2$ be a covering map, i.e., $f$ is continuous, onto, and for each $p \in \mathbb{R}^2$ there exist open neighborhoods $U$ and $W$ of $p$ and $f(p)$, respectively, such that the restriction of $f$ to $U$ is a homeomorphism of $U$ onto $W$. Suppose further that the set of points $q$ such that $|f^{-1}(f(q))| > 1$ is bounded. Then $f$ is a homeomorphism.*

PROOF. The set $S$ of points $q$ such that $|f^{-1}(q)| \geq 2$ is clearly open. Also, the set of points $q$ such that $|f^{-1}(q)| = 1$ is easily proved to be open. This is possible only if $S$ is empty. □

The conclusion of Lemma 2.8.4 can be derived even without the assumption that $\{q \in \mathbb{R}^2 \mid |f^{-1}(f(q))| > 1\}$ is bounded (since $\mathbb{R}^2$ is simply connected, see e.g. Massey [**Ma67**]).

PROPOSITION 2.8.5. *Let $G$ be a 2-connected plane graph with polygonal edges. Let $H$ be a drawing of $G$ in the plane (possibly with edge* *crossings) such that all edges of H are polygonal arcs. Suppose further that:*

   (i) *For each vertex x of G, the edges incident with x in H are pairwise noncrossing and leave x in the same clockwise order as in G.*

  (ii) *Each facial cycle in G corresponds to a simple closed curve in H.*

 (iii) *If C is a facial cycle bounding a bounded face in G, and e is an edge of G leaving C, then the first segment of e is in the exterior of C in H.*

*Then H is a plane representation of G, i.e., H has no edge crossings.*

PROOF. Let $f : G \to H$ be an isomorphism. We extend $f$ to a continuous map of the point set of $G$ onto the point set of $H$ such that $f$ is 1-1 on each edge of $G$. Next we extend $f$ to a continuous map from $\mathbb{R}^2 \to \mathbb{R}^2$ using the Jordan-Schönflies Theorem as follows. Let $C_0$ denote the outer cycle of $G$. For every facial cycle $C \neq C_0$ of $G$ we use (ii) and the Jordan-Schönflies Theorem to extend $f$ to $\mathrm{int}(C)$ such that the restriction of the new map (which we also call $f$) to $\overline{\mathrm{int}}(C)$ is a homeomorphism onto $\overline{\mathrm{int}}(f(C))$. If $p$ is a point in $\mathrm{int}(C_0)$, then by (i) and (iii), $f(p)$ is an interior point in the compact set $f(\overline{\mathrm{int}}(C_0))$. Hence the boundary of $f(\overline{\mathrm{int}}(C_0))$ is a subset of $f(C_0)$. This implies that either $f(\overline{\mathrm{int}}(C_0)) = \overline{\mathrm{int}}(f(C_0))$ or $f(\overline{\mathrm{int}}(C_0)) = \overline{\mathrm{ext}}(f(C_0))$. As $f(\overline{\mathrm{int}}(C_0))$ is compact, the former equality holds. By the Jordan-Schönflies Theorem, $f$ can be extended to $\mathbb{R}^2$ such that the restriction to $\overline{\mathrm{ext}}(C_0)$ is a homeomorphism onto $\overline{\mathrm{ext}}(f(C_0))$. By Lemma 2.8.4, $f$ is a homeomorphism of $\mathbb{R}^2$ onto $\mathbb{R}^2$. In particular, $H$ is a plane representation of $G$. □

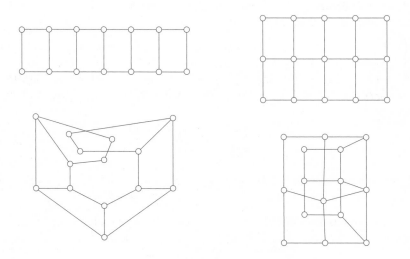

FIGURE 2.12. Locally bad drawings

Figure 2.12 shows that just a slight weakening of conditions (i)–(iii) of Proposition 2.8.5 does not result in the conclusion of that proposition.

LEMMA 2.8.6. *Let $G$ be a 3-connected plane graph and $\Gamma = \Gamma(G)$ its vertex-face graph. If there are positive numbers $r_\nu$, $\nu \in V(\Gamma')$, such that (2.4)–(2.5) are satisfied, then there exists a weak PDCP of $G, G^*$ with radii $r_\nu$, $\nu \in V(\Gamma')$ and with the same local clockwise orientations as in $G$, $G^*$.*

PROOF. For every vertex $\nu \in V(\Gamma')$, we shall determine a point $\bar\nu$ in $\mathbb{R}^2$ such that the circles $C_\nu$ with center $\bar\nu$ and radius $r_\nu$ ($\nu \in V(\Gamma')$) form the desired weak PDCP. We start with a vertex $\tau_0$ of $\Gamma'$ and draw any edge $\tau_0\tau_1$, its length being equal to $\sqrt{r_{\tau_0}^2 + r_{\tau_1}^2}$ (cf. Figure 2.10). Then the position of each neighbor of $\tau_0$ is uniquely determined. So, having drawn $\bar\tau_0$ and $\bar\tau_1$ we consider, for each vertex $\tau$ in $\Gamma'$, a path from $\tau_0$ to $\tau$ and thus get a position for $\bar\tau$. Clearly, the position for $\tau_0$ does not change if we walk from $\tau_0$ to $\tau_0$ along a facial 4-cycle or, more generally, along any cycle $C$ (as can be proved by induction on the number of facial 4-cycles in $\overline{\text{int}}(C)$), or along any closed walk from $\tau_0$ to $\tau_0$. Hence, we get the same point $\bar\tau$ if we consider another path from $\tau_0$ to $\tau$. Also, we get the same drawing of $\Gamma'$ if we start with any other edge instead of $\tau_0\tau_1$ and repeat the construction. It follows that (2.4) is satisfied at every vertex. It only remains to show that the drawing of $\Gamma'$ has no edge crossings. But this follows from Proposition 2.8.5. (Note that we need (2.5) in order to show that the outer facial cycle of $\Gamma'$ satisfies (ii).)                       □

Lemma 2.8.7 below shows that for any 3-connected plane graph $G$, there exist positive real numbers associated with the vertices of $\Gamma = \Gamma(G)$ such that (2.4)–(2.5) are satisfied.

LEMMA 2.8.7. *Let $G$ be a 3-connected plane graph with outer cycle $C = v_1v_2\ldots v_k$. Let $\alpha_1, \alpha_2, \ldots, \alpha_k$ be real numbers such that $0 < \alpha_i < \pi$ ($i = 1, 2, \ldots, k$) and $\alpha_1 + \cdots + \alpha_k = (k-2)\pi$. Then there are positive numbers $r_\nu$, $\nu \in V(\Gamma')$, such that (2.4) holds for $\nu \neq v_1, \ldots, v_k$ and, for each $i = 1, \ldots, k$,*

$$2 \sum_{v_i\tau \in E(\Gamma')} \operatorname{arctg} \frac{r_\tau}{r_{v_i}} = \alpha_i \tag{2.7}$$

*where the summation is taken over all neighbors $\tau$ of $v_i$ in $\Gamma'$. The numbers $r_\nu$, $\nu \in V(\Gamma')$, are unique up to a multiplicative constant.*

PROOF. Suppose that we have a list of positive numbers, $r = (r_\nu \mid \nu \in V(\Gamma'))$. For each $\nu \in V(\Gamma') \setminus \{v_1, \ldots, v_k\}$ we define

$$\vartheta_\nu = \vartheta_\nu(r) = \sum_{\nu\tau \in E(\Gamma)} \operatorname{arctg} \frac{r_\tau}{r_\nu} - \pi.$$

Also, we define for $i = 1, \ldots, k$

$$\vartheta_{v_i} = \vartheta_{v_i}(r) = \sum_{v_i \tau \in E(\Gamma')} \operatorname{arctg} \frac{r_\tau}{r_{v_i}} - \tfrac{1}{2}\alpha_i \,.$$

The number

$$\mu(r) = \sum_{\nu \in V(\Gamma')} \vartheta_\nu^2 \tag{2.8}$$

is a measure for how far $r$ is from a solution. To prove the theorem, it suffices to see that there are positive numbers $r = (r_\nu)$ such that $\mu(r) = 0$.

We claim that

$$\sum_{\nu \in V(\Gamma')} \vartheta_\nu = 0 \,. \tag{2.9}$$

To see this, we expand the sum of $\vartheta_\nu$ as:

$$\sum_{\nu \in V(\Gamma')} \vartheta_\nu = \sum_{\nu\tau \in E(\Gamma')} \left( \operatorname{arctg} \frac{r_\tau}{r_\nu} + \operatorname{arctg} \frac{r_\nu}{r_\tau} \right)$$

$$-\pi(|V(\Gamma')| - k) - \frac{1}{2} \sum_{i=1}^{k} \alpha_i \,. \tag{2.10}$$

Since $\Gamma$ is a quadrangulation, we have by Proposition 2.1.6(b)

$$2|V(\Gamma')| = |E(\Gamma)| + 2 = |E(\Gamma')| + k + 2 \,. \tag{2.11}$$

Since $\operatorname{arctg}(x) + \operatorname{arctg}(1/x) = \pi/2$ for every $x > 0$, (2.10) and (2.11) clearly imply (2.9).

Let $S$ be a proper nonempty subset of $V(\Gamma')$. Denote by $t$ the number of vertices among $v_1, \ldots, v_k$ that are contained in $S$. By applying (2.3) on the set $S \cup \{\omega\} \subset V(\Gamma)$, we see that

$$2|S| - |E(\Gamma(S))| \geq t + 3 \tag{2.12}$$

if $|S| \geq 4$. It is easy to see that (2.12) holds also in cases when $|S| \in \{2,3\}$ and $t = 0$. If $t > 0$ and $|S| \in \{2,3\}$, then we have:

$$2|S| - |E(\Gamma(S))| \geq t + 2 \,. \tag{2.13}$$

Let $\mathfrak{Q}$ be the set of all sequences $r = (r_\nu \mid \nu \in V(\Gamma'))$ (of radii candidates) such that $0 < r_\nu \leq 1$ and $r_\nu = 1$ if $\vartheta_\nu > 0$ $(\nu \in V(\Gamma'))$. Moreover, we require that $r_\nu = 1$ for some $\nu \in V(\Gamma')$. Clearly, $\mathfrak{Q}$ is nonempty since the sequence with $r_\nu = 1$ for each $\nu \in V(\Gamma')$ belongs to $\mathfrak{Q}$.

Let $m = \inf\{\mu(r) \mid r \in \mathfrak{Q}\}$. We claim that the infimum is attained, i.e., $m = \mu(r)$ for some $r \in \mathfrak{Q}$. Let $r^{(1)}, r^{(2)}, r^{(3)}, \ldots$ be a sequence in $\mathfrak{Q}$ such that $\mu(r^{(i)}) \to m$ as $i \to \infty$. By standard arguments, there is a subsequence such that for each $\nu \in V(\Gamma')$ the corresponding numbers $r_\nu^{(i)}$ converge. We may assume that this holds for the sequence $r^{(1)}, r^{(2)}, r^{(3)}, \ldots$.

Let $S \subseteq V(\Gamma')$ be the set of vertices $\nu$ for which $\lim_{i \to \infty} r_\nu^{(i)} \neq 0$. Suppose that $S \neq V(\Gamma')$. By a calculation similar to that in (2.10) we get

$$\sum_{\nu \in S} \vartheta_\nu(r^{(i)}) = \frac{\pi}{2} |E(\Gamma(S))| - \pi(|S| - t)$$

$$- \frac{1}{2} \sum_{v_j \in S} \alpha_j + \sum_{\nu, \tau} \mathrm{arctg} \frac{r_\tau^{(i)}}{r_\nu^{(i)}} \qquad (2.14)$$

where the last sum is taken over all edges $\nu\tau \in E(\Gamma')$ such that $\nu \in S$ and $\tau \notin S$. By definition of $S$, this latter sum tends to 0 as $i \to \infty$. Therefore $\sum_{\nu \in S} \vartheta_\nu(r^{(i)})$ tends to

$$-\frac{\pi}{2} (2|S| - |E(\Gamma(S))| - t - 2) + \frac{1}{2} \sum_{v_j \in S} (\pi - \alpha_j) - \pi \qquad (2.15)$$

as $i \to \infty$. Since $\pi - \alpha_j > 0$ for $j = 1, \ldots, k$ and $\sum_{j=1}^{k} (\pi - \alpha_j) = 2\pi$, (2.15) implies that $\sum_{\nu \in S} \vartheta_\nu(r^{(i)}) < 0$ if $i$ is large enough and if (2.12) holds. The same is true when we have equality in (2.13) since in that case $t < k$. The remaining case when $|S| = 1$ trivially gives the same conclusion. This result and (2.9) imply that

$$\sum_{\nu \notin S} \vartheta_\nu(r^{(i)}) > 0$$

if $i$ is sufficiently large. But $\vartheta_\nu(r^{(i)}) > 0$ implies that $r_\nu^{(i)} = 1$, a contradiction to the definition of $S$. Hence $S = V(\Gamma')$.

Let $r = \lim_{i \to \infty} r^{(i)}$. Since the functions $\vartheta_\nu$ are continuous, $r \in \mathfrak{Q}$. Now we prove that $m = \mu(r) = 0$. Suppose that this is not the case. Let $S'$ be the set of vertices $\nu$ with $\vartheta_\nu(r) < 0$. By (2.9), $S' \neq V(\Gamma')$ and $S' \neq \emptyset$. Let $r_\nu' = r_\nu$ if $\nu \notin S'$ and let $r_\nu' = \alpha r_\nu$ if $\nu \in S'$, where $\alpha < 1$. If $\alpha$ is close enough to 1 (so that no $\vartheta_\nu(r')$, $\nu \in S'$, becomes positive), then $r' \in \mathfrak{Q}$. Using (2.9) and the definition of $\vartheta_\nu$, it is easy to see that $\mu(r') < \mu(r)$ if $\alpha$ is close enough to 1. This contradicts the minimality of $\mu(r)$.

Suppose now that there are distinct solutions $r$ and $r'$ such that $\max\{r_\nu \mid \nu \in V(\Gamma')\} = \max\{r_\nu' \mid \nu \in V(\Gamma')\} = 1$. Then $\vartheta_\nu(r) = \vartheta_\nu(r') = 0$ for all $\nu \in V(\Gamma')$. Assume without loss of generality that the set $S = \{\nu \mid r_\nu > r_\nu'\}$ is nonempty. Clearly, $S \neq V(\Gamma')$. From (2.14) applied to $r$ and $r'$, respectively, we get

$$0 = \sum_{\nu \in S} \vartheta_\nu(r) - \sum_{\nu \in S} \vartheta_\nu(r') = \sum_{\nu, \tau} \left( \mathrm{arctg} \frac{r_\tau}{r_\nu} - \mathrm{arctg} \frac{r_\tau'}{r_\nu'} \right)$$

where the last sum is taken over all edges $\nu\tau \in E(\Gamma')$ such that $\nu \in S$ and $\tau \notin S$. By definition of $S$, the latter sum is negative, a contradiction. The proof is complete. $\qquad \square$

We now apply Lemmas 2.8.6 and 2.8.7 to show that every 3-connected planar graph admits a PDCP.

THEOREM 2.8.8 (Brightwell and Scheinerman [**BS93**]). *Let $G$ be a 3-connected plane graph. Then $G$ admits a PDCP representation. The PDCP of $G$ is unique up to fractional linear transformations and reflections in the plane.*

PROOF. Suppose first that the outer cycle of $G$ is a 3-cycle. Using the notation in Lemma 2.8.7 we let $\alpha_1 = \alpha_2 = \alpha_3 = \pi/3$ and then apply Lemma 2.8.7. The list $r = (r_\nu \mid \nu \in V(\Gamma'))$ satisfies (2.4)–(2.5). By Lemma 2.8.6, there is a weak PDCP of $G$ with these radii. In particular, $r_{v_1} = r_{v_2} = r_{v_3}$. This implies that this weak PDCP can be extended to a PDCP by adding the circle $C_\omega$. Because of the uniqueness of the radii in Lemma 2.8.7, the resulting PDCP is unique once the three circles $C_1, C_2, C_3$ corresponding to $v_1, v_2, v_3$, respectively, have been prescribed.

Suppose next that the outer cycle of $G$ has length greater than 3. Then either $G$ or $G^*$ has a facial 3-cycle by the proof of Proposition 2.1.6. Now redraw $G$ and $G^*$ (using a fractional linear transformation), and interchange the roles of $G$ and $G^*$ if necessary, so that the outer cycle of $G$ in the new embedding is a 3-cycle. By the previous paragraph the new embedding of $G, G^*$ has a PDCP representation. The inverse of the applied fractional linear transformation gives rise to a PDCP representation of the original pair $G, G^*$.

For any PDCP representation of $G$, there exists a fractional linear transformation (possibly followed by reflection) taking the PDCP representation into one using the prescribed circles $V_1, C_2, C_3$ in the previous paragraph. That proves uniqueness.                                              □

Mohar [**Mo97a**] proved that, given a 3-connected planar graph $G$ and an $\varepsilon > 0$, one can determine the centers and radii of the PDCP of $G$ with precision $\varepsilon$ in time that is bounded by a polynomial in $|V(G)|$ and $\max\{\log(1/\varepsilon), 1\}$.

An immediate corollary of Theorem 2.8.8 is a result of Koebe [**Ko36**] which was independently discovered by Andreev [**An70a, An70b**] and Thurston [**Th78**].

COROLLARY 2.8.9 (Koebe–Andreev–Thurston). *Every plane graph admits a circle packing representation.*

Koebe (and also Andreev and Thurston) also proved that circle packing representations of planar triangulations are unique up to fractional linear transformations.

In Theorem 2.3.2 we proved that every 3-connected planar graph has a convex representation. Since every PDCP representation is convex, Theorem 2.8.8 yields a much stronger result:

THEOREM 2.8.10 (Brightwell and Scheinerman [**BS93**]). *If G is a planar 3-connected graph, then G and its dual G\* can be embedded in the
plane with straight edges and with the outer vertex of G\* at infinity such
that they form a geometric dual pair. Both embeddings are convex and
each pair of dual edges is perpendicular.*

Theorem 2.8.10 was conjectured by Tutte [**Tu63**] who proved that every planar 3-connected graph $G$ and its dual $G^*$ have simultaneous straight
line representations so that only dual pairs of edges cross.

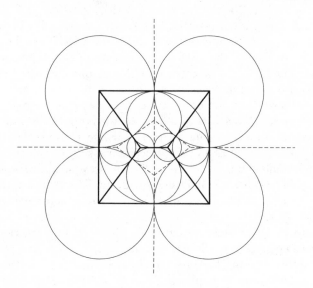

FIGURE 2.13. A PDCP representation of the 3-prism

In Figure 2.13 a PDCP representation of the 3-prism is shown. The
edges of the dual graph are represented by broken lines.

Tutte established a necessary and sufficient condition for a 2-connected
graph $G$ with a given cycle $C$ to have a convex embedding in the plane
such that $C$ bounds the outer face and it is a convex $|V(C)|$-gon. The
condition is that $G$ is a subdivision of a 2-connected graph $H$ such that
every separating set $\{u, v\}$ of $H$ (if any) is contained in the cycle of $H$
corresponding to $C$. This result was generalized by Thomassen [**Th80a**]
to cover also the case where $C$ need not be strictly convex, i.e., it could
be a $k$-gon with $k < |V(C)|$. Thomassen [**Th88**] also extended Tutte's
result of simultaneous straight line representations of a planar graph and
its dual to the 2-connected case. The proof of [**Th80a**] yields a linear time
algorithm for convex drawings of planar graphs. See [**CYN84, NC88**]
for details.

The inverse of the stereographic projection which maps the extended plane onto the unit sphere in $\mathbb{R}^3$ takes every CP in the extended plane into a *spherical circle packing*, a set of circles on the unit sphere in $\mathbb{R}^3$ that has the same properties as are required for circle packings in the plane. Theorem 2.8.8 implies the following geometric result.

THEOREM 2.8.11. *If $G$ is a 3-connected planar graph, then there is a convex polyhedron $Q$ in $\mathbb{R}^3$ whose graph is isomorphic to $G$ such that all edges of $Q$ are tangent to the unit sphere in $\mathbb{R}^3$.*

PROOF. By Theorem 2.8.8, $G$ has a PDCP. The inverse of the stereographic projection maps the circles of the PDCP to circles $\tilde{C}_\nu$, $\nu \in V(\Gamma(G))$, on the unit sphere in $\mathbb{R}^3$. Denote by $\Pi_\nu$ the plane in $\mathbb{R}^3$ that contains $\tilde{C}_\nu$. It is easy to see that the planes $\Pi_\nu$, $\nu \in V(G^*)$, determine a convex polytope whose graph is isomorphic to $G$ and whose edges are tangent to the unit sphere. □

The construction in the above proof gives, at the same time, a convex polyhedron whose graph is $G^*$ and whose edges are tangent to the unit sphere at the same points as their dual edges, and are perpendicular to their dual edges.

We have proved that part of Theorem 2.8.8 implies Theorem 2.8.11. Also, the converse holds as pointed out by Sachs [**Sa94**]. To see this, let $G$ be a 3-connected planar graph and $Q$ a convex polyhedron whose graph is isomorphic to $G$ and whose edges are tangent to the unit sphere $S^2$ in $\mathbb{R}^3$. Then the faces of $Q$ intersect the unit sphere in circles which determine a spherical CP. Let $\tilde{C}_\tau$ be the circle corresponding to the face $\tau$ of $Q$. The stereographic projection maps these circles into a CP representation of the dual graph $G^*$ of $G$ in the plane. At the same time, we get a CP of $G$ as follows. Let $\nu$ be a vertex of $Q$. The cone with apex $\nu$ that is tangent to the unit sphere $S^2$ has a circle $\tilde{C}_\nu$ in common with $S^2$. All such circles $\tilde{C}_\nu$, $\nu \in V(Q) = V(G)$, determine a circle packing of $G$ on $S^2$. Clearly, these circles intersect the corresponding circles $\tilde{C}_\tau$ in the CP of $G^*$ perpendicularly (or not at all). The stereographic projection therefore gives rise to a PDCP of $G$ and $G^*$ in the plane.

Theorem 2.8.11 was conjectured by Grünbaum and Shephard [**GS87**] (cf. also Schulte [**Sch87**]) and independently by Sachs [**Sa94**] (see also Lehel and Sachs [**LS90**]).

Theorem 2.8.11 implies the difficult part of Steinitz' Theorem [**St22**].

**Steinitz' Theorem.** *A graph $G$ is the graph of a convex polytope in $\mathbb{R}^3$ if and only if it is planar and 3-connected.*

For the definition of the graph (1-skeleton) of a polytope and for the easy part of Steinitz' Theorem, the reader is referred to Grünbaum [**Gr67**] or Brøndsted [**Br83**].

The uniqueness of the PDCP in Theorem 2.8.8 implies that every automorphism[4] of a 3-connected planar graph $G$ induces a geometric symmetry of the corresponding polyhedron $Q$ in Theorem 2.8.11. This implies an extension of Steinitz' Theorem.

THEOREM 2.8.12 (Mani [**Ma71**]). *If $G$ is a 3-connected planar graph, then there is a convex polyhedron $Q$ in $\mathbb{R}^3$ whose graph is isomorphic to $G$ such that every automorphism of $G$ induces a symmetry of $Q$.*

Schramm has obtained the following generalizations of (part of) Theorems 2.8.8 and 2.8.11, respectively.

THEOREM 2.8.13 (Schramm [**Sc96**]). *Let $G$ be a planar graph and $(P_v \,;\, v \in V(G))$ a collection of strictly convex compact subsets of the plane with smooth boundary. Then there are numbers $\alpha_v > 0$ $(v \in V(G))$ and points $\rho_v \in \mathbb{R}^2$ $(v \in V(G))$ such that the sets $Q_v = \alpha_v P_v + \rho_v$ $(v \in V(G))$ have pairwise disjoint interiors, and $Q_u, Q_v$ intersect if and only if $u$ and $v$ are adjacent in $G$.*

THEOREM 2.8.14 (Schramm [**Sc92**]). *Let $S$ be a strictly convex compact set in $\mathbb{R}^3$ with nonempty interior. If $G$ is a 3-connected planar graph, then there is a convex polyhedron $P$ in $\mathbb{R}^3$, whose graph is isomorphic to $G$, all of whose edges are tangent to $S$.*

Colin de Verdière [**CV89, CV91**] and Mohar [**Mo97a**] obtained analogues of the PDCP results for graphs on arbitrary surfaces.

H. Harborth (private communication) has raised the following

PROBLEM 2.8.15. *Does every planar graph have a straight line representation such that all edges have integer length?*

Brightwell and Scheinerman [**BS93**] showed that this cannot in general be achieved by a circle packing representation since otherwise it would be possible to trisect an angle of $\pi/3$ by ruler and compass.

## 2.9. The Riemann Mapping Theorem

An open connected set $\Omega \subseteq \mathbb{R}^2$ is *simply connected* if, for every simple closed curve $J \subseteq \Omega$, we have $\mathrm{int}(J) \subseteq \Omega$. Using the proof of the Jordan–Schönflies Theorem presented in Section 2.2, it is not difficult to prove that every open simply connected set $\Omega$ in $\mathbb{R}^2$ is homeomorphic to the open unit disc $\Delta$ in $\mathbb{R}^2$. The Riemann Mapping Theorem says that, if $\Omega$ is bounded, then $\Omega$ is conformally equivalent to $\Delta$, i.e., there exists a homeomorphism $f : \Omega \to \Delta$ which is conformal (analytic). At the International Symposium in Celebration of the Proof of the Bieberbach Conjecture (Purdue University, March 1985), William Thurston conjectured that the conformal mapping of $\Omega$ to the unit disc $\Delta$ can be approximated by manipulating

---

[4]An *automorphism* of a graph is an isomorphism of the graph onto itself.

hexagonal circle packing configurations in $\Omega$. More precisely, let $p$ and $q$ be fixed distinct points in $\Omega$. Consider the standard hexagonal tiling of the plane into hexagons of diameter $1/n$. Let $G_n'$ be the graph which is the union of those hexagons that are in $\Omega$. Let $G_n''$ be the maximal subgraph of $G_n'$ which contains $p$ and $q$ in its hexagons and which has no two adjacent vertices which separate $G_n''$. For $n$ sufficiently large, $G_n''$ exists and is a subdivision of a 3-connected graph $G_n$. By Theorem 2.8.8, the pair $G_n, G_n^*$ has a PDCP such that the vertex of $G_n^*$ in the unbounded face of $G_n$ corresponds to the unit circle (i.e., the boundary of $\Delta$) centered at infinity. We focus on the circle packing of $G_n^*$. An example is indicated in Figure 2.14. It shows a region $\Omega$ with the hexagonal lattice graph in it and the graph $G_n^*$ without its vertex corresponding to the outer face of $G_n$. A circle packing of $G_n^*$ is shown in Figure 2.15.

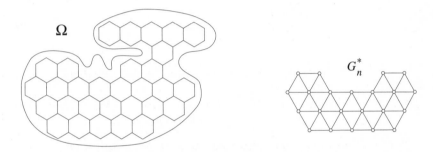

FIGURE 2.14. Discretization of a region $\Omega$

We define a map $f_n$ from the interior of the outer cycle of $G_n''$ into $\Delta$ by mapping all points inside a hexagon to the center of the corresponding circle in $\Delta$. By modifying the circle packing by a Möbius transformation, we may assume that $f_n(p) = 0$, and using a rotation we may assume that $f_n(q)$ is a positive real number. Thurston conjectured that $f_n$ converges to a homeomorphism which is analytic, and this was verified by Rodin and Sullivan [RS87].

The Riemann Mapping Theorem has several other proofs but the proof of Rodin and Sullivan based on circle packings is particularly interesting because of its combinatorial and constructive nature. It can be used for computer experiments on conformal mappings, see Dubejko and Stephenson [DS95b] and Collins and Stephenson [CS98p].

Some other central results in the theory of conformal mappings have been successfully attacked by the use of circle packings, e.g., Schwartz' Lemma (e.g., Beardon and Stephenson [BS91], Dubejko and Stephenson [DS95a], Rodin [Ro87, Ro89]), Koebe uniformization (He and Schramm [HS93, HS95]), etc. (Aharonov [Ah90]). More references can be found in Stephenson's Cumulative bibliography on circle packings [St93].

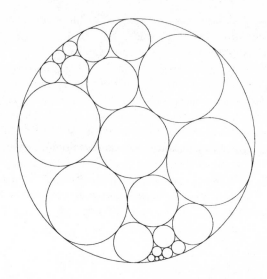

FIGURE 2.15. An approximation to a Riemann mapping

## 2.10. The Jordan Curve Theorem and Kuratowski's Theorem in general topological spaces

We have previously observed that, in Kuratowski's theorem, $K_{3,3}$ is more fundamental in that $K_5$ can be omitted when we restrict Kuratowski's theorem to 3-connected graphs of order at least 6 (cf. Lemma 2.5.5). The fundamental character of $K_{3,3}$ becomes even more clear when we consider more general topological spaces that have the Jordan curve separation property, as demonstrated by the following result that provides a link between the Jordan Curve Theorem and Kuratowski's theorem.

THEOREM 2.10.1 (Thomassen [**Th90a**]). *Let $X$ be an arcwise connected Hausdorff space that cannot be separated by a simple arc. Assume also that $X$ is not homeomorphic to a simple closed curve. Then the following statements are equivalent:*

(a) *Every simple closed curve separates $X$.*
(b) *Every simple closed curve separates $X$ into precisely two arcwise connected components.*
(c) *$K_{3,3}$ cannot be embedded in $X$.*
(d) *Neither $K_{3,3}$ nor $K_5$ can be embedded in $X$.*

PROOF. Clearly (b) $\Rightarrow$ (a) and (d) $\Rightarrow$ (c). It is also easy to prove that (c) $\Rightarrow$ (d). For suppose that $K_5$ is embedded in $X$. Let $v_1, v_2, \ldots, v_5$ be the vertices of $K_5$. Let $p_1, p_2$ be points on the edge $v_1 v_2$ such that $v_1, p_1, p_2, v_2$ are distinct. Now consider an arc $A$ in $K_5$ which connects $p_1$ and $p_2$ and which contains all of $v_1, v_2, v_3, v_4, v_5$. Since $X \setminus A$ is arcwise connected it contains an arc $B$ from a point on the edge $v_1v_2$ to a point on the edge $v_3v_5$. It is now easy to find a $K_{3,3}$ in $K_5 \cup B$.

We shall now assume (c) and prove (b). Let $C$ be any simple closed curve in $X$. Since $X$ is not homeomorphic to a simple closed curve, there is a point $p \in X \setminus C$. Let $A$ be a simple arc from $p$ to $C$ such that $A \cap C$ consists of one point $q$. Let $A_1$ be a segment of $A$ from $q$ to a point $p_1 \neq p$. Since $X \setminus A_1$ is arcwise connected it contains an arc $A_2$ from $A \setminus A_1$ to $C$ having only its ends in common with $A \cup C$. Now $A \cup A_2$ contains a simple arc $A_3$ connecting $q$ with another point $q'$ on $C$ such that $A_3 \cap C = \{q, q'\}$. Let $A_4$, $A_5$ be the two segments of $C$ from $q$ to $q'$. Since $X \setminus A_3$ is arcwise connected, there is a simple arc $A_6$ joining a point $r$ on $A_4 \setminus \{q, q'\}$ with a point $r'$ on $A_5 \setminus \{q, q'\}$ such that $A_6 \cap (A_3 \cup A_4 \cup A_5) = \{r, r'\}$. Now $A_3 \setminus \{q, q'\}$ and $A_6 \setminus \{r, r'\}$ belong to distinct arcwise connected components of $X \setminus C$ since a simple arc from $A_3$ to $A_6$ in $X \setminus C$ (having only its ends in common with $A_3$ and $A_6$) together with $A_3 \cup A_4 \cup A_5 \cup A_6$ would be an embedding of $K_{3,3}$. So, $X \setminus C$ has at least two arcwise connected components.

To see that $X \setminus C$ cannot have more than two arcwise connected components we assume (*reductio ad absurdum*) that $p_1, p_2, p_3$ belong to distinct arcwise connected components of $X \setminus C$. Let $A_1, A_2, A_3$ be pairwise disjoint segments of $C$. Since $X \setminus (C \setminus A_1)$ is arcwise connected, it has a simple arc from $p_1$ to $p_2$. This arc contains an arc $A_{1,1}$ from $p_1$ to a point $q_{1,1}$ on $A_1$ such that $A_{1,1} \cap C = \{q_{1,1}\}$. Similarly, there is a simple arc $A_{i,j}$ from $p_i$ to a point $q_{i,j}$ on $A_j$ such that $C \cap A_{i,j} = \{q_{i,j}\}$ for $i, j = 1, 2, 3$. Clearly, $A_{i,j} \cap A_{k,l} = \emptyset$ or $\{q_{i,j}\}$ when $i \neq k$. We may assume that $A_{i,j} \cap A_{i,l} = \{p_i\}$ when $j \neq l$. For $j = 1, 2, 3$, let $B_j$ be the minimal segment of $A_j$ that contains $q_{1,j}$, $q_{2,j}$, and $q_{3,j}$. Then the nine arcs $A_{i,j}$ ($i, j = 1, 2, 3$) together with $B_1 \cup B_2 \cup B_3$ form an embedding of $K_{3,3}$. This contradiction proves that $X \setminus C$ has precisely two arcwise connected components.

We shall complete the proof by showing that (a) implies (c). Assuming (a), suppose that $X$ contains an embedding of $K_{3,3}$. The graph $K_{3,3}$ has nine 4-cycles and six 6-cycles (and no other cycles). Each of them separates $X$. Consider now a 4-cycle $C$. Since $K_{3,3} \setminus C$ is contained in an arcwise connected component of $X \setminus C$, there is another arcwise connected component which we denote by $Q(C)$. Then $Q(C)$ is also an arcwise connected component of $X \setminus K_{3,3}$. If $A$ is a simple arc from $Q(C)$ to $K_{3,3}$ such that $A$ has only an end $q$ in common with $K_{3,3}$, then we say that $q$ is a *point of attachment* of $Q(C)$ on $K_{3,3}$. All points of attachment of $Q(C)$ are on $C$. Moreover, if $A$ is any segment of $C$, then $X \setminus (C \setminus A)$ is arcwise connected, which implies that $Q(C)$ has a point of attachment on $A$. In other words, the points of attachment of $Q(C)$ on $K_{3,3}$ are dense in $C$.

We may think of $K_{3,3}$ as a 6-cycle $C = v_1v_2\ldots v_6v_1$ with chords $v_iv_{i+3}$
($i = 1, 2, 3$) (where the indices are expressed modulo 6). For each edge
$e_i = v_iv_{i+1}$, we let $C_i$ be the 4-cycle $v_iv_{i+1}v_{i+2}v_{i+3}v_i$, and in $Q(C_i) \cup e_i$
we select an arc $A_i$ joining two distinct points $p_i$ and $q_i$ on $e_i$ such that
$A_i \cap K_{3,3} = \{p_i, q_i\}$. In $C$ we now replace the segment $B_i$ of $e_i$ between
$p_i$ and $q_i$ by $A_i$, $i = 1, 2, 3, 4, 5, 6$. Denote the resulting simple curve by
$C'$. We claim that $X \setminus C'$ is arcwise connected. $K_{3,3}$ has three 4-cycles
containing two diagonals of $C$. For each of these, there is a corresponding
component of $X \setminus K_{3,3}$, and using those we conclude that all arcs $B_i$ and
all diagonals of $C$ (except the ends) are in the same arcwise connected
component $R$ of $X \setminus C'$. In other words, $K_{3,3} \setminus C'$ is contained in $R$.
Now suppose (*reductio ad absurdum*) that $p \in X \setminus (C' \cup R)$. Since $C' \setminus$
$(A_1 \setminus \{p_1, q_1\})$ does not separate $X$, it follows that $X$ has a simple arc $D$
from $p$ to $A_1$. Since $p \notin R$, the arc $D$ does not intersect $K_{3,3}$. But then
$D \subseteq Q(C_1)$. In particular, $p \in Q(C_1)$. By similar reasoning we prove
that $p \in Q(C_2)$. But $Q(C_1)$ and $Q(C_2)$ are disjoint. This contradiction
completes the proof.                                                    □

Every open connected subset of the sphere satisfies all conditions of
Theorem 2.10.1. Since a topological space satisfying condition (d) of Theorem 2.10.1 admits no embedding of nonplanar graphs, by Kuratowski's
theorem, it must be "sphere-like" or "plane-like". This raises the question
of whether every topological space satisfying the conditions of the theorem
can be embedded into the sphere. However, this is not so. Indeed, there
are spaces which do not admit an embedding of the Kuratowski graph $K_{3,3}$
but cannot be embedded in the 2-sphere. For example, if $L$ denotes the
long line,[5] then the Cartesian product $L^2$ is such a space. Another such
space can be obtained from two disjoint copies of $L^2$ by cutting out an
open disc in each of them and identifying the boundaries. This operation
can be performed up to $\aleph_1$ times.

Thomassen proved that condition (a) combined with a slight modification of condition (b) in Theorem 2.10.1 characterize the 2-sphere. More
precisely

THEOREM 2.10.2 (Thomassen [**Th92b**]). *Let $X$ be an arcwise connected compact metric space satisfying* (J1) *and* (J2) *below.*

(J1) *If $J$ is a simple arc in $X$, then $X \setminus J$ is arcwise connected.*
(J2) *If $J$ is a simple closed curve in $X$, then $X \setminus J$ is disconnected in
the topological sense, i.e., $X \setminus J$ contains a proper nonempty subset
which is both open and closed.*

---

[5]The *long line* is obtained from two copies of the *long half-line*. This is a topological space which is a union of uncountably many unit intervals. It can be described
as the Cartesian product $A \times [0, 1)$ where $A$ is the set of all countable ordinals (so $A$ is
uncountable). The product is equipped with the topology induced by the lexicographic
ordering.

*Then $X$ is homeomorphic to the sphere $S^2$.*

Since the 2-sphere satisfies conditions (a) and (b) in Theorem 2.10.1, by Theorem 2.1.11 and the Jordan Curve Theorem, it also satisfies (J1) and (J2). It is perhaps surprising that the converse holds since no local topological property is included explicitly in (J1) or (J2). (Other known characterizations of the 2-sphere usually assume that the space is locally arcwise connected.)

Note that (J1) and (J2) assume different connectivity properties. This seems inevitable. Indeed, "arcwise connected" cannot be replaced by "connected" in (J1). (To see this, take a compact set $Y$ in $\mathbb{R}^2$ which is connected but contains no simple arc. It is well known that such sets exist. Then add a new point $p$ in $\mathbb{R}^3 \backslash \mathbb{R}^2$ and also add all line segments from $p$ to $Y$. The resulting set is arcwise connected but has no simple closed curve.) Also, "connected" cannot be replaced by "arcwise connected" in (J2). (To see this, take the long line $L$, then form the Cartesian product $L^2$, take a one-point-compactification of $L^2$, and identify the new point with any point of $L^2$.)

A short proof of Theorem 2.10.2 based on other topological results is given in [**Th92b**]. We present here the longer but self-contained proof of [**Th90d**].

In the following, $X$ is a Hausdorff topological space. If $J$ is a simple arc in $X$ with ends $p, q$, then we denote by $X - J$ the set $(X \backslash J) \cup \{p, q\}$. If $C$ is a simple arc or a simple closed curve and $Y \subseteq X$, then a $Y$-*segment* of $C$ is a segment of $C$ having only its ends in $Y$. We say that $Y$ *separates* sets $A$ and $B$ if $A$ and $B$ are contained in distinct faces (that is, arcwise connected components) of $Y$.

LEMMA 2.10.3. *If $Y_1, Y_2$ are disjoint closed sets in $X$ and $C$ is a simple arc or a simple closed curve, then $C$ has only finitely many $(Y_1 \cup Y_2)$-segments with one end in $Y_1$ and the other end in $Y_2$.*

Let $G$ and $H$ be graphs embedded in $X$. In this section, the notation $H \subseteq G$ means that $G$ is obtained from $H$ by first inserting vertices of degree 2 on edges and then adding edges. (In particular, $H \subseteq G$ implies that $H$ is a subspace of $G$ when we consider $H$ and $G$ as subspaces of $X$. It does not necessarily imply that $H$ is a subgraph of $G$ when $H$ and $G$ are viewed as graphs.) Also, plane graphs are assumed to be polygonal arc embedded.

If $G$ and $H$ are embedded graphs, we say that a graph isomorphism $\varphi : G \to H$ is *face-preserving* if a subgraph $G'$ of $G$ is a face boundary if and only if $\varphi(G')$ is a face boundary of $H$. (In the case of a 2-connected plane graph, the above definition means that embedded graphs $G$ and $\varphi(G)$ are weakly equivalent.)

Although Proposition 2.10.4 below will be applied to 2-connected graphs only, it is convenient to prove it for more general graphs.

PROPOSITION 2.10.4. *If $G$ is a plane graph and $p, q$ are distinct points in $\mathbb{R}^2$, then precisely one of (a) and (b) below holds.*

(a) *$G$ has a cycle $C$ separating $p$ and $q$ in the plane.*

(b) *There exists a simple polygonal arc $J$ connecting $p$ and $q$ such that $J\backslash\{p, q\}$ is contained in some face of $G$.*

PROOF. If (a) holds, then (b) cannot hold. Now assume that (a) does not hold. We prove, by induction on $|V(G)| + |E(G)|$, that (b) holds. For $|E(G)| = 0$, this is trivial. So assume that $E(G) \neq \emptyset$. If $G$ has a vertex $x$ of degree 0 or 1, we delete $x$ and use induction. (We may assume that the graph $G - x$ satisfies (b) and since we are dealing with simple polygonal arcs, it is easy to see that also $G$ satisfies (b).) So assume that all vertices have degree at least 2. In particular, $G$ contains a cycle $C$. We may assume that $C$ does not separate $p$ and $q$.

Let $e$ be an edge of $C$. If possible, we choose $e$ such that it contains either $p$ or $q$. By the induction hypothesis, there is a simple arc $J$ connecting $p$ and $q$ such that $J\backslash\{p, q\}$ is contained in a face of $G - e$. We may assume that $J\backslash\{p, q\}$ intersects $e$. Let $J_p$ and $J_q$ be the unique minimal segments from $p$ (respectively $q$) to $e$. As $C$ does not separate $p$ and $q$, $J_p$ and $J_q$ hit $e$ on the same side of $e$. Hence it is easy to connect $J_p$ and $J_q$ by a simple arc (close to $e$) not intersecting $e$ (except at $p$ or $q$ in case one or both of $p$, $q$ belong to $e$). □

From now on, let $X$ denote any arcwise connected Hausdorff topological space satisfying the conditions (J1) and (J2). As indicated earlier, $X$ may have a rather complicated structure. We shall show, however, that such spaces share many properties with the sphere.

Theorem 2.10.1 implies:

(1) If $J$ is any simple closed curve in $X$, then $J$ has precisely two faces. Each of them is open and has $J$ as boundary. For each face $F$ of $J$, the set of points on $J$ that are curve-accessible from $F$ is dense on $J$.

PROOF. As $X$ satisfies (J1) and (J2), $X$ also satisfies (a) (and hence (b)) of Theorem 2.10.1. This proves the first statement of (1). The last statement of (1) also follows from (J1). For, if $J'$ is any segment of $J$, then $X\backslash(J - J')$ is arcwise connected. Hence $J'$ contains a point which is curve-accessible from $F$. It remains to be proved that both faces $F_1, F_2$ of $J$ are open. By (J2), $X\backslash J$ is the union of two nonempty disjoint open sets $R_1, R_2$. For each $i \in \{1, 2\}$ there is a $j \in \{1, 2\}$ such that $F_i \subseteq R_j$. Since $X\backslash J = R_1 \cup R_2 = F_1 \cup F_2$, the notation can be chosen such that $R_1 = F_1$ and $R_2 = F_2$. □

We now consider a simple closed curve $J$ in $X$. Let $F, F'$ be the two faces of $J$. Let $p$ and $q$ be two points on $J$ which are curve-accessible from $F'$, and let $K$ be a simple arc from $p$ to $q$ such that $K\backslash\{p, q\} \subseteq F'$. Let

$J_1, J_2$ be the two simple arcs in $J$ connecting $p$ and $q$. Let $F_i$ be the face of $J_i \cup K$ which does not contain $F$, for $i = 1, 2$. With this notation we have:

(2) $J \cup K$ has precisely three faces, namely $F, F_1, F_2$. They are open in $X$. The boundary of $F_i$ is $J_i \cup K$ and the points of $J_i \cup K$ curve-accessible from $F_i$ are dense in $J_i \cup K$ for $i = 1, 2$.

PROOF. Clearly, $F$ is a face of $J \cup K$. The face of $J_i \cup K$ containing $F$ also contains $J_{3-i} \backslash \{p, q\}$, by the last statement of (1). Hence $F_i$ is a face of $J \cup K$ for $i = 1, 2$. It remains to be shown that there are no other faces. Let $s$ be any element in $X \backslash (J \cup K)$. Let $J'$ be a simple arc in $X \backslash K$ from $s$ to a point $t$ on $J$ such that $J' \cap J = \{t\}$. If $t \in J_i$, then $s \in F$ or $s \in F_i$. The last two statements of (2) follow from (1) and the definition of $F, F_1, F_2$.                                    □

(3) Let $G$ be a 2-connected graph with $n$ vertices and $q$ edges embedded in $X$. Then $G$ has $q - n + 2$ faces each of which is open and has a cycle of $G$ as boundary. The set of curve-accessible points from a face is dense in the boundary. Moreover, there exists a plane graph $G'$ and a face-preserving isomorphism $\varphi : G' \to G$.

PROOF. The proof is analogous to the proof of Proposition 2.1.5. By Proposition 1.4.2, $G$ can be obtained from a cycle $C$ by a sequence of path-additions. Each such path-addition takes place in a face. Proposition (2) tells us what happens with that face. Therefore we can draw $C$ in the Euclidean plane and perform the path-addition in the plane so as to obtain the desired $G'$.                                    □

Combining (3) with Proposition 2.10.4 (applied to $G'$) yields:

(4) If $G$ is a 2-connected graph embedded in $X$, and $p, q$ are distinct elements of $X$, then precisely one of the following two statements holds:
    (a) $G$ has a cycle $C$ which separates $p$ and $q$, or
    (b) $G$ has a face $F$ with boundary $C$ such that $p, q \in C \cup F$.

Next we have:

(5) If $p$ and $q$ are distinct elements in $X$, then $X$ has a simple closed curve separating $p$ and $q$.

PROOF. Let $J$ be a simple arc from $p$ to $q$. Let $J'$ be a segment of $J$ not containing $p$ or $q$. Let $J''$ be a simple arc in $X \backslash J'$ connecting the two segments of $J - J'$. Now $J \cup J''$ may be described as the union of a simple closed curve, say $C$, and two disjoint simple arcs $J_p, J_q$ connecting $p$ and $q$ with $C$. Let $p'$ (respectively $q'$) be the end of $J_p$ (respectively $J_q$) on $C$. Possibly $p = q'$ or $q = q'$ (or both). If $p$ and $q$ belong to distinct faces of $C$ we have finished. So assume $F$ is a face of $C$ not intersecting $J_p \cup J_q$. By (1), $X$ has a simple arc $K$ which joins two elements $s, t$ on distinct segments of $C\backslash\{p',q'\}$ such that $K\backslash\{s,t\} \subseteq F$. Let $S$ and $T$ be disjoint simple arcs on $C$ such that $S$ has ends $s,p'$, and $T$ has ends $t,q'$. Now $J_p \cup S \cup K \cup T \cup J_q$ is a simple arc and so $X\backslash(J_p \cup S \cup K \cup T \cup J_q)$ has a simple arc $K'$ joining the two simple arcs of $(C-S)-T$ (and having only its ends $s',t'$ in common with $C$). Thus $C \cup K \cup K'$ has a unique simple closed curve $C'$ which does not contain any of $p',q'$. Then $J_p$ and $J_q$ belong to distinct faces of $C'$. Otherwise $X\backslash C'$ has a simple arc, say $P$, from $J_p$ to $J_q$. Now, $C \cup K \cup K' \cup P$ contains a $K_{3,3}$, contradicting Theorem 2.10.1(c).                                                    □

Now, let $C$ be a simple closed curve in $X$, let $F$ be a face of $C$, and let $p,q$ be distinct elements of $C \cup F$. By (5), there exists a simple closed curve $C'$ separating $p$ and $q$. With this notation we have

(6) If $C'$ has more than one point in common with $C$, then $C'$ contains a $C$-segment $J$ such that $p$ and $q$ belong to distinct arcwise connected components of $(C \cup F)\backslash J$.

PROOF. Assume that $C$ and $C'$ have at least two points in common. Assume also (*reductio ad absurdum*) that no $C$-segment of $C'$ satisfies (6). We define a partial ordering $<$ on the $C$-segments of $C'$ in $F \cup C$. Let $L$ be any $C$-segment of $C'$ in $F \cup C$. By (2), $F$ is partitioned into two faces by $C \cup L$. One of these, say $F''$ has the property that $p$ and $q$ belong to the same arcwise connected component of $(C \cup F'')\backslash L$. Hence the other, say $F'$, has the property that none of $p$, $q$ are in $F'$ nor on the boundary of $F'$. Now if $L'$ is a $C$-segment of $C'$ in $F' \cup C$, then we write $L' < L$. This defines a partial ordering. Suppose that there exists an infinite sequence $L_1 < L_2 < \cdots$. We then obtain a contradiction to Lemma 2.10.3 as follows. If $p$ (or $q$) is in $F$, we let $Y_1$ be a simple arc in $X\backslash C$ from $p$ to $L_1$ and we put $Y_2 = C$. If both $p$ and $q$ are on $C$, then we let $Y_1$ and $Y_2$ denote the disjoint segments of $C$ connecting the ends of $L_1$ with $p$ and $q$. This proves that there exists no infinite sequence $L_1 < L_2 < \cdots$.

Consider now a $C$-segment $L$ of $C'$ which is maximal with respect to $<$. Let $f$ be a homeomorphism of a circle onto $C$. We modify $f$ as follows: $C$ has a unique simple arc $T_L$ which joins the two ends of $L$ such that neither $T_L \cup L$ nor the face $F_L$ of $T_L \cup L$ in $F$ contains any of $p$ or $q$. We now map $f^{-1}(T_L)$ onto $L$ instead of $T_L$ such that the new $f$ is a homeomorphism. We perform this modification for every such maximal $L$. (There may be infinitely many such modifications). The resulting map $f'$ from the circle to $X$ is clearly 1-1. Using Lemma 2.10.3, it is easy to see that $f'$ is continuous and hence the image of $f'$ is a simple closed curve which we denote $C'''$. Let $F'$ be the face of $C$ distinct from $F$. Let $F_1$, $F_2$ be the two faces of $C'''$. We may assume that $F' \subseteq F_1$.

Consider again a $C$-segment $L$ of $C'$ which is maximal with respect to $<$. Let $F_L$ and $T_L$ be as above. Clearly, $F_1$ contains both $F_L$ and $T_L$

(except its ends). We claim that $F_1$ is the union of $F'$ and all the sets $F_L$, $T_L$ where $L$ is maximal with respect to $<$. To

prove this, let $s$ be any element in $F_1$. If $s \in F' \cup C$ there is nothing to prove; so assume $s \notin F' \cup C$. We traverse a simple arc in $F_1$ from $s$ to $F'$ until we hit $C' \cup C$ in a point $t$, say. Either $t$ is on $C$ or on a $C$-segment on $C'$ which is not maximal. In either case $t \in F_L \cup T_L$ for some maximal $L$, and hence $s \in F_L$. This proves the claim. The claim implies in particular that none of $p$, $q$ belong to $F_1$. Hence $p$, $q$ belong to $F_2 \cup C''$. If $p, q \in F_2$, we let $Z$ be a simple arc in $F_2$ joining $p$ and $q$. Otherwise we let $p_n$ and $q_n$ be points on $C''$ curve-accessible from $F_2$ such that $p_n \to p$ and $q_n \to q$ as $n \to \infty$. Let $Z_1$ be a simple arc from $p_1$ to $q_1$ such that $Z_1 \backslash \{p_1, q_2\} \subseteq F_2$. Having constructed $Z_1 \subseteq \cdots \subseteq Z_{n-1}$, we let $Z_n$ denote the union of $Z_{n-1}$, a simple arc from $p_n$ to $Z_{n-1}$, and a simple arc from $q_n$ to $Z_{n-1}$ such that $Z_n \cap C'' = \{p_1, q_1, \ldots, p_n, q_n\}$. Put $Z = \cup_{n=1}^\infty Z_n$. As $p$ and $q$ belong to distinct faces of $C'$, $Z \cap C' \neq \emptyset$. Any element of $Z \cap C'$ is on a $C$-segment $L_1$ of $C'$. Let $L_2$ be a maximal $C$-segment of $C'$ such that $L_1 \leq L_2$. By the definition of $C''$, $L_2 \subseteq C''$. Moreover, the closure of some face of $C \cup L_2$ contains $L_1$ while the closure of some other face contains $p$ and $q$ and hence also $Z$. But then $Z$ cannot intersect $L_1$, a contradiction which proves (6). $\square$

(7) If $G$ is a 2-connected graph embedded in $X$ and $p$, $q$ are distinct elements of $X$, then $G$ can be extended to a 2-connected graph $G'$ containing a cycle which separates $p$ and $q$.

PROOF. By (5), $X$ has a simple closed curve $C'$ separating $p$ and $q$. If $C' \cap G$ contains at most one point, then $C' \cup G$ can be extended to a 2-connected graph $G'$ in which $C'$ is the desired cycle. So assume that $C' \cap G$ has at least two elements. Let $F$ be any face of $G$ and let $C$ be its boundary. If $C \cup F$ contains both $p$ and $q$, then we add a $C$-segment of $C'$ in $F \cup C$ such that the last statement of (6) holds. We do this for every such face $F$. Then the resulting graph $G'$ has no face whose closure contains both $p$ and $q$. By Proposition 2.10.4 and (3), $G'$ has a cycle which separates $p$ and $q$. $\square$

We shall now assume that the space $X$ is also compact and metrizable. We shall use the above results to construct a sequence $G_1 \subseteq G_2 \subseteq \cdots$ of 2-connected graphs embedded in $X$ and a sequence $H_1 \subseteq H_2 \subseteq \cdots$ of graphs embedded in the sphere such that $G_n$ and $H_n$ are isomorphic and such that the maximum face diameter in both $G_n$ and $H_n$ tends to zero as $n \to \infty$. Then the natural homeomorphism of $\cup_{n=1}^\infty G_n$ onto $\cup_{n=1}^\infty H_n$ extends to a homeomorphism of $X$ onto the 2-sphere. We need some additional notation. The *distance* between $p$ and $q$ in $X$ is denoted by $d(p,q)$. If $r > 0$, then

$$D(p,r) = \{q \mid d(p,q) < r\}$$

and

$$\overline{D}(p, r) = \{q \mid d(p, q) \leq r\}.$$

For $p \in X$ and $A \subseteq X$, we put

$$d(p, A) = \inf\{d(p, q) \mid q \in A\}.$$

Let $r_0$ be the diameter of $X$. If $C$ is a simple closed curve with faces $Z$ and $Y$ and $p \in Z$, $q \in Y$, then we write

$$\rho(p, q, C) = \min\{d(q, C \cup Z), d(p, C \cup Y)\}.$$

As $Z$ and $Y$ are open, $C \cup Z$ and $C \cup Y$ are compact and hence $\rho(p, q, C) > 0$. We let $\rho(p, q)$ denote the supremum of $\rho(p, q, C)$ taken over all simple closed curves that separate $p$ and $q$. By (5), $\rho(p, q)$ is defined for each pair of distinct points $p$ and $q$ in $X$. Clearly, $\rho$ is continuous. For $r \leq r_0$, the set $\{(p, q) \in X^2 \mid d(p, q) \geq r\}$ is compact. Let $f(r)$ denote the minimum of $\min\{d(p, q), \rho(p, q)\}$ taken on that set. Then $f(r)$ is increasing and $0 < f(r) \leq r$ for $0 < r \leq r_0$.

PROOF OF THEOREM 2.10.2. Let $M$ be a countable dense set in $X$. Enumerate the elements of $M^2$, say $M^2 = \{(p_1, q_1), (p_2, q_2), \dots\}$. We shall construct a sequence of 2-connected graphs $G_1 \subseteq G_2 \subseteq \cdots$ embedded in $X$, a sequence $H_1 \subseteq H_2 \subseteq \cdots$ of plane graphs, and face-preserving isomorphisms $\varphi_n : G_n \to H_n$ satisfying the following:

(a) $\varphi_{n-1}$ is the restriction of $\varphi_n$ to $G_{n-1}$.

(b) Each bounded face of $H_n$ has diameter $< 1/n$ and each point in the square $[-n, n] \times [-n, n]$ is either in $H_n$ or in a bounded face of $H_n$.

(c) For every $p \in \overline{D}(p_n, \frac{1}{2}f(d(p_n, q_n)))$ and $q \in \overline{D}(q_n, \frac{1}{2}f(d(p_n, q_n)))$, the graph $G_n$ has a cycle $C$ which separates $p$ and $q$.

Assume that we have defined graphs $G_1, \dots, G_{n-1}, H_1, \dots, H_{n-1}$, and isomorphisms $\varphi_1, \dots, \varphi_{n-1}$ where $n \geq 1$. We define $G_n$, $H_n$, and $\varphi_n$ as follows. First we add to $H_{n-1}$ vertical and horizontal straight line segments (containing no vertices of $H_{n-1}$) such that $H_{n-1}$ is extended to a graph $H_n'$ satisfying (b). We add the corresponding simple arcs in $X$ so that $G_{n-1}$ is extended to a graph $G_n'$ which is isomorphic to $H_n'$ with face preserving isomorphism $\varphi_n'$. By the definition of $f$, $X$ has a simple closed curve $J$ such that $J$ separates $A = \overline{D}(p_n, \frac{1}{2}f(d(p_n, q_n)))$ and $B = \overline{D}(q_n, \frac{1}{2}f(d(p_n, q_n)))$. If $J$ has at most one point in common with $G_n'$, it is easy to extend $G_n'$ to the desired $G_n$. So assume that $J$ has at least two points in common with $G_n'$. Let $W$ be any finite collection of $G_n'$-segments of $J$. If we add to $G_n'$ all segments of $W$, we obtain a 2-connected graph $G_W$. Let $O_W$ denote those pairs of points of $X$ which are separated by some cycle in $G_W$. Clearly, $O_W$ is open in $X^2$. Moreover, the proof of (7) implies that the collection of sets of the form $O_W$ is a covering of $A \times B$ which is compact in $X^2$. Hence $A \times B$ is contained in the union of a finite subcovering which is of the form $O_W$. Adding the $G'_n$-segments in $W$ to $G'_n$ results in the desired graph $G_n$. The plane graph $H_n$ and an isomorphism $\varphi_n$ are then defined in the obvious way.

Now (a), (b), and (c) imply:

(d) If $p$ and $q$ are distinct points in $X$, then for $n$ sufficiently large, $G_n$ contains a cycle which separates $p$ and $q$.

To prove (d), let $\delta = d(p, q)$. Choose $n$ such that $p_n \in D(p, \frac{1}{2}f(\delta/2))$ and $q_n \in D(q, \frac{1}{2}f(\delta/2))$. Then $\delta \leq 2d(p_n, q_n)$ and hence $f(\delta/2) \leq f(d(p_n, q_n))$. So $p \in D(p_n, \frac{1}{2}f(d(p_n, q_n)))$ and $q \in D(p_n, \frac{1}{2}f(d(p_n, q_n)))$. Now (d) follows from (c).

If $F_n$ is a face in $G_n$ with boundary $C_n$, then $C_n \cap F_n$ is closed (by (1)) and hence compact. So, if $F_1 \supseteq F_2 \supseteq \cdots$, then $\cap_{n=1}^{\infty}(C_n \cap F_n)$ is nonempty. By (d), $\cap_{n=1}^{\infty}(C_n \cap F_n)$ consists of a single point, say $p$. If $F'_n$ denotes the face of $H_n$ bounded by $\varphi_n(C_n)$, then (b) implies that $\cap_{n=1}^{\infty}(\varphi_n(C_n) \cap F'_n)$ is a single point, which we denote by $\varphi(p)$, unless $F'_n$ is the unbounded face of $H_n$ for all $n \geq 1$. In that case we write $\varphi(p) = \infty$ and regard the sphere $S^2$ as a one-point-compactification of the plane.

It is easy to see that the map $\varphi : X \to S^2$ defined above is 1-1 and onto. We claim that $\varphi$ is a homeomorphism. As $X$ and $S^2$ are compact, it suffices to prove that $\varphi$ is continuous. Let $y \in S^2 \backslash \{\infty\}$ and let $r > 0$. The construction of $H_n$ shows that some $H_n$ contains a cycle $C \subseteq D(y, r)$ such that $y \in \text{int}(C)$. Now $\varphi^{-1}(y)$ is in one of the two faces of $\varphi_n^{-1}(C)$, say $F$. Since $F$ is open and $\varphi(F) = \text{int}(C)$ (by the definition of $\varphi$), it follows that $\varphi$ is continuous at $y$. The case $y = \infty$ is treated similarly. This completes the proof.    $\square$

CHAPTER 3

# Surfaces

In this chapter we establish some basic tools for graphs embedded on
surfaces. We restrict ourselves to compact 2-dimensional surfaces without
boundary. These are compact connected Hausdorff topological spaces in
which every point has a neighborhood that is homeomorphic to the plane
$\mathbb{R}^2$. In Section 3.1 we present a simple rigorous proof (due to Thomassen
[**Th92a**]) of the fact that every surface is homeomorphic to a triangulated
surface. Other available proofs of this result are complicated and appeal to
geometric intuition. Perhaps it is difficult to follow some of the details but
the proof is conceptually very simple since it merely consists of repeated
use of the Jordan–Schönflies Theorem. The surface classification theorem,
which we prove next, says that every surface is homeomorphic to a space
obtained from the sphere by adding handles and crosscaps. One of the
first complete proofs of this fundamental result was given by Kerékjártó
[**Ke23**]. The proof presented in Section 3.1 is very short and follows
[**Th92a**]. It differs from other proofs in the literature in that it uses
no topological results, not even the Jordan Curve Theorem (except for
polygonal simple closed curves in the plane). In particular, it does not
use Euler's formula (which includes the Jordan Curve Theorem). Instead,
Euler's formula is a by-product of the proof.

After obtaining the surface classification theorem, it is shown that
every orientable graph embedding, whose faces are all homeomorphic to
an open disc in the plane, can be described combinatorially, up to homeomorphism, by the local rotations at each vertex. A similar description
exists also for embeddings of graphs into nonorientable surfaces. Again,
this is proved rigorously without appealing to geometric intuition.

Inspired by the possibility of describing embeddings of graphs combinatorially, we *define* in Chapter 4 an embedding as a collection of local
rotations in the orientable case and, more generally, an embedding scheme
in the nonorientable case. In the remaining part of the book we treat embeddings in this purely combinatorial framework.

## 3.1. Classification of surfaces

A *surface* is a connected compact Hausdorff topological space $S$ which is locally homeomorphic to an open disc in the plane[1] (and hence locally homeomorphic to $\mathbb{R}^2$), i.e., each point of $S$ has an open neighborhood homeomorphic to the open unit disc in $\mathbb{R}^2$.

Examples of surfaces can be constructed as follows. Let $\mathcal{F}$ be a finite collection of pairwise disjoint (convex) polygons in the plane (including their interiors) with all sides of length 1. Suppose that all these polygons together have $m$ sides $\sigma_1, \sigma_2, \ldots, \sigma_m$, where $m$ is even. Orient arbitrarily each of the sides $\sigma_i$ by choosing one of its endpoints as the initial point, and choose an arbitrary partition of the sides into pairs. From the disjoint union of polygons in $\mathcal{F}$ form a topological space $S$ by identifying sides in given pairs of our partition in such a way that the orientations agree, i.e., if $\sigma_i, \sigma_j$ is such a pair, then they are identified in such a way that the initial point of $\sigma_i$ is identified with the initial point of $\sigma_j$. In this way we get a compact Hausdorff space $S$ which is locally homeomorphic to the unit disc in the plane. This is obvious for points inside polygons and in the interiors of identified sides. To see what happens with points obtained by identifying corners of the polygons, observe that after each identification of a pair of sides, each such point $x$ has a neighborhood homeomorphic to the closed upper halfplane or to the plane. The latter case occurs only when all of the sides adjacent to corners that have been identified into $x$ have already been processed. This shows that all points of $S$ have open neighborhoods homeomorphic to the plane. Thus, if $S$ is connected (which we shall assume henceforth), then it is a surface. The sides $\sigma_1, \sigma_2, \ldots, \sigma_m$ and their endpoints determine a connected multigraph $G$ embedded in $S$. We say that $G$ is *2-cell embedded* in $S$. The polygons in $\mathcal{F}$ are the *faces* of $G$. If all faces are triangles and $G$ is a graph, then we say that $G$ *triangulates* the surface $S$ and that $S$ is a *triangulated surface*.

In the example of Figure 3.1, four triangles are used to get a surface homeomorphic to the unit *sphere* in $\mathbb{R}^3$ (also called the 2-*sphere* or just the *sphere*). The resulting triangulated surface is the *tetrahedron*. Its graph is $K_4$.

Since every convex polygon in the plane can be triangulated (obtained up to homeomorphism from equilateral triangles by identification of some of their sides), the surfaces with 2-cell embedded multigraphs are all homeomorphic to triangulated surfaces. Theorem 3.1.1 below shows that we can get all surfaces in this way.

THEOREM 3.1.1. *Every surface is homeomorphic to a triangulated surface.*

---

[1]This is sometimes called a *closed surface* in the literature.

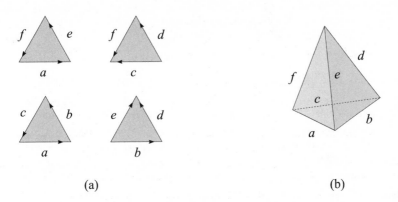

(a)                                    (b)

FIGURE 3.1. The tetrahedron

PROOF (from [**Th92a**]). Since the interior of a convex polygon can be triangulated, it is sufficient to prove that every surface $S$ is homeomorphic to a surface with a 2-cell embedded graph. For each point $p$ on $S$, let $D(p)$ be a disc in the plane which is homeomorphic to a neighborhood of $p$ on $S$. (Instead of specifying a homeomorphism we shall use the same notation for a point in $D(p)$ and the corresponding point on $S$.) In $D(p)$ we draw two quadrangles $Q_1(p)$ and $Q_2(p)$ such that $p \in \text{int}(Q_1(p))$ and $Q_1(p) \subset \text{int}(Q_2(p))$. Since $S$ is compact, it has a finite number of points $p_1, p_2, \ldots, p_n$ such that $S = \overset{n}{\underset{i=1}{\cup}} \text{int}(Q_1(p_i))$. Viewed as subsets in the plane, $D(p_1), \ldots, D(p_n)$ can be assumed to be pairwise disjoint. In what follows we are going to keep $D(p_1), \ldots, D(p_n)$ fixed in the plane (keeping in mind, though, that they also correspond to subsets of $S$). However, we shall modify the homeomorphisms between $D(p_i)$ and the corresponding set in $S$ and consider new quadrangles $Q_1(p_i)$. More precisely, we shall show that $Q_1(p_1), \ldots, Q_1(p_n)$ can be chosen such that they form a 2-cell embedding of $S$.

Suppose, by induction on $k$, that they have been chosen such that any two of $Q_1(p_1), Q_1(p_2), \ldots, Q_1(p_{k-1})$ have only a finite number of points in common on $S$. We now focus on $Q_2(p_k)$. We define a *bad segment* as a segment $P$ of some $Q_1(p_j)$ $(1 \leq j \leq k-1)$ which joins two points of $Q_2(p_k)$ and which has all other points in $\text{int}(Q_2(p_k))$. Let $Q_3(p_k)$ be a square between $Q_1(p_k)$ and $Q_2(p_k)$. We say that a bad segment inside $Q_2(p_k)$ is *very bad* if it intersects $Q_3(p_k)$. There may be infinitely many bad segments but only finitely many very bad ones. The very bad ones together with $Q_2(p_k)$ form a 2-connected graph $\Gamma$. We redraw $\Gamma$ inside $Q_2(p_k)$ such that we get a graph $\Gamma'$ which is isomorphic to $\Gamma$ such that all edges of $\Gamma'$ are simple polygonal arcs and such that the facial cycles in $\Gamma$ and $\Gamma'$ are the same. This can be done using Proposition 1.4.2. Now we apply Theorem 2.2.3 to extend the isomorphism from $\Gamma$ to $\Gamma'$ to a homeomorphism of $\overline{\text{int}}(Q_2(p_k))$ keeping $Q_2(p_k)$ fixed. This transforms $Q_1(p_k)$ and $Q_3(p_k)$ into simple closed curves $Q_1'$ and $Q_3'$ such that $p_k \in \text{int}(Q_1') \subseteq \text{int}(Q_3')$. We consider a closed polygonal curve $Q_3''$ in $\text{int}(Q_2(p_k))$ such that $Q_1' \subseteq \text{int}(Q_3'')$ and such that $Q_3''$ intersects no bad segments except the very bad ones (which are now polygonal arcs). (The existence of $Q_3''$ can be established as follows: For every point $p$ on $Q_3'$ we let $R(p)$ be a square with $p$ as the center such that $R(p)$ intersects neither $Q_1'$ nor any bad segment which is not very bad. We consider a (minimal) finite covering of $Q_3'$ by such squares. The plane union of those squares determines a 2-connected plane graph whose outer cycle can play the role of $Q_3''$.) By redrawing $\Gamma' \cup Q_3''$ (which is either 2-connected or has two blocks) and using Theorem 2.2.3 once more, we may assume that $Q_3''$ is in fact a quadrangle having $Q_1'$ in its interior. If we let $Q_3''$ be the new choice of $Q_1(p_k)$, then any two of $Q_1(p_1), \ldots, Q_1(p_k)$ have only finite intersection. The inductive hypothesis is hence proved for all $k = 1, 2, \ldots, n$.

Thus we may assume that there are only finitely many very bad segments inside each $Q_2(p_k)$ and that those segments are simple polygonal curves forming a 2-connected plane graph. The union $\overset{n}{\underset{i=1}{\cup}} Q_1(p_i)$ may be thought of as a graph $\Gamma$ drawn on $S$. Each region of $S \backslash \Gamma$ is bounded by a cycle $C$ in $\Gamma$. (We may think of $C$ as a simple closed polygonal curve inside some $Q_2(p_i)$.) Now we draw a convex polygon $C'$ with sides of length 1 such that the corners of $C'$ correspond to the vertices of $C$. After appropriate identifications of the sides of the polygons $C'$ corresponding to the faces of $\Gamma$ in $S$ we get a surface $S'$ with a 2-cell embedded graph $\Gamma'$ which is isomorphic to $\Gamma$. The isomorphism of $\Gamma$ to $\Gamma'$ can be extended to a homeomorphism $f$ of the point set of $\Gamma$ on $S$ onto the point set of $\Gamma'$ on $S'$. In particular, the restriction of $f$ to the above cycle $C$ is a homeomorphism onto $C'$. By the Jordan-Schönflies Theorem, $f$ can be extended to a homeomorphism of $\overline{\text{int}}(C)$ to $\overline{\text{int}}(C')$. This defines a homeomorphism of $S$ onto $S'$.                                                                                         □

Theorem 3.1.1 implies in particular that every surface is metrizable.

Consider now two disjoint triangles $T_1, T_2$ (such that all six sides have the same length) in a face $F$ of a surface $S$ with a 2-cell embedded multigraph $G$. We form a new surface $S'$ by deleting from $F$ the interior of $T_1$ and $T_2$ and identifying $T_1$ with $T_2$ such that the clockwise orientations[2] around $T_1$ and $T_2$ disagree, see Figure 3.2(b). We say that the surface $S'$ is obtained from $S$ by *adding a handle*. There is another possibility of identifying pairs of edges obtained after removing $T_1$ and $T_2$ as indicated in Figure 3.2(c). In this case, the orientations of $T_1$ and $T_2$ agree. We say that the resulting surface $S''$ has been obtained from $S$ by *adding a*

---

[2]We recall that $S$ consists of polygons and their interiors in the plane. So, when we speak of clockwise orientation we simply refer to the plane. We do not yet discuss orientability of surfaces.

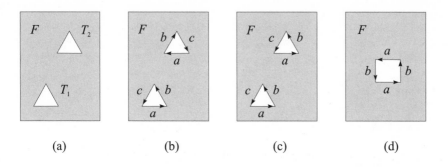

(a)          (b)          (c)          (d)

FIGURE 3.2. Adding a handle, a twisted handle, or a crosscap

*twisted handle.* Finally, let $T$ be a quadrangle (with equilateral sides) in $F$. We let $S'''$ denote the surface obtained by deleting the interior of $T$ and identifying diametrically opposite points of the quadrangle as shown in Figure 3.2(d). Then $S'''$ is said to be obtained by *adding a crosscap.* It is easy to extend $G$ to a 2-cell embedded graph in $S', S'', S'''$, respectively. Also, it is an easy exercise to show that $S', S''$ and $S'''$ are independent (up to homeomorphism) of where in $F$ the triangles $T_1$ and $T_2$ (the square $T$, respectively) are located since it is easy to continuously deform a pair of triangles into another pair of triangles inside the given face $F$. Similarly, $S', S'', S'''$ are independent of the choice of the face $F$. This is easy to see for faces sharing an edge, and then one can apply the assumption that $S$ is connected. The triangles $T_1$ and $T_2$ may also belong to distinct faces. In that case it is clear that we add either a handle or a twisted handle but we may (at this stage) have difficulties distinguishing between a handle and a twisted handle. When adding a crosscap, it is sufficient that $T$ is a simple closed polygonal curve, which can be continuously deformed into a point (and hence to a quadrangle in a face).

We shall now consider all surfaces obtained from the sphere $\mathbb{S}_0$ (which we here think of as a tetrahedron) by adding handles, twisted handles and crosscaps. If we add $h$ handles to $\mathbb{S}_0$, we obtain the surface $\mathbb{S}_h$ which we refer to as the *orientable surface* of *genus* $h$ ($h \geq 0$). If we add $h$ crosscaps to $\mathbb{S}_0$, we get the *nonorientable surface* of *genus* $h$. This surface will be denoted by $\mathbb{N}_h$. The surfaces $\mathbb{S}_1, \mathbb{S}_2, \mathbb{N}_1, \mathbb{N}_2$ are also called the *torus*, the *double torus*, the *projective plane*, and the *Klein bottle*, respectively.

Let $S$ be a surface obtained from $\mathbb{S}_0$ by adding some handles, twisted handles and crosscaps. By the above remarks, the order in which we add the handles and crosscaps is not important. The resulting surface is always the same (up to homeomorphism). Figure 3.3 shows that the addition of a twisted handle can be replaced by adding two crosscaps. Moreover, if we have already added a crosscap, then adding a handle amounts to the same, up to homeomorphism, as adding a twisted handle. To verify this

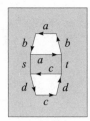 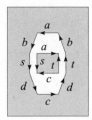 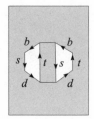 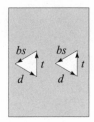

FIGURE 3.3. Adding two crosscaps

statement, it suffices to consider the case when we add a crosscap and then a handle or a twisted handle inside the same face of the surface. This can be done in a similar way as proving that adding a twisted handle is the same as adding two crosscaps. (We leave details to the reader.) This implies the following result:

PROPOSITION 3.1.2. *Let $S$ be the surface obtained from the sphere by adding $h$ handles, $t$ twisted handles and $c$ crosscaps. If $t = c = 0$, then $S = \mathbb{S}_h$. Otherwise, $S = \mathbb{N}_{2h+2t+c}$.*

We now show that every surface is homeomorphic to one of $\mathbb{S}_h$ ($h \geq 0$), or $\mathbb{N}_k$ ($k \geq 1$). The proof is from [**Th92a**].

THEOREM 3.1.3 (Classification of Surfaces). *Every surface is homeomorphic to precisely one of the surfaces $\mathbb{S}_h$ ($h \geq 0$), or $\mathbb{N}_k$ ($k \geq 1$).*

By Theorem 3.1.1 it suffices to show that no two of the surfaces $\mathbb{S}_0, \mathbb{N}_1,$ $\mathbb{S}_1, \mathbb{N}_2, \mathbb{S}_2, \mathbb{N}_3, \ldots$ are homeomorphic (which we shall prove later) and to establish the following lemma:

LEMMA 3.1.4. *Let $S$ be a surface and $G$ a multigraph that is 2-cell embedded in $S$, with $n$ vertices, $q$ edges and $f$ faces. Then $S$ is homeomorphic to either $\mathbb{S}_h$ or $\mathbb{N}_k$ where $h$ and $k$ are defined by the equations*

$$n - q + f = 2 - 2h = 2 - k . \tag{3.1}$$

PROOF. Since $S$ is connected, also $G$ is connected. We first show that $n - q + f \leq 2$. For this we successively delete edges from $G$ until we get a spanning tree $H$ of $G$. For each edge deletion, the number of faces (which are now not necessarily 2-cells) is unchanged or decreased by 1. It is easy to see that $H$ has only one face. Since $H$ has $n$ vertices and $n - 1$ edges, it follows that $n - q + f \leq 2$.

We next extend $G$ to a triangulation of $S$ as follows. If $G$ is not a graph, we subdivide its loops and parallel edges to get a graph. Clearly, the lemma holds for the resulting graph if and only if it holds for the multigraph we started with. For each face $F$ of $G$ which is a convex polygon with corners $v_1, v_2, \ldots, v_r$, where $r \geq 4$ and their indices are expressed modulo $r$, we add new vertices $u, u_1, \dots, u_r$ in $F$ and we add the edges $u_i v_i, u_i v_{i+1}, u_i u_{i+1}, u_i u$ for $i = 1, 2, \dots, r$. Denote the obtained graph by $G'$, and let $n'$, $q'$, $f'$ be the number of vertices, edges and faces, respectively, of $G'$. Clearly, $n' - q' + f' = n - q + f$. Thus it is sufficient to prove the lemma in the case where $G$ is a graph triangulating $S$ which we now assume.

Suppose (reductio ad absurdum) that $S$, $G$ are a counterexample to Lemma 3.1.4 such that $G$ is a graph triangulating $S$, with at least four vertices and such that

(1) $2 - n + q - f$ is as small as possible,

(2) subject to (1), $n$ is minimum and

(3) the minimum degree $m$ of $G$ is as small as possible
    subject to (1) and (2).

Let $v$ be a vertex of minimum degree in $G$. Let $v_1, v_2, \dots, v_m$ be the neighbors of $v$ such that the 3-cycles $vv_1v_2, vv_2v_3, \dots, vv_mv_1$ bound the faces incident with $v$ and the indices are expressed modulo $m$. Since $G$ is a graph, we have $m \geq 3$. If $m = 3$, then $G - v$ is a triangulation of $S$ unless $n = 4$ in which case $S$ is the tetrahedron. In either case this contradicts (2) or the assumption that $S, G$ are a counterexample. So $m \geq 4$.

If, for some $i = 1, 2, \dots, m$, the vertex $v_i$ is not joined to $v_{i+2}$ by an edge, then we let $G'$ be obtained from $G$ by deleting the edge $vv_{i+1}$ and adding the edge $v_i v_{i+2}$ instead. Clearly, $G'$ triangulates $S$, contradicting (3). So we may assume that $G$ contains all edges $v_i v_{i+2}$, $i = 1, 2, \dots, m$, when $v$ is a vertex of minimum degree.

Intuitively, we complete the proof by cutting the surface along the triangle $T = vv_1v_3$. This transforms $T$ into either two triangles $T_1$ and $T_2$ or into a hexagon $H$ (in case $S$ has a Möbius strip that contains $T$). We get a new surface $S'$ by adding two new triangles (and their interior) or a hexagon (and its interior which we triangulate) and identify their sides with sides of $T_1$ and $T_2$ or $H$, respectively. Then $S'$ is a triangulated surface with smaller value of $2 - n + q - f$ than $S$. By the minimality of this parameter, $S'$ is of the form $\mathbb{S}_g$ or $\mathbb{N}_g$. Then $S$ is of that form, too.

Formally, we argue as follows. Recall that $S$ is a triangulated surface, i.e., $S$ is obtained by identifying sides of pairwise disjoint triangles in the plane. Let $M$ denote the topological space which is formed by using the same triangles and the same identifications of sides, except that those six sides that correspond to the edges $vv_1$, $v_1v_3$, $v_3v$ are not identified with each other. Let us call those six sides the *boundary sides* of $M$. Let $G'$ be the graph whose vertices are the corners of the triangles of $M$ and whose edges are the sides of the triangles. It is easy to see that $G'$ has precisely six vertices which are incident with boundary sides and that each of these six vertices is incident with exactly two boundary sides. Thus the boundary sides determine a 2-regular subgraph $C$ of $G'$. Clearly, $C$ is either a 6-cycle or the union of two 3-cycles. If $C$ is two disjoint 3-cycles,

then we add to $M$ two disjoint triangles (and their interior) in the plane
and identify their sides with $C$ such that we obtain a new surface which is
triangulated by $G'$. If $C$ is a 6-cycle, then we add to $M$ a hexagon in the
plane together with its interior (which we triangulate) and then identify
the sides of this hexagon with the edges of $C$. In this way $M$ is extended
to a surface $S''$ and $G'$ is extended to a graph $G''$ which triangulates $S''$.
(Note that $G'$ is obtained from $G$ by "cutting" the triangle $vv_1v_3$. Then
$G'$ is connected because of the edge $v_2v_4$. Hence the spaces $M, S', S''$ are
also connected.) Thus we have transformed $G$ and $S$ into a triangulation
$G'$ with $n'$ vertices, $q'$ edges and $f'$ faces of a surface $S'$, or a triangulation
$G''$ with $n''$ vertices, $q''$ edges and $f''$ faces of a surface $S''$. In the former
case we have

$$q' - n' + f' = q - n + f + 2. \tag{3.2}$$

In the latter case we have

$$q'' - n'' + f'' = q - n + f + 1. \tag{3.3}$$

By (1), $S'$ or $S''$ is homeomorphic to the surface $\mathbb{S}_{h'}$ or $\mathbb{N}_{k'}$ where $2h'$
or $k'$ is equal to $2 - n' + q' - f'$ or $2 - n'' + q'' - f''$, respectively. If
$C$ consists of two triangles, then clearly $S$ is obtained from $S'$ by adding
a handle or a twisted handle. If $C$ is a 6-cycle, then in $S''$, $C$ can be
continuously deformed into a point, and hence $S$ is obtained from $S''$ by
adding a crosscap (see the discussion preceding Proposition 3.1.2). Then $S$
is homeomorphic to $\mathbb{N}_{k'+1}$ or $\mathbb{N}_{2h'+1}$ (by Proposition 3.1.2). By (3.3), this
contradicts the assumption that $S$ and $G$ are a counterexample. Similarly,
if $C$ is the union of two 3-cycles, then $S$ is homeomorphic to either $\mathbb{N}_{k'+2}$ or
$\mathbb{S}_{h'+1}$ or $\mathbb{N}_{2h'+2}$, and again we obtain a contradiction. This finally proves
the lemma.                                                                    $\square$

PROOF OF THEOREM 3.1.3. To complete the proof of Theorem 3.1.3
we indicate a way to see that no two of the surfaces $\mathbb{S}_0, \mathbb{S}_1, \ldots, \mathbb{N}_1, \mathbb{N}_2, \ldots$
are homeomorphic. In this discussion, however, many details will be left
to the reader.

Consider any connected graph $G$ with $n$ vertices and $q$ edges embedded
in $\mathbb{S}_h$. Suppose that $\mathbb{S}_h$ is triangulated and that $G'$ is the graph of the
triangulation. Using Lemma 2.1.1, we can obtain (another) embedding
of $G$ in $\mathbb{S}_h$ such that all edges are simple polygonal arcs. Let $f$ be the
number of faces of this embedding. We may assume that $G \cap G' \neq \emptyset$.
Then $G \sqcup G'$ (the "union" $\sqcup$ is defined after Lemma 2.1.9) determines a
2-cell embedded (multi)graph satisfying Euler's formula and containing a
subdivision of $G$. By successively deleting edges (and isolated vertices)
from $G \sqcup G'$ until we get a subdivision of $G$, we conclude that

$$n - q + f \geq 2 - 2h . \tag{3.4}$$

Since $3f \leq 2q$, we conclude that

$$q \leq 3n - 6 + 6h \tag{3.5}$$

with equality if and only if $G$ is a triangulation of $\mathbb{S}_h$. Thus a triangulation of $\mathbb{S}_h$ has too many edges in order to be drawn on $\mathbb{S}_{h'}$ when $h' < h$, and hence $\mathbb{S}_h$ and $\mathbb{S}_{h'}$ are not homeomorphic if $h' \neq h$. More generally, this argument shows that $\mathbb{S}_0, \mathbb{S}_1 \ldots, \mathbb{N}_1, \mathbb{N}_2, \ldots$ are pairwise nonhomeomorphic except that $\mathbb{S}_h$ and $\mathbb{N}_{2h}$ might be homeomorphic. We sketch an argument which shows that they are not.

It is easy to describe a simple closed polygonal curve $C$ in $\mathbb{N}_{2h}$ such that, when we traverse $C$, left and right interchange. Also, it is easy (though a little tedious) to show that $\mathbb{S}_h$ contains no such simple closed polygonal curve $C'$. (It is convenient to consider a 2-cell embedded graph $G$ that contains no such curve $C'$ and then extend the argument to an arbitrary simple polygonal curve $C'$ in the surface.) So it suffices to show the following: If there exists a homeomorphism $f : \mathbb{N}_{2h} \to \mathbb{S}_h$, then there exists a homeomorphism $f' : \mathbb{N}_{2h} \to \mathbb{S}_h$ such that $f'(C)$ is a simple closed polygonal curve. To see this, we let $G$ be 2-cell embedded in $\mathbb{N}_{2h}$. Then also $G \sqcup C$ may be regarded as a graph which is 2-cell embedded in $\mathbb{N}_{2h}$, and $G \sqcup C$ can be extended to a triangulation $H$ of $\mathbb{N}_{2h}$. We construct $H$ such that it has no triangles other than the face boundaries. Then $f(H)$ is a graph drawn on $\mathbb{S}_h$ and we redraw $f(H)$, resulting in a graph $H'$, such that all edges are simple polygonal arcs (using the argument from the proof of Lemma 2.1.1). Since $H'$ and $H$ are isomorphic and $H$ is a triangulation of $\mathbb{N}_{2h}$, it follows from (3.5) that $H'$ is a triangulation of $\mathbb{S}_h$. Since all the triangles of $H$ are facial, the face boundaries of $H'$ are the same as the face boundaries of $H$. So, any isomorphism $H \to H'$ can be extended into a homeomorphism $f' : \mathbb{N}_{2h} \to \mathbb{S}_h$ taking $C$ into a closed polygonal curve. □

Note that so far we have not referred to orientability of surfaces or used *Euler's formula* (3.1) in the proof of Theorem 3.1.3. To make it explicit, the *Euler characteristic* $\chi(S)$ of a surface $S$ is defined as

$$\chi(S) = \begin{cases} 2 - 2h, & S = \mathbb{S}_h \\ 2 - k, & S = \mathbb{N}_k. \end{cases} \tag{3.6}$$

Then Theorem 3.1.3 and Lemma 3.1.4 imply the following.

**Euler's Formula.** *Let $G$ be a multigraph which is 2-cell embedded in a surface $S$. If $G$ has $n$ vertices, $q$ edges, and $f$ faces in $S$, then*

$$n - q + f = \chi(S) . \tag{3.7}$$

The proof of Theorem 3.1.3 shows that in any representation of $\mathbb{N}_k$ ($k \geq 1$) as a finite union of polygons (together with their interiors), there is a simple closed curve such that right and left interchange along that curve. We have also noted that no $\mathbb{S}_h$ $(h \geq 0)$ has that property. We say that $\mathbb{S}_h$ $(h \geq 0)$ is *orientable* and that $\mathbb{N}_k$ $(k \geq 1)$ is *nonorientable*.

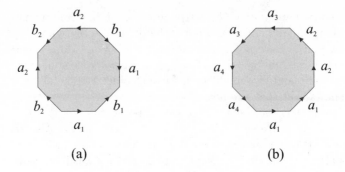

(a)                                          (b)

FIGURE 3.4. The surfaces $\mathbb{S}_2$ and $\mathbb{N}_4$

Let us consider some examples of surfaces obtained from only one polygon (called the *fundamental polygon* of the corresponding surface). Let $g \geq 1$ be an integer, and let $F$ be the regular polygon with $4g$ sides $a_1$, $b_1$, $a_1'$, $b_1'$, $a_2$, $b_2$, $a_2'$, $b_2'$, ..., $a_g$, $b_g$, $a_g'$, $b_g'$ of unit length, enumerated in the same order as they appear on the boundary of $F$. By identifying $a_i$ with $a_i'$, and $b_i$ with $b_i'$ for $i = 1, 2, \ldots, g$, so that the sides $s$ and $s'$ ($s = a_i$ or $s = b_i$) are oriented differently (see Figure 3.4(a)), we get a surface. It is easy to see that the surface is orientable. The multigraph in this example has one vertex and $2g$ loops. So, by Euler's formula, the surface is $\mathbb{S}_g$.

A similar example is obtained by taking a $2g$-sided polygon[3] with sides $a_1, a_1', a_2, a_2', \ldots, a_g, a_g'$ and by identifying $a_i$ with $a_i'$ $(i = 1, \ldots, g)$ having the same orientation (see Figure 3.4(b)). The resulting surface is $\mathbb{N}_g$.

FIGURE 3.5. The torus and the double torus in the 3-space

Figure 3.5 shows the torus $\mathbb{S}_1$ and the double torus $\mathbb{S}_2$ as surfaces represented in the 3-space. It can be shown that no nonorientable surface admits such a representation in the 3-space.

---

[3]If $g = 1$, we use polygonal lines for the sides of the polygon.

## 3.2. Rotation systems

In this section we focus on 2-cell embeddings in orientable surfaces and define their local clockwise ordering. First we observe that 2-cell embeddings can be formed without requesting that the sides of the polygons are of unit length. It is sufficient that the sides are simple polygonal arcs. Identifying such sides pair by pair also produces surfaces with 2-cell embedded multigraphs. Note that this generalized construction allows also polygons with just one or just two sides.

We now point out how any connected multigraph with at least one edge can be 2-cell embedded. This idea, which is attributed to Heffter [**He891**] and Edmonds [**Ed60**], will play a crucial role in the next chapters. Let $G$ be a connected multigraph with at least one edge. Suppose that we have, for each $v \in V(G)$, a cyclic permutation $\pi_v$ of edges incident with $v$. Let us consider an edge $e_1 = v_1 v_2$ and the closed walk $W = v_1 e_1 v_2 e_2 v_3 \ldots v_k e_k v_1$ which is determined by the requirement that, for $i = 1, \ldots, k$, we have $\pi_{v_{i+1}}(e_i) = e_{i+1}$ where $e_{k+1} = e_1$ (and $k$ is minimal). We should explain why there exists an integer $k$ such that $e_{k+1} = e_1$. Since $G$ is finite, $W$ cannot be infinite without repetition of an edge in the same direction. It is easy to see that the first edge repeated in the same direction is $e_1$. (Note however, that some edges $e_i$ may occur in $W$ traversed also in the direction from $v_{i+1}$ to $v_i$.) We shall not distinguish between $W$ and its cyclic shifts. If $\pi = \{\pi_v \mid v \in V(G)\}$, then we call $W$ a $\pi$-*walk*. For each $\pi$-walk we take a polygon in the plane with $k$ sides (where $k$ is the length of the walk) so that it is disjoint from the other polygons, and we call it a $\pi$-*polygon*. Now we take all $\pi$-polygons. Each edge of $G$ appears exactly twice in $\pi$-walks and this determines orientations of the sides of the $\pi$-polygons. By identifying each side with its mate we obtain a 2-cell embedding whose multigraph is isomorphic to $G$.

We claim that the resulting surface $S$ is orientable. It suffices to show that the surface does not contain a simple closed polygonal curve $C$ such that, when we traverse $C$, left and right interchange. Since we are working with $\pi$-polygons in the plane, it makes sense to speak of a point "close to" a closed polygonal arc $C$ on the left side of $C$ when a positive direction is assigned to $C$. Now we only have to check that walking "close to $P$" on $S$ will never take us from the left side of $P$ to the right side. We leave the details to the reader.

This construction shows that every connected multigraph with at least one edge admits a 2-cell embedding in some orientable surface.

An embedding[4] of $G$ in $S$ is *cellular* if every face of $G$ is homeomorphic to an open disc in $\mathbb{R}^2$. It is clear that every 2-cell embedding is cellular, and we shall prove below (Theorem 3.2.4) that also the converse is true

---

[4]Embeddings of graphs in topological spaces and the notion of a face of an embedding are defined in Section 2.1.

in the sense that every cellular embedding is homeomorphic to a 2-cell embedding. To prove that, we need some basic results on the plane $\mathbb{R}^2$.

Let $C$ be a simple closed polygonal curve in $\mathbb{R}^2$, let $q_1, \ldots, q_k \in C$ ($k \geq 1$) and $p \in \text{int}(C)$. Let $P_1, \ldots, P_k$ be simple polygonal arcs in $\overline{\text{int}}(C)$ such that $P_i$ connects $p$ and $q_i$ for $i = 1, \ldots, k$, and $P_i \cap P_j = \{p\}$ for $1 \leq i < j \leq k$ (see Figure 3.6).

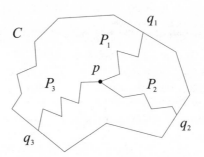

FIGURE 3.6. Clockwise order around a vertex

Assume that $q_1, \ldots, q_k$ appear in that cyclic order when we traverse $C$ in the clockwise direction (meaning that $\text{int}(C)$ is on the right hand side). By Proposition 2.1.5, $P_1 \cup \cdots \cup P_k$ divides $\text{int}(C)$ into $k$ faces. One is bounded by the cycle obtained by walking on the segment $R$ of $C$ from $q_1$ to $q_2$ in the clockwise direction and then turning sharp right. That face is bounded by $R \cup P_2 \cup P_1$. Similarly for the other bounded faces. It follows that the initial straight line segments of $P_1, \ldots, P_k$ occur in that clockwise order around $p$. This simple observation leads to the following.

LEMMA 3.2.1. *Let $C'$ be a simple closed polygonal curve in $\overline{\text{int}}(C)$ such that $p \in \text{int}(C') \subseteq \text{int}(C)$. Let $P_i'$, be the segment of $P_i$ from $p$ to the point $q_i'$ on $C'$ such that $P_i' \cap C' = \{q_i'\}$, $i = 1, \ldots, k$. Then $q_1', \ldots, q_k'$ occur in that clockwise order on $C'$.*

Now let $Q_1, \ldots, Q_k$ be simple arcs (not necessarily polygonal) such that $Q_i \cap Q_j = \{p\}$ for $1 \leq i < j \leq k$. Let $C$ be a simple closed polygonal curve such that $p \in \text{int}(C)$ and $Q_i \cap C \neq \emptyset$ for $i = 1, \ldots, k$. Let $Q_i^\circ$ be the segment of $Q_i$ from $p$ to, say, $q_i$, on $C$ such that $C \cap Q_i^\circ = \{q_i\}$, $i = 1, \ldots, k$. We define the *clockwise order* of $Q_1, \ldots, Q_k$ around $p$ as the clockwise order of $q_1, \ldots, q_k$ on $C$.

LEMMA 3.2.2. *The clockwise order of $Q_1, \ldots, Q_k$ is independent of $C$.*

PROOF. Assume $C'$ is a simple polygonal closed curve such that $p \in \text{int}(C') \subseteq \text{int}(C)$. Let $Q_i'$ be the segment of $Q_i$ from $p$ to the point $q_i'$ on $C'$ such that $Q_i' \cap C' = \{q_i'\}$ for $i = 1, \ldots, k$. As in the proof of Lemma

2.1.1, there are polygonal arcs $P_1, \ldots, P_k$ in $\overline{\text{int}}(C)$ such that $P_i$ connects $p$ and $q_i$, and such that $q_i' \in P_i$, and $P_i \cap P_j = \{p\}$ for $1 \leq i < j \leq k$. By Lemma 3.2.1, the clockwise order of $q_1, \ldots, q_k$ on $C$ is the same as the clockwise order of $q_1', \ldots, q_k'$ on $C'$.

The proof is now completed by observing that for any simple closed curves $C, C''$ containing $p$ in their interior, there is a simple polygonal closed curve $C'$ such that $p \in \text{int}(C') \subseteq \text{int}(C) \cap \text{int}(C'')$. $\qquad \square$

Let $C, Q_i, q_i, Q_i^\circ$ be as above and let $P_i$ be a simple polygonal arc in $\overline{\text{int}}(C)$ from $p$ to $q_i$ such that $P_i \cap C = \{q_i\}$, $i = 1, \ldots, k$, and such that $P_i \cap P_j = \{p\}$, $1 \leq i < j \leq k$. From the Jordan–Schönflies theorem (combined with Theorem 2.2.6 if $k = 1$) we have:

LEMMA 3.2.3. *There exists a homeomorphism $f$ of $\mathbb{R}^2$ onto $\mathbb{R}^2$ keeping $C \cup \text{ext}(C)$ fixed such that $f(P_i) = Q_i^\circ$, $i = 1, \ldots, k$.*

Let $P$ be a simple arc in an open disc $D$ in $\mathbb{R}^2$. As $D$ is homeomorphic to $\mathbb{R}^2$, it follows from Theorem 2.1.11 that $D \setminus P$ is arcwise connected. Also, it is easy to find a simple closed polygonal curve $C$ in $D$ such that $P \subseteq \text{int}(C)$. We apply this observation to a connected graph $G$ embedded in an orientable triangulated surface $S$. (In particular, we think of $S$ as a union of triangles in $\mathbb{R}^2$, so that we can speak of polygonal arcs in $S$.) For every vertex $v$ in $G$ we let $D$ be an open disc containing $v$ but intersecting only edges incident with $v$. We may assume that $D = \text{int}(C_v)$ where $C_v$ is a simple closed polygonal curve. By a simple compactness argument, there exists, for every edge $e$ of $G$ (which is not a loop), a simple closed polygonal curve $C_e$ such that $e$ is in $\text{int}(C_e)$ and $\text{int}(C_e) \cap \text{int}(C_f) = \emptyset$ if $e, f$ are nonadjacent edges, and $\text{int}(C_e) \cap \text{int}(C_f) = \text{int}(C_v)$ if both $e$ and $f$ are incident with the vertex $v$. (If $e$ is a loop, then $\overline{C_e}$ is a cylinder containing $e$. Note that, in order to ensure that $\text{int}(C_e) \cap \text{int}(C_f) = \text{int}(C_v)$ we may have to replace the original $C_v$ by a smaller one. This is an easy exercise on plane graphs with polygonal edges.) Moreover, we can assume that the union of all $C_e$ and $C_v$ ($e \in E(G)$, $v \in V(G)$) form a cubic graph, see Figure 3.7.

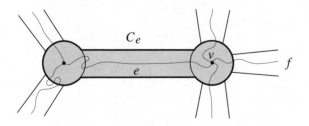

FIGURE 3.7. Discs around vertices and edges

Assume now that all vertices of $G$ have degree different from 2. Lemma 3.2.2 enables us to speak of the clockwise ordering of the edges incident with $v$ except that we have two possibilities (since we can interchange between clockwise and anticlockwise). We fix one of these as being clockwise. Using $\mathrm{int}(C_e)$ we then obtain a clockwise ordering of the edges incident with the other end of $e$. Continuing like that and using the facts that $G$ is connected and that $S$ does not contain a Möbius strip, we obtain a clockwise ordering around each vertex $v$ of $G$. In order to make this precise, the following observation is useful. If $C$ is a cycle in $G$ with edges $e_1, e_2, \ldots, e_r$, then $\overline{\mathrm{int}}(C_{e_1}) \cup \overline{\mathrm{int}}(C_{e_2}) \cup \cdots \cup \overline{\mathrm{int}}(C_{e_r})$ is homeomorphic to an annulus (i.e., a cylinder) in $\mathbb{R}^2$ because $S$ does not contain a Möbius strip. This observation (the proof of which we leave for the reader) shows that clockwise does not change to anticlockwise when we traverse $C$.

If $G$ contains vertices of degree two, we suppress them to get a homeomorphic graph $G'$ without such vertices. The above method determines clockwise ordering around each vertex of $G'$ and hence also around each vertex of $G$.

Assume that $G$ is cellularly embedded in $S$. Let $\pi = \{\pi_v \mid v \in V(G)\}$ where $\pi_v$ is the cyclic permutation of the edges incident with the vertex $v$ such that $\pi_v(e)$ is the successor of $e$ in the clockwise ordering around $v$. The cyclic permutation $\pi_v$ is called the *local rotation* at $v$, and the set $\pi$ is the *rotation system* of the given embedding of $G$ in $S$.

As noted at the beginning of this section, we can use the rotation system $\pi$ to define a surface $S'$ which is formed by pasting pairwise disjoint $\pi$-polygons in $\mathbb{R}^2$ together. With this notation we have:

THEOREM 3.2.4. *Suppose that $G$ is a connected multigraph with at least one edge that is cellularly embedded in an orientable surface $S$. Let $\pi$ be the rotation system of this embedding, and let $S'$ be the surface of the corresponding 2-cell embedding of $G$. Then there exists a homeomorphism of $S$ onto $S'$ taking $G$ in $S$ onto $G$ in $S'$ (in such a way that we induce the identity map from $G$ onto its copy in $S'$). In particular, every cellular embedding of a graph $G$ in an orientable surface is uniquely determined, up to homeomorphism, by its rotation system.*

PROOF. As previously noted and indicated in Figure 3.7, we can define a cubic graph, which we call $H$, which is the union of the simple closed curves $C_e$ ($e \in E(G)$). We draw the corresponding graph $H'$ on $S'$ using polygonal arcs. By iterated use of Lemma 3.2.3 there exists a homeomorphism $f$ of $\bigcup_{e \in E(G)} \overline{\mathrm{int}}(C_e)$ onto the corresponding subset of $S'$ such that $f(H) = H'$ and $f$ takes $G$ (in $S$) onto $G$ (in $S'$). The part of $S$ where $f^{-1}$ is not defined are faces of $H'$ bounded by cycles in $H'$. (To see this, focus on a face $F$ of $G$ in $S'$ bounded by the polygon $P$ which is a $\pi$-walk in $G$. Now $P$ together with the part of $H'$ inside $P$ forms a 2-connected graph with precisely one facial cycle that does not intersect $P$.) By the

Jordan–Schönflies Theorem, we can extend $f^{-1}$ to a homeomorphism of $S'$ onto $S$. There is only one detail which needs discussion. If $F_1'$ is face in $S'$ on which $f^{-1}$ is undefined, then there is a corresponding face $F_1$ in $S$. We only have to show that, if $F_2'$ is another such face in $S'$, then the corresponding face $F_2$ in $S$ is distinct from $F_1$. So assume that $F_1 = F_2$. Let $Q$ be a simple arc in $F_1$ from $f^{-1}(\partial F_1')$ to $f^{-1}(\partial F_2')$ where $\partial F_i'$ is the boundary of $F_i'$ for $i = 1, 2$, such that all points of $Q$, except the ends, are in $F_1$. It is easy to see that $Q$ does not separate $F_1$. This contradicts Corollary 2.1.4 (applied to $\overline{F_1}$). □

The last sentence of Theorem 3.2.4 is often called the *Heffter–Edmonds–Ringel rotation principle*. The idea is implicitly used by Heffter [**He891**]. It was made explicit by Edmonds [**Ed60**], and Ringel [**Ri74**] demonstrated its importance in the proof of the Heawood conjecture (see Section 8.3).

Rotation systems $\pi = \{\pi_v \mid v \in V(G)\}$ and $\pi' = \{\pi_v' \mid v \in V(G)\}$ of $G$ are *equivalent* if they are either the same or for each $v \in V(G)$ we have $\pi_v' = \pi_v^{-1}$. A simple corollary of Theorem 3.2.4 is:

COROLLARY 3.2.5. *Suppose that we have cellular embeddings of a connected multigraph $G$ in orientable surfaces $S$ and $S'$ with rotation systems $\pi$ and $\pi'$, respectively. Then there is a homeomorphism $S \to S'$ whose restriction to $G$ induces the identity if and only if $\pi$ and $\pi'$ are equivalent.*

We conclude this section with two examples.

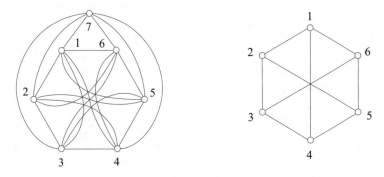

FIGURE 3.8. Rotation systems of $K_7$ and $K_{3,3}$

In Figure 3.8 we have drawings of the graphs $K_7$ and $K_{3,3}$ in the plane. The local rotations of these drawings determine embeddings in the torus. They are shown in Figure 3.9 (where the surface of the torus is represented as a square with opposite sides identified).

## 3.3. Embedding schemes

In Section 3.2 we showed how a cellular embedding of a connected graph in an orientable surface can be described as a 2-cell embedding by

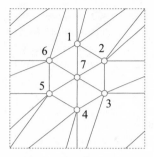   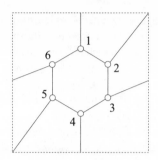

FIGURE 3.9. Embeddings of $K_7$ and $K_{3,3}$ in the torus

its rotation system. In this section we extend the notion of the rotation system in order to include embeddings in nonorientable surfaces. Let $G$ be a connected multigraph cellularly embedded in a surface $S$. Assume first that $G$ has no vertices of degree 2. Lemma 3.2.2 enables us to speak of clockwise ordering of edges incident with a vertex. For each vertex we have two possible choices, and we choose one of them to be clockwise. Define curves $C_v$, $v \in V(G)$, and $C_e$, $e \in E(G)$, as in Section 3.2. For each edge $e = uv$ of $G$ we check if the clockwise orderings at $v$ and $u$ agree in the disc $\operatorname{int}(C_e)$. If this is the case, then we set $\lambda(e) = 1$, otherwise $\lambda(e) = -1$. This defines a mapping $\lambda : E(G) \to \{1, -1\}$, called a *signature*. The pair $\Pi = (\pi, \lambda)$, where $\pi = \{\pi_v \mid v \in V(G)\}$ is the set of chosen local rotations, is an *embedding scheme* for the given embedding of $G$. If $\lambda(e) = 1$ for each edge $e$, then $S$ is orientable and the embedding coincides with the 2-cell embedding obtained from the rotation system $\pi$.

If $G$ has vertices of degree 2, we suppress them to get a multigraph $G'$ on which we can define an embedding scheme $\Pi' = (\pi', \lambda')$ as described above. Now, for each edge $e'$ of $G'$, which corresponds to a path $P$ in $G$, we replace $\lambda'(e')$ by the signature $\lambda$ on $E(P)$ such that the first edge of $P$ has signature $\lambda'(e')$, and all other edges have positive signature. This defines an embedding scheme for $G$.

The embedding scheme corresponding to an embedding of $G$ is not uniquely determined. For example, if we change clockwise ordering at a vertex $v$ to anticlockwise, then $\pi_v$ is replaced by $\pi_v^{-1}$, and for each edge $e$ incident with $v$ we change $\lambda(e)$ to $-\lambda(e)$. If we select a spanning tree $T$ of $G$, then clearly the local rotations $\pi$ can be chosen in such a way that $\lambda(e) = 1$ for each edge $e \in E(T)$.

Two embedding schemes $\Pi$ and $\Pi'$ of $G$ are *equivalent* if $\Pi'$ can be obtained from $\Pi$ by a sequence of operations, each one involving a change of clockwise to anticlockwise at a vertex $v$ and the corresponding change of the signs of the edges incident with $v$.

Having an embedding scheme $\Pi$ of $G$, one can define a 2-cell embedding of $G$ in a surface $S'$ generalizing the method at the beginning of Section 3.2. The only difference is that the $\Pi$-*facial walks* and $\Pi$-*polygons* are determined by the following generalized process, called the *face traversal procedure*. We start with an arbitrary vertex $v$ and an edge $e = vu$ incident with $v$. Traverse the edge $e$ from $v$ to $u$. We continue the walk along the edge $e' = \pi_u(e)$. We repeat this procedure as in the orientable case, except that, when we traverse an edge with signature $-1$, the $\pi$-anticlockwise rotation is used to determine the next edge of the walk. (This can happen already at $u$ if $\lambda(e) = -1$.) We continue using $\pi$-anticlockwise ordering until the next edge with signature $-1$ is traversed, and so forth. The walk is completed when the initial edge $e$ is encountered in the same direction from $v$ to $u$ and we are in the same mode (the $\pi$-clockwise ordering) with which we started. The other $\Pi$-facial walks are determined in the same way by starting with other edges. At the beginning of Section 3.2 we described how to obtain a surface from a rotation system. In the present more general context we must argue why every edge appears precisely twice in $\Pi$-facial walks. We leave this to the reader. Clearly, two embedding schemes are equivalent if and only if they have the same set of facial walks.

It is easy to see that the surface $S'$ is nonorientable if and only if $G$ contains a cycle $C$ which has odd number of edges $e$ with $\lambda(e) = -1$. For, if $C$ exists, then along $C$ left and right interchange, hence $S'$ contains a Möbius strip. If such a cycle does not exist, then we modify $\Pi$ to an equivalent embedding scheme such that the signature is positive on a spanning tree of $G$, and hence positive everywhere. It follows that every connected multigraph with at least one cycle has a 2-cell embedding in some nonorientable surface.

The proof of Theorem 3.2.4 easily extends to the following.

THEOREM 3.3.1. *Suppose that $G$ is a connected multigraph (with at least one edge) that is cellularly embedded in a surface $S$. Let $\Pi$ be the corresponding embedding scheme, and let $S'$ be the surface of the 2-cell embedding of $G$ corresponding to $\Pi$. Then there exists a homeomorphism of $S$ onto $S'$ taking $G$ in $S$ onto $G$ in $S'$ whose restriction to $G$ induces the identity on $G$. In particular, every cellular embedding of a graph $G$ in some surface is uniquely determined, up to homeomorphism, by its embedding scheme.*

Theorem 3.3.1 was first made explicit by Ringel [**Ri77a**] and by Stahl [**St78**]. Hoffman and Richter [**HR84**] presented a combinatorial description of embeddings which are not necessarily cellular.

A simple corollary of Theorem 3.3.1 is:

COROLLARY 3.3.2. *Let $\Pi$ and $\Pi'$ be embedding schemes corresponding to cellular embeddings of a connected multigraph $G$ in surfaces $S$ and*

*S'*, respectively. Then there is a homeomorphism of *S* to *S'* whose restriction to *G* induces the identity if and only if Π and Π' are equivalent.

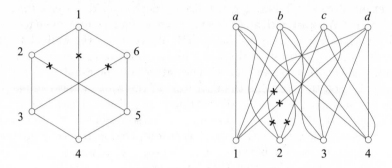

FIGURE 3.10. Embedding schemes of $K_{3,3}$ and $K_{4,4}$

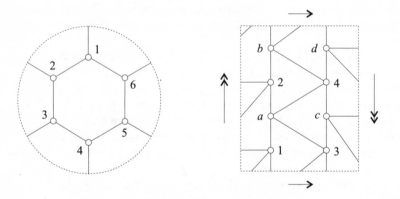

FIGURE 3.11. Embeddings of $K_{3,3}$ and $K_{4,4}$

Consider, for example, the drawings of $K_{3,3}$ and $K_{4,4}$ in Figure 3.10 with local rotations as indicated in the picture. A negative signature of an edge is marked by a cross. The corresponding 2-cell embeddings in the projective plane and in the Klein bottle, respectively, are shown in Figure 3.11 (where the projective plane is represented as a disk *D* with every pair of opposite points on the boundary of *D* identified, while the Klein bottle is represented by a rectangle whose sides are identified as shown in the figure).

### 3.4. The genus of a graph

We define the *genus* $\mathbf{g}(G)$ and the *nonorientable genus* $\widetilde{\mathbf{g}}(G)$ of a graph *G* as the minimum *h* and the minimum *k*, respectively, such that *G* has an embedding into the surface $\mathbb{S}_h$, respectively into $\mathbb{N}_k$. Embeddings of $G$ into $\mathbb{S}_{\mathbf{g}(G)}$ are *minimum genus embeddings*. Similarly, a *nonorientable minimum genus embedding* of a graph $G$ is an embedding into $\mathbb{N}_{\widetilde{\mathbf{g}}(G)}$.

The following observation of Youngs [**Yo63**] shows that the problem of determining the genus of a graph $G$ can be reduced to 2-cell embeddings of $G$. By the results of the previous section, the genus calculation is thus a finite problem which can be solved by considering all rotation systems of $G$.

PROPOSITION 3.4.1. *Every minimum genus embedding of a connected graph $G$ is cellular.*

PROOF. Denote by $n$ and $q$ the number of vertices and edges of $G$, respectively. Suppose that we have an embedding of $G$ into $\mathbb{S}_{\mathbf{g}(G)}$ that is not cellular. If $f$ is the number of faces of this embedding, then a proof similar to the proof of (3.4) shows that

$$n - q + f > 2 - 2\mathbf{g}(G).$$

Let $\pi$ be the rotation system of the embedding and let $r$ be the number of $\pi$-polygons. Then $r \geq f$ which implies that

$$n - q + r > 2 - 2\mathbf{g}(G).$$

The corresponding 2-cell embedding of $G$ has genus $g' = 1 - (n-q+r)/2 < \mathbf{g}(G)$, a contradiction. □

For the nonorientable case Parsons, Pica, Pisanski, and Ventre [**PPPV87**] proved the following.

PROPOSITION 3.4.2. *Let $G$ be a connected graph. If $\widetilde{\mathbf{g}}(G) < 2\mathbf{g}(G) + 1$, then every nonorientable minimum genus embedding of $G$ is cellular. If $\widetilde{\mathbf{g}}(G) = 2\mathbf{g}(G) + 1$ and $G$ is not a tree, then $G$ has both a cellular and a noncellular embedding in $\mathbb{N}_{\widetilde{\mathbf{g}}(G)}$.*

It is easy to see that $\widetilde{\mathbf{g}}(G) \leq 2\mathbf{g}(G) + 1$, cf. Proposition 4.4.1.

## 3.5. Classification of noncompact surfaces

Recall that a surface is a connected compact Hausdorff topological space $S$ which is locally homeomorphic to an open disc in the plane. If we drop the requirement that the surface is compact, we can get complicated topological spaces, see Radó [**Ra25**]. If we add the additional requirement that such a space is metrizable (or equivalently, paracompact or triangulable) but not compact, we get the notion of *noncompact surfaces*. There is a classification theorem for noncompact surfaces (see Kerékjártó [**Ke23**] and Richards [**Ri63**]) which we present in the sequel.

Let $S$ be a noncompact surface. A *boundary component* of $S$ is a decreasing sequence $R_1 \supset R_2 \supset R_3 \supset \cdots$ of connected noncompact regions in $S$ such that the boundary of each $R_i$ in $S$ is compact and for each compact subset $A \subseteq S$ there is an index $i$ such that $R_i \cap A = \emptyset$. Two boundary components $R_1 \supset R_2 \supset R_3 \supset \cdots$ and $R_1' \supset R_2' \supset R_3' \supset \cdots$ are *equivalent* if for each $n$ there is an integer $N$ such that $R_n \subset R_N'$ and $R_n' \subset R_N$. The set of equivalence classes of boundary components is called the set of *ends* of $S$, and is denoted by $\beta(S)$. It is a totally disconnected, compact metric space, as a subspace of $S^* = S \cup \beta(S)$ endowed with the following topology. The points of $S$ have the same basis of topology as in $S$, and a point $\zeta \in \beta(S)$ has the basis consisting of all sets $U \subset S^*$ such that:

(a) The boundary of $U \cap S$ is compact in $S$.
(b) For some (and hence any) boundary component $R_1 \supset R_2 \supset R_3 \supset \cdots$ of the end $\zeta$, $U \cap S$ contains all $R_n$ for $n$ sufficiently large.
(c) $U \cap \beta(S)$ contains precisely those ends $\zeta'$ which satisfy condition (b) above (with $\zeta'$ instead of $\zeta$).

One can show that the space $S^*$ is a metrizable compact space.

The ends are of three types. An end, represented by a boundary component $R_1 \supset R_2 \supset R_3 \supset \cdots$ is *orientable* if some $R_n$ is orientable, and is *nonorientable* otherwise. An orientable end is *planar* if some $R_n$ is homeomorphic to a subset of the plane, and *nonplanar* otherwise. For example, if we take a point $\zeta$ (or more generally a totally disconnected closed set $X$) from a compact surface $S'$, then $S = S' \backslash \{\zeta\}$ (or $S' \backslash X$) is a noncompact surface with a planar end $\zeta$ (respectively, the set of planar ends homeomorphic to $X$), and $S^* = S'$. The one way infinite connected sum of tori $\mathbb{S}_\infty$ has an orientable nonplanar end. Note that $\mathbb{S}_\infty^*$ is not a surface. Similarly, an infinite connected sum of projective planes has one or more nonorientable ends. Let $\beta'(S)$ be the set of orientable ends, and let $\beta''(S)$ be the set of planar ends of $S$. Then $\beta(S) \supseteq \beta'(S) \supseteq \beta''(S)$. The sets $\beta'(S)$ and $\beta''(S)$ are open in $\beta(S)$.

The *genus* of $S$ is the maximum genus of all compact subsurfaces (with boundary) in $S$. (A *compact surface with boundary* is obtained from a compact surface $S$ by deleting the interiors of a finite number of pairwise disjoint contractible simple closed curves on $S$.) The genus may be infinite as well. There are four *orientability types* of noncompact surfaces: *orientable* (each compact subsurface is orientable), *infinitely nonorientable* (for no compact subsurface $A$, $S \backslash A$ is orientable), *finitely nonorientable* of either *even* or *odd nonorientability type* (every sufficiently large compact subsurface of $S$ is of even or odd nonorientable genus). The following theorems classify noncompact surfaces [**Ke23, Ri63**].

THEOREM 3.5.1. *Let $S_1$ and $S_2$ be noncompact surfaces having the same genus and orientability type. Then $S_1$ and $S_2$ are homeomorphic if and only if there is a homeomorphism $\phi : \beta(S_1) \to \beta(S_2)$ such that $\phi(\beta'(S_1)) = \beta'(S_2)$ and $\phi(\beta''(S_1)) = \beta''(S_2)$.*

THEOREM 3.5.2. *Let $X$ be a nonempty totally disconnected compact metric space, and let $A, B$ be open subsets of $X$ such that $A \supseteq B$. Then there exists a noncompact surface $S$ such that the triple $(\beta(S), \beta'(S), \beta''(S))$ is homeomorphic to $(X, A, B)$.*

Basic results about embeddings of infinite graphs in noncompact surfaces have been derived by Mohar [**Mo88b**]. There is a vast literature on infinite graphs in the plane.

# CHAPTER 4

# Embeddings combinatorially, contractibility of cycles, and the genus problem

By Theorems 3.2.4 and 3.3.1, a cellular embedding is uniquely determined by its rotation system (or embedding scheme). Motivated by this result we now *define* an embedding of a connected graph to be a rotation system (or an embedding scheme) of the graph. So from now on, an embedding is a purely combinatorial object and it makes sense to speak of computational aspects of embeddings. We derive some basic properties of embeddings and describe techniques which enable us still to use geometric intuition. We then treat some classical problems mentioned below.

Heawood [**He890**] introduced the problem of determining the smallest number $H(S)$ such that every map on the surface $S$ can be colored in $H(S)$ colors in such a way that no two neighboring countries receive the same color. This problem which we shall treat in more detail in Chapter 8 can be reduced to the problem of deciding which complete graphs can be embedded in $S$. This is a special case of the *genus problem*, one of the fundamental problems listed by Garey and Johnson [**GJ79**]: Given a graph $G$, determine the minimum genus of a surface in which $G$ can be embedded. The solution of the Heawood problem obtained by Ringel and Youngs [**RY68, Ri74**] involves the calculation of the genera of the complete graphs.

Embeddings and the genus of graphs are studied in Section 4.4. It is shown that the genus problem is as difficult as many other well-known hard combinatorial problems. More precisely, the decision version of the genus problem is **NP**-complete as shown by Thomassen [**Th89**]. In the last section we treat the maximum genus of graphs which is better understood than the (minimum) genus.

The calculation of genera of various classes of graphs are extensively treated in other books on topological graph theory. The reader is referred to the monographs by Ringel [**Ri74**], White [**Wh73**], and Gross and Tucker [**GT87**].

## 4.1. Embeddings combinatorially

Let $G$ be a connected multigraph. An *embedding* of $G$ is a pair $\Pi = (\pi, \lambda)$ where $\pi = \{\pi_v \mid v \in V(G)\}$ is a *rotation system* (which means that for each vertex $v$, $\pi_v$ is a cyclic permutation of the edges incident with $v$) and $\lambda$ is a *signature* mapping which assigns to each edge $e \in E(G)$ a *sign* $\lambda(e) \in \{-1, 1\}$. If $e$ is an edge incident with $v \in V(G)$, then the cyclic sequence $e, \pi_v(e), \pi_v^2(e), \ldots$ is called the $\Pi$-*clockwise ordering* around $v$ (or the *local rotation* at $v$). Given an embedding $\Pi$ of $G$ we say that $G$ is $\Pi$-*embedded*.

We define the $\Pi$-*facial walks* (which we shall also refer to as $\Pi$-*faces*) as the closed walks in $G$ that are determined by the face traversal procedure described in Chapter 3. If $G = K_1$, then we define the walk of length 0 to be the facial walk. Facial walks that differ only by a cyclic shift are considered to be the same. Clearly, each edge of $G$ either appears twice on the same facial walk or appears in exactly two facial walks. The edges that are contained (twice) in only one facial walk are called *singular*. Similarly, if $v$ is a vertex that appears more than once on a $\Pi$-facial walk $W$, then $v$ is *singular* in $W$. If $W$ contains no singular vertices or edges, then $W$ is a cycle and we call it a $\Pi$-*facial cycle*.

Edges $e$ and $f$ incident with vertex $v$ are $\Pi$-*consecutive* if $\pi_v(e) = f$ or $\pi_v(f) = e$. Every such pair $\{e, f\}$ of edges forms a $\Pi$-*angle*.

A *subwalk* of a walk is defined in the obvious way. Two facial walks can have a subwalk in common. Also, a walk $W$ can appear twice as a subwalk of a facial walk. However, this occurs only in special cases as the next lemma shows.

LEMMA 4.1.1. *If* $W$ *is a walk that appears as a (proper) subwalk in two* $\Pi$-*facial walks or appears twice as a subwalk of the same* $\Pi$-*facial walk, then all intermediate vertices of* $W$ *are of degree 2. If* $C$ *is a cycle that appears twice as a subwalk of* $\Pi$-*facial walks, then* $C$ *has at most two vertices of degree distinct from 2.*

PROOF. Let $W = v_0 e_1 v_1 e_2 \ldots e_k v_k$. Under the assumptions of the lemma, each $\Pi$-angle $\{e_i, e_{i+1}\}$ $(i = 1, \ldots, k-1)$ appears twice among the angles of the embedding. This implies that $e_{i+1}$ is both the successor and the predecessor of $e_i$ in the $\Pi$-clockwise ordering around $v_i$. Therefore $v_i$ has degree 2 in $G$. A cycle can appear twice as a subwalk with different initial vertices $v, v'$. All vertices different from $v$ and $v'$ are intermediate vertices of common subwalks and hence of degree 2 by the above.          $\square$

A *local change* of the embedding $\Pi = (\pi, \lambda)$ changes the clockwise ordering to anticlockwise at some vertex $v \in V(G)$, i.e., $\pi_v$ is replaced by its inverse $\pi_v^{-1}$, and $\lambda(e)$ is replaced by $-\lambda(e)$ for all edges $e$ that are incident with $v$. Two embeddings are *equivalent* if one can be obtained from the other by a sequence of local changes. It is easy to see that for an arbitrary spanning tree $T$ of $G$ there is an embedding equivalent to $\Pi$ such that the signs of the edges in $T$ are all positive.

LEMMA 4.1.2. *Two embeddings of $G$ are equivalent if and only if they have the same set of facial walks.*

PROOF. It is clear that local changes preserve facial walks. Hence equivalent embeddings have the same facial walks. Conversely, if $\Pi = (\pi, \lambda)$ and $\Pi'$ have the same facial walks, then their local rotations are the same up to a change of clockwise with anticlockwise. So, $\Pi'$ is equivalent to an embedding $(\pi, \lambda'')$. The face traversal procedure shows that $\lambda = \lambda''$. □

By Corollary 3.3.2, equivalent embeddings are topologically the same. Therefore, all important properties of embeddings depend only on the equivalence class of an embedding. Lemma 4.1.2 shows that all invariants defined by using facial walks satisfy this condition.

The following property is immediate from the definition of facial walks.

LEMMA 4.1.3. *If $W$ is a $\Pi$-facial walk, then the number of (appearances of) edges on $W$ with negative signature is even.*

A cycle $C$ of a $\Pi$-embedded graph $G$ is $\Pi$-*onesided* if it has an odd number of edges with negative sign. Otherwise $C$ is $\Pi$-*twosided*. If $G$ contains a $\Pi$-onesided cycle, then the embedding $\Pi$ is *nonorientable*. Otherwise $\Pi$ is *orientable*. With this notation, the remark preceding Lemma 4.1.2 implies

LEMMA 4.1.4. *An embedding is orientable if and only if it is equivalent to an embedding with positive signature.*

Suppose that a $\Pi$-embedded connected multigraph $G$ has $n$ vertices, $q$ edges and that there are $f$ $\Pi$-facial walks. The number

$$\chi(\Pi) = n - q + f \tag{4.1}$$

is called the *Euler characteristic* of the embedding $\Pi$. (Although (4.1) is now just the definition of $\chi(\Pi)$ we still call it *Euler's formula*.) If $\Pi$ is an orientable embedding, then we define the *genus* of $\Pi$ (or the $\Pi$-*genus* of $G$) as

$$\mathbf{g}(\Pi) = 1 - \tfrac{1}{2}\chi(\Pi). \tag{4.2}$$

If $\Pi$ is nonorientable, then its *genus* (which is sometimes also referred to as the *nonorientable genus* or the *crosscap number*) is

$$\widetilde{\mathbf{g}}(\Pi) = 2 - \chi(\Pi). \tag{4.3}$$

It is also convenient to introduce the *Euler genus* $\mathbf{eg}(\Pi)$ of an embedding $\Pi$ which is defined as

$$\mathbf{eg}(\Pi) = 2 - \chi(\Pi). \tag{4.4}$$

If $\Pi$ is an orientable embedding of genus $g$, then we also say that $\Pi$ is an embedding of $G$ in $\mathbb{S}_g$, the orientable surface of genus $g$. For example, we say that an orientable embedding $\Pi$ is an embedding in the torus if $\mathbf{g}(\Pi) = 1$. The same terminology is used for nonorientable embeddings. Thus we speak of embeddings in the projective plane $\mathbb{N}_1$, the Klein bottle $\mathbb{N}_2$, etc.

Let $G$ be a $\Pi$-embedded graph and $H$ a connected subgraph of $G$. Then the *induced embedding* of $H$ is obtained from $\Pi$ by ignoring all edges of $E(G)\backslash E(H)$. More precisely, we restrict the signature to $E(H)$, and if $e \in E(H)$ is an edge incident with $v$, then the successor of $e$ in the clockwise ordering around $v$ in $H$ is the first edge of $H$ in the sequence $\pi_v(e), \pi_v^2(e), \ldots$. We denote the induced embedding by $\Pi$ as well.

Suppose that $e \in E(G)$ is an edge such that $G - e$ is connected. If $e$ belongs to two $\Pi$-facial walks, then the number of faces of the induced embedding is one less than the number of faces of $G$, and so $G - e$ has the same Euler characteristic as $G$. Using Lemmas 4.1.3 and 4.1.4 it is not difficult to see that the orientability of the embedding remains unchanged. Putting the edge $e$ back to $G - e$ results in splitting a facial walk of $G - e$ in two walks. We say that $e$ is *embedded in a face* of $G - e$. On the other hand, if $e$ appears twice on the same facial walk, then $G - e$ either has the same number of facial walks or one more than $G$. Hence the Euler characteristic increases by one (which can happen only when the embedding of $G$ is nonorientable) or two.

If $v$ is a vertex of degree 1 in $G$, then $G$ and $G - v$ have the same genus and orientability. (There is only one facial walk of $G$ containing $v$. It gives rise to a $\Pi$-face $F$ of $G - v$, and we say that $v$ is *embedded in $F$*.) Since every connected subgraph of $G$ can be obtained from $G$ by successively deleting edges or vertices of degree 1, it follows that the $\Pi$-genus of any connected subgraph of $G$ is smaller than or equal to the $\Pi$-genus of $G$. Since the $\Pi$-genus of a single vertex of $G$ is 0 and the Euler characteristic changes by one only in the nonorientable case, it follows that if $\Pi$ is an orientable embedding, then its genus is a nonnegative integer.

The above argument shows that going from an embedded multigraph to the induced embedding of a subgraph, never increases the genus and never changes an orientable embedding into a nonorientable one.

PROPOSITION 4.1.5. *Let $G'$ be a connected subgraph of a $\Pi$-embedded graph and let $\Pi'$ be the induced embedding of $G'$. Then $\mathbf{eg}(\Pi') \le \mathbf{eg}(\Pi)$, and if $\mathbf{eg}(\Pi') = \mathbf{eg}(\Pi)$, then $\Pi$ and $\Pi'$ are embeddings of the same orientability type.*

Let $G$ be a connected multigraph and $K$ a connected subgraph of $G$. Let $\Pi$ be an embedding of $K$. An embedding $\tilde{\Pi}$ of $G$ is an *extension* of $\Pi$ to $G$ if $\Pi$ is the induced embedding of $\tilde{\Pi}$, and $\Pi$ and $\tilde{\Pi}$ are embeddings in the same surface.

Let $e = uv$ be an edge of $G$ which is not a loop, and let $w$ be the vertex of $G/e$ obtained after the contraction of $e$. If $\Pi = (\pi, \lambda)$ is an embedding of $G$ with $\lambda(e) = 1$ and $\pi_v = (e, e_1, \ldots, e_p)$, $\pi_u = (e, e'_1, \ldots, e'_r)$, let $\pi_w = (e_1, \ldots, e_p, e'_1, \ldots, e'_r)$. This defines an embedding of $G/e$ which we call the *embedding induced* by $\Pi$. The facial walks in $G/e$ are the same as the facial walks of $G$ except that in the walk(s) containing $e$, the edge $e$ is suppressed. Therefore $G/e$ is embedded in the same surface as $G$.

Suppose now that $e = uv$ is an edge of $G$ which is not a loop, and that $\lambda(e) = -1$. To define the *induced embedding* of $G/e$, we first make the local change at $u$ (or, equivalently, at $v$) to obtain an equivalent embedding of $G$ where $e$ has positive signature. After that, the induced embedding of $G/e$ is defined as above. If we make the local change at $v$ instead of $u$, we obtain an equivalent induced embedding of $G/e$.

Since a connected minor of $G$ can be obtained from $G$ by successively deleting edges and contracting edges that are not loops, we have now defined the *induced embedding* $\Pi'$ of an arbitrary connected minor $G'$ of $G$. Observe that $\Pi'$ is unique up to equivalence of embeddings. We say that the $\Pi'$-embedded minor $G'$ of $G$ is a *surface minor* of the $\Pi$-embedded graph $G$. Note that the order of edge deletions and edge contractions used to get a surface minor is not important.

Given an embedding $\Pi = (\pi, \lambda)$ of a connected multigraph $G$, we define the *geometric dual multigraph* (or the $\Pi$-*dual multigraph*) $G^*$ and its embedding $\Pi^* = (\pi^*, \lambda^*)$, called the *dual embedding* of $\Pi$, as follows. The vertices of $G^*$ correspond to the $\Pi$-facial walks. The edges of $G^*$ are in bijective correspondence $e \mapsto e^*$ with the edges of $G$, and the edge $e^*$ joins the vertices corresponding to the $\Pi$-facial walks containing $e$. (If $e$ is singular, then $e^*$ is a loop.) If $W = e_1 \ldots e_k$ is a $\Pi$-facial walk and $w$ its vertex of $G^*$, then $\pi_w^* = (e_1^*, \ldots, e_k^*)$. For $e^* = ww'$ we set $\lambda^*(e^*) = 1$ if the $\Pi$-facial walks $W$ and $W'$ used to define $\pi_w^*$ and $\pi_{w'}^*$ traverse the edge $e$ in opposite direction; otherwise $\lambda^*(e^*) = -1$. Clearly, $\Pi^*$ is orientable if $\Pi$ is orientable. It is easy to see that $G$ is the geometric dual of $G^*$. In particular, $\Pi$ is orientable if $\Pi^*$ is orientable. By Euler's formula, $\Pi^*$ is an embedding in the same surface as $\Pi$.

Proposition 2.6.3 extends from plane graphs to arbitrary embeddings.

PROPOSITION 4.1.6. *Let $G$ be a connected $\Pi$-embedded multigraph and $G^*$ its geometric dual multigraph. If $e$ is an edge of $G$ that is not a loop, then $(G/e)^* = G^* - e^*$. If $e$ is contained in two facial walks, then $(G - e)^* = G^*/e^*$.*

The easy proof is omitted.

## 4.2. Cycles of embedded graphs

For simplicity we now focus on graphs although most of the results also hold for multigraphs. (Note that an embedded multigraph can be modified to a graph by inserting vertices of degree 2 on the loops and multiple edges.)

Let $C = v_0 e_1 v_1 e_2 \ldots v_{l-1} e_l v_l$ be a $\Pi$-twosided cycle of a $\Pi$-embedded graph $G$. Suppose that the signature of $\Pi$ is positive on $C$. We define the *left graph* and the *right graph* of $C$ as follows. For $i = 1, \ldots, l$, if $e_{i+1} = \pi_{v_i}^{k_i}(e_i)$, then all edges $\pi_{v_i}(e_i), \pi_{v_i}^2(e_i), \ldots, \pi_{v_i}^{k_i-1}(e_i)$ are said to be *on the left side* of $C$. Now, the *left graph* of $C$, denoted by $G_l(C, \Pi)$ (or just $G_l(C)$), is defined as the union of all $C$-bridges that contain an edge on the left side of $C$. The *right graph* $G_r(C, \Pi)$ (or just $G_r(C)$) is defined analogously. If the signature is not positive on $C$, then there is an embedding $\Pi'$ equivalent to $\Pi$ whose signature is positive on $C$ (since $C$ is $\Pi$-twosided). Now we define $G_l(C, \Pi)$ and $G_r(C, \Pi)$ as the left and the right graph of $C$ with respect to the embedding $\Pi'$. Note that a different choice of $\Pi'$ gives rise to the same pair $\{G_l(C), G_r(C)\}$ but the left and the right graphs may interchange.

A cycle $C$ of a $\Pi$-embedded graph $G$ is $\Pi$-*separating* (or *surface separating*) if $C$ is $\Pi$-twosided and $G_l(C)$ and $G_r(C)$ have no edges in common.

PROPOSITION 4.2.1. *Let $C$ be a surface separating cycle in a $\Pi$-embedded graph $G$. Then the Euler genus of $\Pi$ equals the sum of the Euler genera of the induced embeddings of $G_l(C) \cup C$ and $G_r(C) \cup C$.*

PROOF. Each $\Pi$-facial walk of $G$ is either in $G_l(C) \cup C$ or in $G_r(C) \cup C$. In addition, $C$ is a facial cycle in both graphs and there are no other facial walks in either of the graphs. Now Euler's formula proves the proposition. $\square$

If one of $G_l(C) \cup C$ or $G_r(C) \cup C$ (say the former) in Proposition 4.2.1 has $\Pi$-genus zero, then we say that $C$ is $\Pi$-*contractible* and we call $G_l(C)$ the $\Pi$-*interior* of $C$ and denote it by $\text{int}(C, \Pi)$ (or just $\text{int}(C)$). We put

$$\text{Int}(C, \Pi) = \text{int}(C, \Pi) \cup C.$$

We define the $\Pi$-*exterior* $\text{ext}(C)$ as $G_r(C)$ and put $\text{Ext}(C) = \text{ext}(C) \cup C$. It is easy to see that $\text{Ext}(C)$ is $\Pi$-embedded in the same surface as $G$. We say that $G$ is obtained from $\text{Ext}(C)$ by *embedding* $\text{Int}(C)$ in a face of $\text{Ext}(C)$. Note that every $\Pi$-facial cycle $C$ is $\Pi$-contractible (and hence also surface separating). In that case $\text{int}(C) = \emptyset$.

Figure 4.1 shows a graph on the double torus $\mathbb{S}_2$ where the cycle $C_0$ is contractible, $C_1, C_2, C_3$ are noncontractible, $C_1$ is surface separating, and neither of $C_2, C_3$ is surface separating.

PROPOSITION 4.2.2. *If $H$ is a connected subgraph of a $\Pi$-embedded graph $G$ such that $G$ and $H$ have the same $\Pi$-genus, then each $\Pi$-facial cycle in $H$ is $\Pi$-contractible in $G$. If all $\Pi$-facial walks of $H$ are cycles, then $G$ is obtained from $H$ by embedding subgraphs of genus 0 in faces of $H$.*

PROOF. Let $C$ be a facial cycle in $H$. By Lemma 4.1.3, $C$ is twosided in $G$. If $G_l(C)$ and $G_r(C)$ have an edge in common, then there exists a

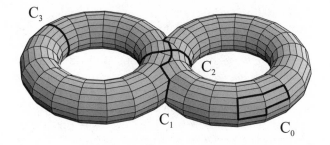

FIGURE 4.1. Cycles in an embedded graph

path $P$ (or a cycle) in $G$ starting with an edge $e$ on the left side of $C$, ending with an edge on the right side of $C$ and such that no intermediate vertex of $P$ is on $C$. All edges of $H - E(C)$ incident with $C$ are on the same side of $C$, say on the right. Let $P'$ be the shortest subpath of $P$ starting with $e$ and ending in $H$. Now $H \cup P'$ has fewer $\Pi$-facial walks than $H$ and hence $H \cup P'$ has larger $\Pi$-genus than $H$. This contradicts the assumption that $H$ and $G$ (and hence also every connected subgraph of $G$ containing $H$) have the same $\Pi$-genus. Hence $C$ is $\Pi$-separating.

By Proposition 4.2.1, the $\Pi$-genus of $G$ equals the sum of the $\Pi$-genera of $G_l(C) \cup C$ and $G_r(C) \cup C$. Since the latter contains $H$, we conclude that the former graph has $\Pi$-genus zero, and hence $C$ is $\Pi$-contractible in $G$. $\qquad\qquad\qquad\qquad\qquad\qquad\qquad\qquad\qquad\qquad\qquad\qquad\qquad$ □

COROLLARY 4.2.3. *Let $H$ be a connected subgraph of a $\Pi$-embedded graph $G$ such that the $\Pi$-genera of $G$ and $H$ are equal. If all $\Pi$-facial walks of $H$ are cycles, then a cycle $C$ in $H$ is $\Pi$-contractible in $H$ if and only if it is $\Pi$-contractible in $G$.*

PROOF. A cycle which is $\Pi$-contractible in $G$ is clearly $\Pi$-contractible in every connected subgraph which contains the cycle. Suppose now that $C$ is $\Pi$-contractible in $H$. By Proposition 4.2.2, $G$ is obtained from $H$ by adding subgraphs of genus zero in faces of $H$. Hence $\mathrm{Int}(C)$ also has genus 0 in $G$. $\qquad\qquad\qquad\qquad\qquad\qquad\qquad\qquad\qquad\qquad\qquad\qquad\qquad$ □

Suppose now that $C$ is a surface nonseparating cycle of a $\Pi$-embedded graph $G$. If $C$ is $\Pi$-twosided, let $\overline{G}$ be the graph obtained from $G$ by replacing $C$ with two copies of $C$ such that all edges on the left side of $C$ are incident with one copy of $C$ and all edges on the right side of $C$ are incident with the other copy of $C$. We say that $\overline{G}$ is obtained from $G$ by *cutting* (or $\Pi$-*cutting*) *along* the cycle $C$. The embedding $\Pi$ defines an embedding of $\overline{G}$ (which we also denote by $\Pi$). The only difference between the two embeddings is that two edges of $C$ which are incident with the same vertex $v$ of $C$ are consecutive in the new rotation around $v$.

Similarly we define *cutting along* a $\Pi$-onesided cycle. If $C = v_0 e_1 v_1 e_2 \ldots e_k v_0$ is such a cycle, we first define edges on the left side of $C$ (and edges on the right side) at each vertex of $C$ in a similar way as in the case of twosided cycles: By taking an equivalent embedding if necessary, we may assume that the signature $\lambda$ of $\Pi$ satisfies: $\lambda(e_i) = 1$ for $1 \leq i < k$, and $\lambda(e_k) = -1$. Then we use pairs of consecutive edges $e_i, e_{i+1}$ on $C$ $(i = 1, \ldots, k - 1)$ to define edges on the left side of $C$ incident with the vertex $v_i$ incident with $e_i$ and $e_{i+1}$. Then we construct the graph $\overline{G}$ by replacing $C$ in $G$ by the cycle $\overline{C} = v_0 e_1 v_1 \ldots e_k \bar{v}_0 \bar{e}_1 \bar{v}_1 \ldots \bar{e}_k v_0$ whose length is twice the length of $C$. The edges on the left side of $C$ are incident to $v_0, v_1, \ldots, v_{k-1}$ and the edges on the right side of $C$ are incident to $\bar{v}_0, \bar{v}_1, \ldots, \bar{v}_{k-1}$. We extend $\lambda$ by putting $\lambda(\bar{e}_1) = \cdots = \lambda(\bar{e}_{k-1}) = 1$ and $\lambda(\bar{e}_k) = -1$ and hence obtain an embedding $\overline{\Pi}$ of $\overline{G}$.

LEMMA 4.2.4. *Suppose that $C$ is a $\Pi$-nonseparating cycle of a $\Pi$-embedded graph $G$. Let $\overline{G}$ be the graph obtained by cutting along $C$ and let $\overline{\Pi}$ be its embedding. Then all $\Pi$-facial walks are $\overline{\Pi}$-facial walks in $\overline{G}$, where edges of $C$ are replaced by their copies in $\overline{G}$. If $C$ is $\Pi$-twosided, then $\mathbf{eg}(\overline{\Pi}) = \mathbf{eg}(\Pi) - 2$ and the two copies of $C$ are the new $\overline{\Pi}$-facial cycles. If $C$ is $\Pi$-onesided, then $\mathbf{eg}(\overline{\Pi}) = \mathbf{eg}(\Pi) - 1$ and $\overline{C}$ is a facial cycle in $\overline{G}$.*

PROOF. The claims about $\overline{\Pi}$-facial walks in the graph $\overline{G}$ are easy consequences of the definitions. Now the lemma follows from Euler's formula. □

Lemma 4.2.4 can be used in proofs by induction on the genus. In the nonorientable case the new embedding may become orientable, though.

Every embedding $\Pi$ of positive genus contains $\Pi$-nonseparating (and hence also $\Pi$-noncontractible) cycles. They can be found by the following method. We may assume that the signature of $\Pi$ is positive on a spanning tree $T$ of $G$. For each edge $e \in E(G) \backslash E(T)$ we let $C(e, T)$ be the fundamental cycle of $e$ with respect to $T$ (i.e., the unique cycle in $T + e$). If $\Pi$ is nonorientable, then at least one of the fundamental cycles $C(e, T)$, $e \in E(G) \backslash E(T)$, is $\Pi$-onesided. We say that edges $e, f \in E(G) \backslash E(T)$ $\Pi$-*overlap* (with respect to $T$) if the $\Pi$-genus of $T + e + f$ is nonzero. If $e$ and $f$ $\Pi$-overlap and $\Pi$ is orientable, then it is easy to see that $f \in G_l(C(e, T)) \cap G_r(C(e, T))$ and $e \in G_l(C(f, T)) \cap G_r(C(f, T))$. In particular, $C(e, T)$ and $C(f, T)$ are both $\Pi$-nonseparating cycles. If there are no $\Pi$-overlapping pairs of edges in $E(G) \backslash E(T)$, then it is easy to see that the embedding has genus zero, hence the claim.

Overlapping pairs of edges in $F = E(G) \backslash E(T)$ determine a symmetric 01-matrix $A(\Pi, T)$ whose rows and columns are indexed by the edges in $F$ and whose $(e, f)$-entry is equal to 1 if $e$ and $f$ $\Pi$-overlap with respect to $T$, and is 0 otherwise. In particular, the diagonal entry corresponding to the edge $e$ is 1 if and only if $C(e, T)$ is $\Pi$-onesided. Then the rank over

$GF(2)$ of $A(\Pi, T)$ is equal to the Euler genus of $\Pi$ as proved by Mohar [**Mo89**]. Analogous versions were proved by Brahana [**Br21**], Cohn and Lempel [**CL72**], Goldstein and Turner [**GT79**], and Marx [**Ma81**].

Now we introduce a classification of cycles into free homotopy classes which refines their partition into contractible and noncontractible cycles. First we extend the definition of *cutting along a cycle* also to surface separating cycles. In that case, the cutting gives rise to two graphs $G_l(C, \Pi) \cup C$ and $G_r(C, \Pi) \cup C$. More generally, if $\mathcal{C} = \{C_1, \ldots, C_k\}$ is a set of disjoint cycles of $G$, we define *cutting along* $\mathcal{C}$ as cutting first along $C_1$, then along $C_2$, etc. Clearly, the order of cutting along cycles in $\mathcal{C}$ is not important.

We say that two disjoint cycles $C, C'$ are (*freely*) $\Pi$-*homotopic* if either $C$ and $C'$ are both $\Pi$-contractible, or $C$ and $C'$ are $\Pi$-twosided and cutting along $\{C, C'\}$ results in a graph which has a component $D$ which contains precisely one copy of $C$ and one copy of $C'$ and whose $\Pi$-genus is zero. In the latter case we write $D = \text{Int}(C \cup C', \Pi)$.

LEMMA 4.2.5. *Suppose that $C_1, C_2, C_3$ are pairwise disjoint cycles and that $C_1, C_2$ and $C_2, C_3$ are $\Pi$-homotopic (respectively). Then also $C_1$ and $C_3$ are $\Pi$-homotopic. If one of the cycles is $\Pi$-noncontractible, then so are the other two and there is a pair $\{r, s\} \subset \{1, 2, 3\}$ such that for $1 \leq i < j \leq 3$*

$$\text{Int}(C_i \cup C_j, \Pi) \subseteq \text{Int}(C_r \cup C_s, \Pi).$$

PROOF. If one of $C_1, C_2, C_3$ is $\Pi$-contractible, then so are the other two. (We leave the details to the reader.) So, let us assume that none of the cycles are contractible. Let $D_{ij} = \text{Int}(C_i \cup C_j, \Pi)$ and let $\Pi_{ij}$ be the induced embedding of $D_{ij}$, $(i, j) \in \{(1, 2), (2, 3)\}$. Suppose that $C_3 \subseteq D_{12}$. Since $C_3$ is $\Pi_{12}$-contractible and is not $\Pi$-contractible, the notation can be chosen such that $C_1 \subseteq \text{int}(C_3, \Pi_{12})$ and $C_2 \subseteq \text{ext}(C_3, \Pi_{12})$. Then we may take $r = 1$, $s = 2$, and the claims of the lemma follow easily. The same is true if $C_1 \subseteq D_{23}$. Otherwise, $D_{12} \cap D_{23} = C_2$ and $D_{12}$ contains all edges on the left side of $C_2$ (say), and $D_{23}$ contains all edges on the right side of $C_2$. Then $D_{12} \cup D_{23}$ is one of the components obtained by cutting along $\{C_1, C_3\}$ and, clearly, its $\Pi$-genus is 0. Therefore $C_1$ and $C_3$ are $\Pi$-homotopic and $D_{12} \cup D_{23}$ can be taken as $\text{Int}(C_1 \cup C_3, \Pi)$. This completes the proof. $\square$

In the next chapter we shall use the following propositions.

PROPOSITION 4.2.6 (Malnič and Mohar [**MM92**]). *Let $G$ be a $\Pi$-embedded graph and $C_1, \ldots, C_k$ pairwise disjoint, $\Pi$-noncontractible and pairwise $\Pi$-nonhomotopic cycles of $G$. Let $g = \mathbf{g}(\Pi)$ or $g = \widetilde{\mathbf{g}}(\Pi)$ if $\Pi$ is orientable or nonorientable, respectively. Then*

$$k \leq \begin{cases} g, & \text{if } g \leq 1 \\ 3g - 3, & \text{if } g \geq 2. \end{cases} \tag{4.5}$$

PROOF. Using Lemma 4.2.4, (4.5) is easy to verify when $g \leq 1$. Suppose now that $g \geq 2$, and let $\mathcal{C}_1$ and $\mathcal{C}_2$ be the partition of $\mathcal{C} = \{C_1, \dots, C_k\}$ into the $\Pi$-onesided and $\Pi$-twosided cycles, respectively. Let $G_1, \dots, G_r$ be the components and $\Pi_1, \dots, \Pi_r$ (respectively) their induced embeddings, obtained after cutting along $\mathcal{C}$. By Proposition 4.2.1 and Lemma 4.2.4,

$$\sum_{i=1}^{r} \mathbf{eg}(\Pi_i) = \mathbf{eg}(\Pi) - 2(|\mathcal{C}_2| - (r-1)) - |\mathcal{C}_1|. \qquad (4.6)$$

Let $\psi(\Pi_i) = \mathbf{eg}(\Pi_i) - 2 + b_i$ where $b_i$ is the number of copies of cycles from $\mathcal{C}$ contained in $G_i$ $(i = 1, \dots, r)$. Then (4.6) implies

$$\sum_{i=1}^{r} \psi(\Pi_i) = \mathbf{eg}(\Pi) - 2. \qquad (4.7)$$

Since the left hand side of (4.6) is nonnegative, we also get

$$2|\mathcal{C}| - \mathbf{eg}(\Pi) + 2 \leq 2r + |\mathcal{C}_1|. \qquad (4.8)$$

Suppose first that $\Pi$ is orientable. Since $g \geq 2$ and no cycle in $\mathcal{C}$ is $\Pi$-contractible and no two cycles are $\Pi$-homotopic, $\psi(\Pi_i) \geq 1$ for $i = 1, \dots, r$. Now (4.7) implies $r \leq \mathbf{eg}(\Pi) - 2$. Inserting this inequality in (4.8) and using the assumption that $|\mathcal{C}_1| = 0$ gives (4.5).

If $\Pi$ is nonorientable, then $|\mathcal{C}_1| \leq \mathbf{eg}(\Pi)$ by Lemma 4.2.4. For the same reason as above we have $\psi(\Pi_i) \geq 0$, and if $\psi(\Pi_i) = 0$, then either $\mathbf{eg}(\Pi_i) = 0$ and $b_i = 2$ (in which case one of the cycles from $\mathcal{C}$ in $G_i$ corresponds to a $\Pi$-onesided cycle), or $\mathbf{eg}(\Pi_i) = 1$ and $b_i = 1$. By Proposition 4.2.1, the number of components $G_i$ for which $\psi(\Pi_i) = 0$ is at most $\mathbf{eg}(\Pi)$. Now (4.7) implies $r \leq 2\mathbf{eg}(\Pi) - 2$. By combining this bound and the bound on $\mathcal{C}_1$ with (4.8), we get (4.5). $\qquad \square$

It is easy to give examples showing that the inequalities of (4.5) are sharp.

If $a, b$ are distinct vertices of a $\Pi$-embedded graph $G$, and $P_1, P_2$ are internally disjoint paths from $a$ to $b$ in $G$, then $P_1$ and $P_2$ are said to be $\Pi$-*homotopic* if the cycle $C = P_1 \cup P_2$ is $\Pi$-contractible. This definition extends to the case when $a = b$ and $P_1, P_2$ are cycles with $P_1 \cap P_2 = \{a\}$ as follows: If the $\Pi$-clockwise (or the $\Pi$-anticlockwise) ordering around $a$ is $e_1, f_1, \dots, f_k, e_2, g_1, \dots, g_l$ where $e_i$ is an edge of $P_i$ $(i = 1, 2)$ and none of the edges $f_1, \dots, f_k$ are in $P_1 \cup P_2$, then we replace the vertex $a$ by two adjacent vertices $a', a''$ such that $a'$ is incident with $e_1, f_1, \dots, f_k, e_2$ and $a''$ is incident with $g_1, \dots, g_l$. The embedding $\Pi$ determines an embedding $\Pi'$ of the resulting graph $G'$. Then we say that $P_1$ and $P_2$ are $\Pi$-*homotopic* if they give rise to $\Pi'$-homotopic paths $P_1'$ and $P_2'$ between $a'$ and $a''$ in $G'$; see Figure 4.2(a). If the edges of $P_1$ and $P_2$ incident to $a$ interlace as indicated in Figure 4.2(b), then we consider each of the splittings in that

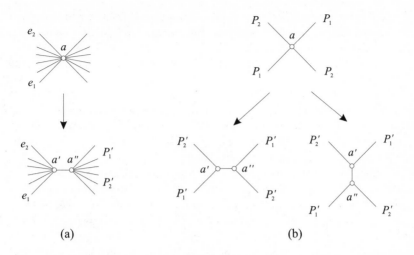

FIGURE 4.2. Splitting the vertex $a$

figure and check if $P_1'$ and $P_2'$ are homotopic in one of the two new graphs. (It is easy to see that this may happen only in the nonorientable case.) If $P_1', P_2'$ are homotopic and $D = \operatorname{int}(P_1' \cup P_2', \Pi')$, we also say that the closed walk $P_1 \cup P_2$ is $\Pi$-*contractible* and write $\operatorname{int}(P_1 \cup P_2, \Pi) = D$ and $\operatorname{Int}(P_1 \cup P_2, \Pi) = D \cup P_1 \cup P_2$.

PROPOSITION 4.2.7 (Malnič and Mohar [**MM92**]). *Let $G$ be a $\Pi$-embedded graph and $a, b$ vertices of $G$ (possibly $a = b$). If $P_0, P_1, \ldots, P_k$ are pairwise internally disjoint paths (or cycles) from $a$ to $b$ such that no two of them are $\Pi$-homotopic, then*

$$k \le \begin{cases} \mathbf{eg}(\Pi), & \text{if } \mathbf{eg}(\Pi) \le 1 \\ 3\,\mathbf{eg}(\Pi) - 3, & \text{if } \mathbf{eg}(\Pi) \ge 2. \end{cases} \qquad (4.9)$$

PROOF. If $\mathbf{eg}(\Pi) \le 1$, the bound follows from Proposition 4.2.1 and Lemma 4.2.4.

Suppose now that $\mathbf{eg}(\Pi) \ge 2$. For $i = 0, \ldots, k$, let $e_i$ be an edge of $P_i$. Let $T$ be a spanning tree of $G$ containing $P_i - e_i$ for $i = 0, \ldots, k$ and if $a \ne b$, also $e_0 \in E(T)$. (If $a = b$, then at most one of $P_0, \ldots, P_k$ is contractible, so we may assume that $P_1, \ldots, P_k$ are all noncontractible if $a = b$.) After contracting $E(T)$, the resulting multigraph $G/E(T)$ is embedded in the same surface as $G$. It contains only one vertex, the loops $e_1, \ldots, e_k$ and some additional loops. Let $G'$ be the largest subgraph of $G/E(T)$ (and $\Pi'$ its induced embedding) such that no edge in $G'$ distinct from $e_1, \ldots, e_k$ is contained in two $\Pi'$-facial walks. Then $\Pi'$ is an embedding in the same surface as $\Pi$. If $W$ is a $\Pi'$-facial walk containing an edge $f$ distinct from $e_1, \ldots, e_k$, then $f$ occurs twice in $W$ and hence the length of $W$ is at least 3. If $W$ consists of loops from $\{e_1, \ldots, e_k\}$ only, then $W$ is not a single loop $e_i$ since otherwise $P_0 \cup P_i$ would be $\Pi$-contractible and hence $P_0$ and $P_i$ would be $\Pi$-homotopic. Similarly, if $W = e_i e_j$, then $P_i$ and $P_j$ are $\Pi$-homotopic. Therefore the length of $W$ is at least 3. It follows that $2|E(G')| \geq 3t$ where $t$ is the number of $\Pi'$-facial walks. Now, Euler's formula implies (4.9).                                                         □

The upper bound on $k$ in (4.9) is best possible when $a = b$. If $a \neq b$, then at most two $\Pi'$-facial walks $W$ in the proof of Proposition 4.2.7 may have length 3. All others have length at least 4. This implies that $k \leq 2\mathbf{eg}(\Pi) - 1$ (and this bound is best possible).

Schrijver [**Sch91a**] proved a min-max characterization for existence of disjoint cycles of prescribed homotopies in an embedded graph $G$. Schrijver's characterization yields a polynomial time algorithm for finding such cycles or proving their nonexistence. A related unsolved problem is to find edge-disjoint cycles of prescribed homotopies. This problem has a close connection to multicommodity flow problems and has been successfully attacked by several authors, cf. Lins [**Li81**], Schrijver [**Sch90, Sch91b**], and de Graaf [**Gr94**].

### 4.3. The 3-path-condition

Let $\mathcal{K}$ be a family of cycles in $G$. Following [**Th90b**] we say that $\mathcal{K}$ satisfies the *3-path-condition* if it has the following property. If $x, y$ are vertices of $G$ and $P_1, P_2, P_3$ are internally disjoint paths joining $x$ and $y$, and if two of the three cycles $C_{i,j} = P_i \cup P_j$ $(1 \leq i < j \leq 3)$ are not in $\mathcal{K}$, then also the third cycle is not in $\mathcal{K}$. An example of such a family is the set of all cycles of odd length in $G$. More generally, the set of all cycles in $G$ that have odd intersection with a prescribed edge set satisfies the 3-path-condition. For embedded graphs we have:

PROPOSITION 4.3.1. *The following families of cycles of a $\Pi$-embedded graph $G$ satisfy the 3-path-condition:*

(a) *The family of $\Pi$-noncontractible cycles.*
(b) *The family of $\Pi$-nonseparating cycles.*
(c) *The family of $\Pi$-onesided cycles.*

PROOF. We shall use the notation from the definition of the 3-path-condition. To prove (a), suppose that $C_{1,2}$ and $C_{2,3}$ are contractible. If $P_3 \subseteq \mathrm{int}(C_{1,2})$ or $P_1 \subseteq \mathrm{int}(C_{2,3})$ we have finished. So assume that $P_3 \subseteq \mathrm{ext}(C_{1,2})$ and $P_1 \subseteq \mathrm{ext}(C_{2,3})$. Now choose the orientation of $C_{i,j}$ $(1 \leq i < j \leq 3)$ such that we first walk from $x$ to $y$ along $P_i$ and then return along $P_j$. Without loss of generality we may assume that $\mathrm{int}(C_{1,2}) = G_r(C_{1,2})$. Since $P_3 \subseteq \mathrm{ext}(C_{1,2}) = G_l(C_{1,2})$, we have $P_1 \subseteq G_l(C_{2,3})$. Since $P_1 \subseteq \mathrm{ext}(C_{2,3})$, we have $G_l(C_{2,3}) = \mathrm{ext}(C_{2,3})$. Now it follows that $G_r(C_{1,3}) = G_r(C_{1,2}) \cup G_r(C_{2,3}) \cup P_2$. Since $C_{1,2}$ and $C_{2,3}$ are contractible,

an easy count shows that $G_r(C_{1,3}) \cup C_{1,3}$ has $\Pi$-genus zero. This proves
(a). The proofs of (b) and (c) are similar. $\square$

In Section 4.2 we observed that every embedding of positive genus has
surface nonseparating (and hence noncontractible) cycles and that every
nonorientable embedding has onesided cycles. Below we present a general
algorithm from [**Th90b**] which we call the *fundamental cycle method* that
finds a shortest cycle in any set $\mathcal{C}$ of cycles satisfying the 3-path-condition.
If we can decide membership in $\mathcal{C}$ in polynomial time, then the worst case
time complexity of this method is polynomial.

Suppose that $\mathcal{C}$ is a collection of cycles satisfying the 3-path-condition
in a connected graph $G$. For each vertex $v$ of $G$ we let $T_v$ be a breadth-
first-search spanning tree[1] rooted at $v$, i.e., a tree such that for each vertex
$u$, the distance between $v$ and $u$ in $T_v$ is equal to the distance between $v$
and $u$ in $G$. For each edge $e \in E(G)\backslash E(T_v)$ we let $C(e,T_v)$ be the unique
cycle in $T_v + e$. The fundamental cycle method constructs a tree $T_v$, for
each $v \in V(G)$, and among the fundamental cycles $C(e,T_v)$, $v \in V(G)$,
$e \in E(G)\backslash E(T_v)$, the algorithm selects a shortest one that is in $\mathcal{C}$. We
claim that this method returns a shortest cycle in $\mathcal{C}$. To prove this, let $C$
be a shortest cycle in $\mathcal{C}$ and let $v_0$ be a vertex of $C$. Suppose that $C$ and
$v_0$ are selected in such a way that $|E(C)\backslash E(T_{v_0})|$ is as small as possible.
Let $k = \lfloor |V(C)|/2 \rfloor$ and let $P_1 = v_0 v_1 \ldots v_k$ and $P_2 = v_0 u_1 \ldots u_k$ be the
two paths in $C$ starting at $v_0$.

We first claim that $v_i$ and $u_i$ have distance $i$ to $v_0$ in $G$ for $i = 1, \ldots, k$.
For otherwise, $T_{v_0}$ would have a path $P$ such that $P \cap C = \{x_j, x_l\}$ where
$0 \le j < l \le k$ and $x_j \in \{v_j, u_j\}$ and $x_l \in \{v_l, u_l\}$ and $P$ has length less
than $|j - l|$. By the minimality of $C$, two of the cycles in $C \cup P$ are not
in $\mathcal{C}$. Since $C \in \mathcal{C}$, this contradicts the assumption that $\mathcal{C}$ satisfies the
3-path-condition.

We next claim that $T_{v_0}$ contains both paths $P_1, P_2$ if $v_k \ne u_k$ and
that $T_{v_0}$ contains one of these paths, say $P_1$, and $P_2 - u_k$ if $v_k = u_k$. For
otherwise, $T_{v_0}$ would contain a path $P$ as in the previous paragraph except
that now $P$ has length $|j - l|$. We obtain a contradiction as in the previous
paragraph unless the ends $x_j$ and $x_l$ are on the same path, say on $P_1$. Since
$\mathcal{C}$ satisfies the 3-path-condition, the cycle $C'$ obtained from $C$ by replacing
the segment of $P_1$ from $v_j$ to $v_l$ by $P$ is in $\mathcal{C}$. Now $|E(C')\backslash E(T_{v_0})| <
|E(C)\backslash E(T_{v_0})|$. This contradiction shows that $E(C)\backslash E(T_{v_0})$ consists of
an edge $e$ incident with $u_k$. Hence $C = C(e,T_{v_0})$. This completes the
proof.

It is easy to test if a given cycle is $\Pi$-noncontractible, $\Pi$-nonseparating,
or $\Pi$-onesided. The fundamental cycle method combined with Proposition
4.3.1 therefore implies the following result.

---

[1]Such a tree can be found by the breadth-first search algorithm, see, e.g.,
[**AHU74**].

THEOREM 4.3.2 (Thomassen [**Th90b**]). *There is a polynomially boun-ded algorithm which, for a given* $\Pi$*-embedded graph* $G$*, finds a shortest* $\Pi$*-noncontractible cycle, a shortest* $\Pi$*-nonseparating cycle, and a shortest* $\Pi$*-onesided cycle of* $G$ *whenever such a cycle exists.*

More generally, the fundamental cycle method applies to an arbitrary family $\mathcal{C}$ of cycles of a graph with the property that the cycles not in $\mathcal{C}$ generate a subspace of the cycle space that contains no cycle from $\mathcal{C}$. In other words, we can find, in polynomial time, a shortest cycle outside a given subspace of the cycle space whenever the membership in that subspace can be tested in polynomial time.

In contrast, to find a shortest cycle in a given subspace of the cycle space is not that easy. For example, the following problems are unsolved:

PROBLEM 4.3.3.     (a) *Is there a polynomially bounded algorithm that finds a shortest* $\Pi$*-contractible cycle of a* $\Pi$*-embedded graph* $G$*?*

(b) *Is there a polynomially bounded algorithm that finds a shortest sur-face separating cycle of a* $\Pi$*-embedded graph* $G$*?*

(c) *Is there a polynomially bounded algorithm that finds a shortest* $\Pi$*-twosided cycle of a* $\Pi$*-embedded graph* $G$*?*

Problem 4.3.3(a) was raised by Thomassen [**Th90b**] who pointed out that the answer is simple if there are no vertices of degree less than three: We may assume that $G$ is 2-connected. If one of the cycles of length at most 5 is $\Pi$-contractible, then a shortest $\Pi$-contractible cycle of $G$ is within this set. Otherwise, a shortest $\Pi$-contractible cycle is one of the $\Pi$-facial cycles. This yields an answer in polynomial time.

## 4.4. The genus of a graph

If $G$ is a connected graph, then its *genus* $\mathbf{g}(G)$ is the minimum genus $\mathbf{g}(\Pi)$ of all orientable embeddings $\Pi$ of $G$. In particular, a graph has genus 0 if and only if it is planar. Similarly, the *Euler genus* $\mathbf{eg}(G)$ of $G$ is the minimum Euler genus of all embeddings of $G$. The *nonorientable genus* $\widetilde{\mathbf{g}}(G)$ of a connected graph $G$ with at least one cycle is the minimum $\widetilde{\mathbf{g}}(\Pi)$ taken over all nonorientable embeddings $\Pi$ of $G$. Clearly, $\mathbf{eg}(G) = \min\{2\mathbf{g}(G), \widetilde{\mathbf{g}}(G)\}$. The following simple observation relates the genus and the nonorientable genus of an arbitrary graph.

PROPOSITION 4.4.1. *For every connected graph* $G$ *which is not a tree,*

$$\widetilde{\mathbf{g}}(G) \leq 2\mathbf{g}(G) + 1\,.$$

PROOF. Let $\Pi$ be a minimum genus embedding of $G$ with positive signature. Choose an arbitrary edge $e_0 \in E(G)$ which belongs to a cycle in $G$ and change the signature of $e_0$. This determines a nonorientable embedding $\Pi'$ of $G$. Clearly, $\Pi$-facial walks that do not contain $e_0$ are also $\Pi'$-facial. By Euler's formula, the genus of this embedding is at most $2\mathbf{g}(G) + 1$. □

If $\widetilde{\mathbf{g}}(G) = 2\mathbf{g}(G) + 1$, then $G$ is said to be *orientably simple*. Every connected planar graph $G$ which is not a tree is orientably simple since $\mathbf{g}(G) = 0$ and $\widetilde{\mathbf{g}}(G) = 1$. Corollaries 5.1.3 and 5.10.9 show that for every integer $g \geq 0$, there are infinitely many orientably simple graphs of genus $g$.

Auslander, Brown, and Youngs [**ABY63**] observed that there is no upper bound for $\mathbf{g}(G)$ in terms of $\widetilde{\mathbf{g}}(G)$. We return to this question in Section 5.8.

The following result reduces the genus problem to 2-connected graphs.

THEOREM 4.4.2 (Battle, Harary, Kodama, and Youngs [**BHKY62**]). *The genus of a connected graph is the sum of the genera of its blocks.*

PROOF. It is sufficient to show that, if $G = G_1 \cup G_2$ where $G_1$ and $G_2$ are connected subgraphs of $G$ whose intersection is a vertex $p \in V(G)$, then $\mathbf{g}(G) = \mathbf{g}(G_1) + \mathbf{g}(G_2)$.

Let $\Pi$ be a minimum genus embedding of $G$. A $\Pi$-angle $\{e_1, e_2\}$ at $p$ is said to be *mixed* if $e_i \in E(G_i)$ for $i = 1, 2$. We claim that $\Pi$ can be modified into an embedding of the same genus that has only two mixed angles at $p$. Let $W$ be a $\Pi$-facial walk with edges from $G_1$ and $G_2$. Then $W$ contains at least two mixed angles, say $\{e_1, e_2\}$ and $\{e_1', e_2'\}$. Suppose that there is another mixed angle containing the edge $e_2''$ of $G_2$. We may assume that the $\Pi$-clockwise ordering around $p$ is $\pi_p = (e_1 e_2 W_1 e_2' e_1' W_2 f e_2'' W_3)$ (possibly $f = e_1'$). Now we modify $\pi_p$ into $\pi_p' = (e_1 e_1' W_2 f e_2 W_1 e_2' e_2'' W_3)$. It is easy to verify that the number of facial walks does not decrease when we replace $\pi_p$ by $\pi_p'$. Hence the genus of the embedding does not increase. (By the minimality of $\Pi$, it does not decrease either.) Moreover, this change decreases the number of mixed angles, thus verifying our claim.

If an embedding $\Pi$ of $G$ has only two mixed angles, then it is easy to see that $\mathbf{g}(\Pi) = \mathbf{g}(\Pi_1) + \mathbf{g}(\Pi_2)$. This proves that $\mathbf{g}(G) \geq \mathbf{g}(G_1) + \mathbf{g}(G_2)$.

If $\Pi_i$ is a minimum genus embedding of $G_i$ $(i = 1, 2)$, then there is an embedding $\Pi$ of $G$ with only two mixed angles which induces the embedding $\Pi_i$ on $G_i$, $i = 1, 2$. The above remark implies that $\mathbf{g}(G) \leq \mathbf{g}(\Pi) = \mathbf{g}(G_1) + \mathbf{g}(G_2)$. This completes the proof. □

The above proof also works for the Euler genus and yields the result of Stahl and Beineke.

THEOREM 4.4.3 (Stahl and Beineke [**SB77**]). *If $G$ is a connected graph and $B_1, \ldots, B_s$ are the blocks of $G$, then $\mathbf{eg}(G) = \mathbf{eg}(B_1) + \ldots + \mathbf{eg}(B_s)$.*

Theorem 4.4.3 can be used to compare $\widetilde{\mathbf{g}}(G)$ and the nonorientable genera of the blocks of $G$. In particular, if $G = G_1 \cup G_2$ where $G_1 \cap G_2$ is a single vertex of $G$, then, by Theorem 4.4.3, $\mathbf{eg}(G) = \mathbf{eg}(G_1) + \mathbf{eg}(G_2)$. Proposition 4.4.1 implies that $\widetilde{\mathbf{g}}(G) = \mathbf{eg}(G)$ or $\widetilde{\mathbf{g}}(G) = \mathbf{eg}(G) + 1$. Now,

Theorem 4.4.3 implies that $\widetilde{\mathbf{g}}(G) = \widetilde{\mathbf{g}}(G_1) + \widetilde{\mathbf{g}}(G_2) - \delta$ where $-1 \le \delta \le 2$. A more careful analysis shows that $\delta = 0$ if neither $G_1$ nor $G_2$ is orientably simple, and $\delta = 1$ otherwise. See [**SB77**].

The genus of $G = G_1 \cup G_2$ is also close to the sum of genera of $G_1$ and $G_2$ if $G_1$ and $G_2$ are edge-disjoint graphs which intersect in two vertices. It was shown by Decker, Glover, and Huneke [**DGH81**] and independently by Stahl [**St80**] that

$$\mathbf{g}(G) = \mathbf{g}(G_1) + \mathbf{g}(G_2) - \delta \qquad (4.10)$$

where $\delta \in \{-1, 0, 1\}$. Richter [**Ri87a**] proved the nonorientable version of (4.10):

$$\widetilde{\mathbf{g}}(G) = \widetilde{\mathbf{g}}(G_1) + \widetilde{\mathbf{g}}(G_2) - \tilde{\delta} \qquad (4.11)$$

where $\tilde{\delta} \in \{-2, -1, 0, 1\}$. Related results are given by Miller [**Mi87**], Richter [**Ri87b**], and Stahl [**St80**]. Archdeacon [**Ar86a**] showed that, if $G = G_1 \cup G_2$ where $G_1, G_2$ are edge-disjoint and $G_1 \cap G_2$ consists of $k$ vertices, then

$$|\widetilde{\mathbf{g}}(G) - \widetilde{\mathbf{g}}(G_1) - \widetilde{\mathbf{g}}(G_2)| \le k^2 - 2\,.$$

It is easy to prove that

$$\mathbf{g}(G) \le \mathbf{g}(G_1) + \mathbf{g}(G_2) + k - 1$$

by first letting $G_1$ and $G_2$ be disjoint and then successively identifying vertices until we obtain $G$. However, Archdeacon [**Ar86b**] showed that there is no lower bound for $\mathbf{g}(G) - \mathbf{g}(G_1) - \mathbf{g}(G_2)$ depending only on $k$. Stahl [**St81**] obtained a lower bound on $\mathbf{g}(G) - \mathbf{g}(G_1) - \mathbf{g}(G_2)$ which depends on the degrees of the $k$ vertices in $G_1 \cap G_2$.

Euler's formula implies a simple lower bound on the genus.

PROPOSITION 4.4.4. *Let $G$ be a connected graph with $n \ge 3$ vertices and $q$ edges. Then*

$$\mathbf{g}(G) \ge \left\lceil \frac{q}{6} - \frac{n}{2} + 1 \right\rceil \quad \text{and} \quad \widetilde{\mathbf{g}}(G) \ge \left\lceil \frac{q}{3} - n + 2 \right\rceil. \qquad (4.12)$$

*Moreover, $\mathbf{g}(G) = q/6 - n/2 + 1$ (respectively, $\widetilde{\mathbf{g}}(G) = q/3 - n + 2$) if and only if $G$ triangulates an orientable (respectively, a nonorientable) surface. If $G$ contains no cycle of length 3, then*

$$\mathbf{g}(G) \ge \left\lceil \frac{q}{4} - \frac{n}{2} + 1 \right\rceil \quad \text{and} \quad \widetilde{\mathbf{g}}(G) \ge \left\lceil \frac{q}{2} - n + 2 \right\rceil. \qquad (4.13)$$

*Moreover, $\mathbf{g}(G) = q/4 - n/2 + 1$ (respectively, $\widetilde{\mathbf{g}}(G) = q/2 - n + 2$) if and only if $G$ admits an embedding in an orientable (respectively, a nonorientable) surface such that all facial walks are of length four.*

PROOF. Let $\Pi$ be an embedding of $G$, and let $f$ denote the number of $\Pi$-facial walks. The sum of the lengths of all facial walks is equal to $2q$.

Since $n \geq 3$, each facial walk has length 3 or more. Consequently $2q \geq 3f$. By Euler's formula,

$$3\chi(\Pi) = 3n - 3q + 3f \leq 3n - q \; .$$

(4.12) now follows from (4.2), (4.3), and the fact that $\mathbf{g}(G)$ and $\widetilde{\mathbf{g}}(G)$ are integers.

If $G$ has no 3-cycles, then $2q \geq 4f$. This yields (4.13). $\qquad\square$

For the complete graph $K_n$, it follows from (4.12) that

$$\mathbf{g}(K_n) \geq \left\lceil \frac{(n-3)(n-4)}{12} \right\rceil \quad \text{and} \quad \widetilde{\mathbf{g}}(K_n) \geq \left\lceil \frac{(n-3)(n-4)}{6} \right\rceil$$

if $n \geq 3$. It was conjectured by Heawood in 1890 [**He890**] that these inequalities are actually equalities. In a series of papers, mainly by Ringel, special cases of the Heawood conjecture were settled, and in 1968 Ringel and Youngs [**RY68**] announced the final solution of the Heawood problem.

THEOREM 4.4.5 (Ringel and Youngs [**Ri74**]). *If $n \geq 3$ then*

$$\mathbf{g}(K_n) = \left\lceil \frac{(n-3)(n-4)}{12} \right\rceil \; . \tag{4.14}$$

*If $n \geq 3$ and $n \neq 7$ then*

$$\widetilde{\mathbf{g}}(K_n) = \left\lceil \frac{(n-3)(n-4)}{6} \right\rceil \; . \tag{4.15}$$

A complete proof of Theorem 4.4.5 is presented in the monograph by Ringel [**Ri74**]. The proof is split into 12 cases according to the value of $n$ modulo 12. Some of the cases were later slightly simplified but, for the most complicated ones, no short proofs are known.

PROOF OF (4.14) FOR $n \equiv 7 \,(\text{MOD } 12)$. Let $n = 12k + 7$ ($k \geq 0$). We shall construct an embedding of $K_n$ of genus $12k^2 + 7k + 1$. Since such an embedding yields equality (without the ceiling) in (4.12) if and only if each facial walk is a 3-cycle, it suffices to show that $K_n$ triangulates some orientable surface.

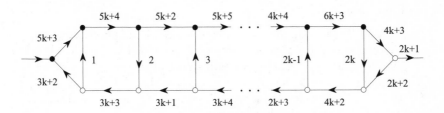

FIGURE 4.3. The graph $G_k$ with labels from $\mathbb{Z}_{12k+7}$

Consider now the cubic graph $G_k$ of Figure 4.3. It has $4k+2$ vertices. Half of them are black, the others are white. Now we define a rotation system of $G_k$ as the clockwise order (shown in Figure 4.3) for white vertices and the anticlockwise order for black vertices. It is easy to see that this rotation system determines an embedding of $G_k$ with exactly one facial walk. Each edge of $G_k$ appears precisely twice on this walk, once in each direction. The edges of $G_k$ are labeled such that each element of $\mathbb{Z}_n\backslash\{0\}$ appears exactly once as a label of an edge or as the negative value (modulo $n$) of an edge label. Also, at every vertex, the sum of incoming labels equals the sum of outgoing labels, and the labels of the vertical edges are $1, 2, \ldots, 2k$. Thus, the facial walk of $G_k$ determines a cyclic permutation $\nu_0$ of $\mathbb{Z}_n\backslash\{0\}$,

$$\nu_0 = (1, 5k+4, 2, 9k+6, 3, 5k+5, 4, 9k+7, 5, 5k+6, 6, 6, \ldots).$$

Now we regard $V(K_n)$ as $\mathbb{Z}_n$. For each vertex $j$ we consider the cyclic permutation $\nu_j$ of the vertices adjacent with $j$ which is obtained from $\nu_0$ by adding $j$ to each term. That is,

$$\begin{aligned} \nu_j = \quad & (1+j, 5k+4+j, 2+j, 9k+6+j, 3+j, 5k+5+j, \\ & 4+j, 9k+7+j, 5+j, 5k+6+j, 6+j, \ldots). \end{aligned}$$

These cyclic permutations determine a rotation system $\Pi$ of $K_n$ such that the $\Pi$-clockwise successor of the edge $jl$ around vertex $j$ is the edge $j\nu_j(l)$. For each vertex $v$ of $G_k$, the sum of the labels on (directed) edges entering $v$ is equal modulo $n$ to the sum of the labels on edges leaving $v$. This implies that each $\Pi$-facial walk is a 3-cycle of $K_n$ and hence $\Pi$ is a triangulation. $\square$

The graph $G_k$ of Figure 4.3 with its edge labels is called a *current graph*. Instead of using $\mathbb{Z}_n$ as labels one may use any Abelian group. Current graphs play an important role in the solution of the Heawood conjecture. For more information on current graphs and the dual concept of *voltage graphs* and their use in embedding theory, the reader is referred to Gross and Tucker [**GT87**].

Next we give a simple proof of the fact that the nonorientable genus of $K_7$ is an exception for the formula (4.15) of Theorem 4.4.5.

THEOREM 4.4.6. $\widetilde{\mathbf{g}}(K_7) = 3$.

PROOF. By (4.14), $\mathbf{g}(K_7) = 1$. Proposition 4.4.1 implies that $\widetilde{\mathbf{g}}(K_7) \leq 3$. So it suffices to prove that $K_7$ cannot be embedded in the Klein bottle.

Suppose that $\Pi = (\pi, \lambda)$ is an embedding of $K_7$ in the Klein bottle. Choose a vertex $x \in V(K_7)$. We may assume that $\lambda(xx') = 1$ for each $x' \in V(K_7 - x)$. By Euler's formula, this embedding has 14 facial walks all of which are 3-cycles. This implies that there is a facial 3-cycle $T = wyz$ that is edge disjoint from the six facial triangles containing $x$. By Lemma 4.1.3, one of the edges of $T$ has positive signature, say $\lambda(yz) = 1$. Then the

3-cycle $C = xyz$ is $\Pi$-twosided and $\Pi$-nonfacial. Moreover, it is induced and nonseparating. By Lemma 4.2.4, cutting along $C$ gives an embedding $\Pi'$ of a graph $G'$ in a surface of Euler genus 0, i.e., the sphere.

Denote by $T_1$ and $T_2$ the copies of $C$ in $G'$, and let $x_i$ be the copy of $x$ in $T_i$, $i = 1, 2$. Note that $G' - T_1 - T_2 = K_4$ and that every vertex of this subgraph is adjacent either to $x_1$ or $x_2$. Thus, $T_1$ and $T_2$ are embedded in distinct faces $F_1, F_2$ (respectively) of $K_4$. By Proposition 2.1.6, $F_i$ contains $T_i$ and precisely six additional edges ($i = 1, 2$). Moreover, the vertex of $F_1 \backslash F_2$ is joined to all vertices of $T_1$, and the vertex of $F_2 \backslash F_1$ is joined to all vertices of $T_2$. Hence the embedding of $G'$ in the sphere is that of Figure 4.4 (without labels) where $T_1$ and $T_2$ are shaded.

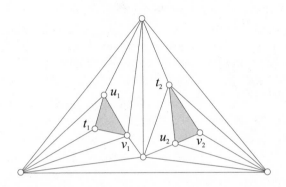

FIGURE 4.4. The plane embedding of $G'$

Now $T_1$ and $T_2$ are identified such that the resulting graph is $K_7$. As the degrees of the vertices in $T_1$ and in $T_2$ are 3, 4, and 5, there is only one way (indicated by labels) of identifying $T_1$ and $T_2$ (in order to get $K_7$) and the identification results in $K_7$ embedded in the torus. $\square$

The inequalities (4.13) of Proposition 4.4.4 applied on the complete bipartite graph $K_{m,n}$ show that $\mathbf{g}(K_{m,n}) \geq \lceil (m - 2)(n - 2)/4 \rceil$ and $\widetilde{\mathbf{g}}(K_{m,n}) \geq \lceil (m - 2)(n - 2)/2 \rceil$. Again, these inequalities are equalities.

THEOREM 4.4.7 (Ringel [**Ri65a, Ri65b**]). *If $m \geq 2$ and $n \geq 2$, then*

$$\mathbf{g}(K_{m,n}) = \left\lceil \frac{(m - 2)(n - 2)}{4} \right\rceil, \tag{4.16}$$

*and if $m \geq 3$, $n \geq 3$, then*

$$\widetilde{\mathbf{g}}(K_{m,n}) = \left\lceil \frac{(m - 2)(n - 2)}{2} \right\rceil. \tag{4.17}$$

PROOF. We present a short proof due to Bouchet [**Bo78a**]. Since the case $m = 2$ or $n = 2$ is trivial, we shall assume $m \geq n \geq 3$. We consider only the orientable case as the proof in the nonorientable case is similar.

Suppose that we have orientable embeddings $\Pi'$ of $K_{p,r}$ and $\Pi''$ of $K_{q,r}$. Choose vertices $v \in V(K_{p,r})$ and $u \in V(K_{q,r})$ that are of degree $r$. Suppose that the clockwise ordering around $v$ and $u$ is $\pi'_v = (vv_1, vv_2, \ldots, vv_r)$ and $\pi''_u = (uu_1, uu_2, \ldots, uu_r)$, respectively. Let $K$ be the graph obtained from $K_{p,r} \cup K_{q,r}$ by deleting $v$ and $u$ and, for $i = 1, \ldots, r$, replacing the edges $vv_i$ and $uu_i$ by an edge $e_i$ joining $v_i$ with $u_i$. Let $\Pi$ be the embedding of $K$ that agrees with $\Pi'$ and $\Pi''$, except that the new edges replace the edges $vv_i$ and $uu_i$ in the local rotations around $v_i$ and $u_i$, respectively ($i = 1, \ldots, r$). After contracting the edges $v_i u_i$, $i = 1, \ldots, r$, we get a graph isomorphic to $K_{p+q-2,r}$. A simple count shows that for the resulting embedding $\Pi$ of this graph

$$\mathbf{g}(\Pi) = \mathbf{g}(\Pi') + \mathbf{g}(\Pi'').  \tag{4.18}$$

Suppose now that $\Pi'$ and $\Pi''$ are minimum genus embeddings such that (4.16) holds, and that $\Pi''$ is a quadrangulation, i.e., $(q-2)(r-2)$ is divisible by 4. Then (4.18) implies that the genus of the embedding $\Pi$ of $K_{p+q-2,r}$ also satisfies (4.16). We shall use the notation $(p,r)\Diamond(q,r) \to (p+q-2, r)$ for this construction.

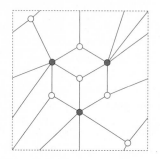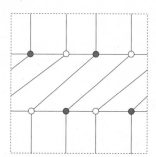

FIGURE 4.5. Embeddings of $K_{6,3}$ and $K_{4,4}$ in the torus

We now prove (4.16) by induction on $m + n$. We first dispose of the cases $K_{m,3}$ ($m \in \{3, 4, 5, 6\}$) and $K_{4,4}$. Embeddings of $K_{6,3}$ and $K_{4,4}$ in the torus are shown in Figure 4.5, and for $K_{3,3}$, $K_{4,3}$, and $K_{5,3}$ we can use the embedding of $K_{6,3}$. If $m = 6$, then we apply the construction $(n-1, 6)\Diamond(3,6) \to (n, 6)$. If $m > 6$, then we use $(m-4, n)\Diamond(6, n) \to (m, n)$. The only remaining cases are $K_{5,4}$ and $K_{5,5}$ for which we can use the embeddings of $K_{6,4}$ and $K_{6,5}$, respectively.    $\square$

Mohar, Parsons, and Pisanski [**MPP85**] generalized the above construction of genus embeddings of complete bipartite graphs to determine the genera of all graphs obtained from $K_{m,n}$ by removing a set of $k \leq \min\{m, n\}$ independent edges. The analogous nonorientable case is treated in [**Mo88a**].

The bounds of Proposition 4.4.4 are sharp for several other families of graphs, in particular several infinite families of complete multipartite graphs, see White [**Wh69**], Gross and Alpert [**GA73**], Jungerman [**Ju75**], Stahl and White [**SW76**], Garman [**Ga78**], Jungerman and Ringel [**JR78**], Bouchet [**Bo78b, Bo82**], and Jackson [**Ja80**]. However, the complete genus classification, even for complete tripartite graphs, is still open. The genus of the $d$-cube was determined by Beineke and Harary [**BH65**] and Ringel [**Ri65c**]. (The $d$-cube is the Cartesian product of $d$ copies of the graph $K_2$.) Ringel [**Ri77b**] determined the genus of $K_2 \square K_n$. Pisanski [**Pi80, Pi82**] generalized the technique of White [**Wh70**] by describing a powerful construction of quadrilateral embeddings of Cartesian products of regular graphs. (If the graphs are triangle-free, then such embeddings are necessarily minimum genus embeddings.)

THEOREM 4.4.8 (Pisanski [**Pi80**]). *Let $G$ and $H$ be connected $r$-regular bipartite graphs. Then the Cartesian product $G \square H$ of $G$ and $H$ has genus $1 + pm(r - 2)/4$ where $p$ and $m$ are the number of vertices of $G$ and $H$, respectively.*

The genus of the lexicographic product has been considered by White [**Wh72**] and Bouchet [**Bo78b**], and the tensor product by Bouchet and Mohar [**BM90**]. There are also some results about genera of Cayley graphs of groups. Such results are related to group actions on surfaces, see [**GT87**].

For a fixed surface $S$, the problem of embedding a graph in $S$ can be settled in polynomial time since there is a characterization of the graphs embeddable in $S$ in terms of a finite number of forbidden subgraphs. (This will be made precise in Chapter 6.) However, if $S$ is not fixed, the problem is difficult as shown by the next result.

THEOREM 4.4.9 (Thomassen [**Th93a**]). *The following problems are* **NP***-complete:*

(a) *Given a graph $G$, does $G$ triangulate a surface?*

(b) *Given a graph $G$, does $G$ triangulate an orientable surface?*

(c) *Given a graph $G$, does $G$ triangulate a nonorientable surface?*

PROOF. All three problems are in **NP** since it is easy to check in polynomial time if a given embedding is triangular or not and if the underlying surface is orientable or not. To prove **NP**-completeness[2] of any of (a), (b), (c), it suffices to find a polynomial time reduction of some known **NP**-complete problem to this problem.

---

[2]For the theory of **NP**-completeness the reader is referred to the monograph of Garey and Johnson [**GJ79**].

The following problem is **NP**-complete [**Th93a**]: Does a given connected cubic bipartite graph contain two Hamilton cycles[3] whose intersection is a perfect matching. This problem can be reduced to either (a) or (b) as follows. Let $G$ be a connected cubic bipartite graph. Let $G'$ be a copy of $G$. Form the disjoint union $G \cup G'$ and join each vertex $v \in V(G)$ with its copy $v'$ in $V(G')$. Add four new vertices $v_1, v_2, v_1', v_2'$ and all edges from $\{v_1, v_2\}$ to $G$ and all edges from $\{v_1', v_2'\}$ to $G'$. Call the resulting graph $M$. Let $A$ be a bipartite class of vertices of $G$ and contract all edges $vv'$, $v \in A$. Denote the resulting graph by $Q$.

We claim that $Q$ triangulates a surface if and only if $G$ has two Hamilton cycles with precisely a matching in common. Moreover, whenever $Q$ triangulates a surface, this surface can be chosen to be orientable. Suppose first that $\Pi$ is a triangular embedding of $Q$. Then the local rotation $\pi_{v_1} = (v_1 u_1, \ldots, v_1 u_n)$ around $v_1$ determines a Hamilton cycle $C_1 = u_1 u_2 \ldots u_n$ of $G$. Similarly, $v_2$ determines a Hamilton cycle $C_2$ of $G$. Since $G$ is cubic, $C_2$ cannot have two consecutive edges in $E(G) \backslash E(C_1)$. Since any two consecutive edges on $C_2$ are $\Pi$-consecutive in the induced embedding of $Q - v_2$, they cannot be consecutive on $C_1$. This implies that $C_1$ and $C_2$ have precisely a perfect matching in common.

Suppose, conversely, that $G$ has two Hamilton cycles $C_1, C_2$ whose intersection is a perfect matching. If $v$ is a vertex in $A$, then we let $e_1, e_2, e_3$ be the edges incident with $v$ such that $e_i \in E(C_i)$ for $i = 1, 2$, and $e_3 \in E(C_1) \cap E(C_2)$. Let

$$\pi_v = (e_1, vv_1, e_3, vv_2, e_2, vv'). \qquad (4.19)$$

be the local rotation around $v$ in $A$. The local rotations in $V(G') \backslash A'$ are defined analogously, and so are the local rotations in $A'$ and in $V(G) \backslash A$ except that here we interchange between 1 and 2 in (4.19). This defines an embedding of $M$ in an orientable surface such that all facial walks containing edges between $G$ and $G'$ are 4-cycles, and all other facial walks are 3-cycles. Contracting the edges between $A$ and $A'$ results in a triangular embedding of $Q$ in an orientable surface.

The above proof shows that problems (a) and (b) are **NP**-complete. To prove (c) we first note that deciding whether the given cubic bipartite graph $J$ with a specified edge $e \in E(J)$ possesses two Hamilton cycles whose intersection is a matching containing the edge $e$ is an **NP**-complete problem [**Th93a**]. Let us therefore consider a bipartite cubic graph $J$ and let $e_1$ be an edge of $J$ through which we want to find two Hamilton cycles intersecting in a matching. Let $u$ be an end of $e_1$ and let $e_2, e_3$ be the other edges incident with $u$. Remove $u$, add a 3-cycle $u_1 u_2 u_3$, and now let $e_i$ be incident with $u_i$, $1 \leq i \leq 3$. The resulting nonbipartite cubic graph $G$ has two Hamilton cycles intersecting in a matching and containing $e_1$ if and only if $J$ does. Let $A$ be the partite class of $J$ not

---

[3]A cycle of the graph $G$ is called a *Hamilton cycle* if it contains all vertices of $G$.

containing $u$. Then $A \subseteq V(G)$, and using it we can form the graphs $M$ and $Q$ as in the first part of the proof. Furthermore, we add to $Q$ two new vertices $z_1, z_2$ and all edges from $z_1$ to $u_1, u_2, u_1', u_2'$ and all edges from $z_2$ to $u_1, u_3, u_1', u_3'$. Call the resulting graph $R$. We claim that $R$ triangulates a nonorientable surface if and only if $G$ contains two Hamilton cycles through $e_1$ intersecting in a matching.

Suppose first that $R$ triangulates some surface. Let $C_1$ and $C_2$ be the two Hamilton cycles corresponding to $v_1$ and $v_2$. As above, $C_1$ and $C_2$ intersect in a matching. Some edge of the 3-cycle $u_1 u_2 u_3$ is in both cycles. This edge must be $u_2 u_3$. For if $u_1 u_2$, say, is in $E(C_1) \cap E(C_2)$, then $u_2 u_3$ is in only one of the cycles. Since $u_2 u_3$ is contained in two facial 3-cycles, and only one of them contains $v_1$ or $v_2$, the 3-cycle $u_1 u_2 u_3$ is facial. But then $u_1 u_2$ is in three facial cycles, a contradiction. So, $C_1$ and $C_2$ both contain $u_2 u_3$ and hence they both contain $e_1$ as well.

Suppose next that $G$ contains Hamilton cycles $C_1, C_2$ whose intersection is a matching containing $e_1$. Then we may assume that $C_1$ contains $u_1 u_3$, $u_3 u_2$, and $e_2$. We form an embedding of the graph $M$ in some surface as follows. Every triangle containing the vertex $v_i$ and an edge of $C_i$ (or the vertex $v_i'$ and an edge of $C_i'$), $i = 1, 2$, will be facial. Furthermore, if $xy \in E(G)$ is contained in only one of $C_1, C_2$, then the quadrilateral $xyy'x'$ will be facial. This determines an embedding of $M$ which is nonorientable since the six triangles containing the edges $e_1, u_1 u_2, u_2 u_3, u_3 u_1$ and one of $v_1, v_2$ form a Möbius strip. When we contract all edges $vv'$, where $v \in A$, then $M$ is transformed into $Q$ which triangulates the same surface except that there are two facial 4-cycles $u_1 u_2 u_2' u_1'$ and $u_1 u_3 u_3' u_1'$. But then $R$ triangulates the same surface. The proof is complete. □

The problem of finding necessary and sufficient conditions for a graph to triangulate a given surface was raised by Ringel [**Ri78**]. Theorem 4.4.9 shows that Ringel's problem does not have a nice solution (if $\mathbf{P} \neq \mathbf{NP}$). We return to Ringel's question in Section 5.3.

The genus problem may be hard even for small graphs. Let $K_{3,3,3,3}$ denote the complete four-partite graph (obtained from $K_{12}$ by removing the edges of four disjoint 3-cycles). White proved in the second edition of his book [**Wh73**, Example 3, p. 169] that $\mathbf{g}(K_{3,3,3,3}) \leq 5$. Jungerman [**Ju75**] reports that the genus of this graph is at least 5. By (4.12), this implies that $K_{3,3,3,3}$ does not triangulate the orientable surface of genus 4.

PROBLEM 4.4.10. *Does there exist a number $c$, $0 < c < 1$, such that every graph with $n$ vertices, whose minimum degree is at least $cn$ and whose number of edges is divisible by 3, triangulates an orientable surface?*

Theorem 4.4.9 also applies to the computational complexity of the *genus problem*: "Given a graph $G$ and a natural number $k$, is $\mathbf{g}(G) \leq k$?" If we replace $\mathbf{g}(G)$ by $\tilde{\mathbf{g}}(G)$, we obtain the *nonorientable genus problem*.

COROLLARY 4.4.11 (Thomassen [**Th89, Th93a**]). *The genus problem and the nonorientable genus problem are* **NP**-*complete*.

PROOF. The decision problem (b) of Theorem 4.4.9 is easily reduced to the genus problem by Proposition 4.4.4. Also, it is clear that the genus problem is in **NP**. Hence it is **NP**-complete.

The **NP**-completeness of the nonorientable genus problem is shown similarly by reducing the problem (c) of Theorem 4.4.9 to it.    □

Originally, Thomassen [**Th89**] proved Corollary 4.4.11 by reducing the independence number problem to the genus problem. He used a modified reduction to prove:

THEOREM 4.4.12 (Thomassen [**Th97a**]). *The genus problem and the nonorientable genus problem for cubic graphs are* **NP**-*complete*.

A graph $G$ is an *apex graph* if it contains a vertex $v$ such that $G - v$ is planar. Neil Robertson asked if it is difficult to determine the genus of apex graphs. If $\mathbf{P} \neq \mathbf{NP}$, the answer is negative:

THEOREM 4.4.13 (Mohar [**Mo98p**]). *The genus problem is* **NP**-*complete also when the input is restricted to apex graphs*.

## 4.5. The maximum genus of a graph

If $G$ is a connected graph, then the *maximum genus* $\mathbf{g}_M(G)$ of $G$ is the largest integer $g$ such that $G$ has an orientable embedding of genus $g$. If $G$ is connected and not a tree, then the *maximum nonorientable genus* $\widetilde{\mathbf{g}}_M(G)$ of $G$ is the maximum integer $h$ such that $G$ has a nonorientable embedding of genus $h$. If $G$ is $\Pi$-embedded with $f$ $\Pi$-facial walks, then Euler's formula states that $n - q + f = 2 - 2\mathbf{g}(\Pi)$, where $n = |V(G)|$, $q = |E(G)|$. Since $f \geq 1$,

$$\mathbf{g}_M(G) \leq \left\lfloor \frac{q - n + 1}{2} \right\rfloor . \tag{4.20}$$

Equality holds if and only if there exists an embedding of $G$ with exactly one facial walk (if $q - n + 1$ is even), or with exactly two facial walks (if $q - n + 1$ is odd). Similarly, $\widetilde{\mathbf{g}}_M(G) \leq q - n + 1$. Unlike (4.20), where strict inequality may occur, Ringel [**Ri77a**] and Stahl [**St78**] observed the following.

THEOREM 4.5.1 (Ringel [**Ri77a**], Stahl [**St78**]). *Let $G$ be a connected graph with $n$ vertices and $q$ edges. If $G$ is not a tree, then*

$$\widetilde{\mathbf{g}}_M(G) = q - n + 1 .$$

Theorem 4.5.1 follows easily from Theorem 4.5.2 below.

THEOREM 4.5.2. *Let $G$ be a connected graph which is not a tree, and let $\Pi = (\pi, \lambda)$ be a nonorientable embedding of genus $k$. Suppose that $\lambda$ is positive on a spanning tree $T$ of $G$. Then for any integer $h$, $k \leq h \leq |E(G)| - |V(G)| + 1$, there is a signature $\lambda'$ such that $\lambda'$ is positive on $T$ and such that $(\pi, \lambda')$ is an embedding into $\mathbb{N}_h$.*

PROOF. Let $E(G) \backslash E(T) = \{e_1, \ldots, e_m\}$. Then $m = |E(G)| - |V(G)| + 1$. Let $G_i = T \cup \{e_1, \ldots, e_i\}$, $i = 0, 1, \ldots, m$. Let $\Pi_0 = (\pi_0, \lambda_0)$ be the induced embedding of $T$. For $i = 1, \ldots, m$ we define a nonorientable embedding $\Pi_i = (\pi_i, \lambda_i)$ of $G_i$ such that the rotation system $\pi_i$ is induced by $\pi$, such that $\lambda_i$ is an extension of the signature $\lambda_{i-1}$, and such that the nonorientable genus of this embedding is equal to $i$, i.e., there is only one $\Pi_i$-facial walk. For $i = 1$, this is achieved by taking $\lambda_1(e_1) = -1$. For $i > 1$ we have two choices for $\lambda_i(e_i)$. One of them gives an embedding with only one facial walk, and we choose that one. The embedding $\Pi_m = (\pi, \lambda_m)$ of $G$ has only one facial walk. $\Pi_m$ can be obtained from $\Pi$ by successively changing the signature of an edge. Such a transformation changes the number of facial walks by at most one. By first changing the signature of $e_1$ to $-1$, if necessary, all intermediate embeddings are nonorientable and their genus takes all values between $k$ and $m$. The proof is complete. $\square$

Theorem 4.5.2 implies that $G$ has nonorientable embeddings in all surfaces $\mathbb{N}_h$ where $\widetilde{\mathbf{g}}(G) \leq h \leq q - n + 1$. The orientable counterpart to this result is known as *Duke's Interpolation Theorem*.

THEOREM 4.5.3 (Duke [**Du66**]). *A connected graph $G$ has embeddings in all surfaces $\mathbb{S}_g$, where $\mathbf{g}(G) \leq g \leq \mathbf{g}_M(G)$.*

PROOF. Every rotation system $\Pi'$ can be obtained from any other rotation system $\Pi$ of $G$ by successively exchanging the order of two consecutive edges in the local rotation of a vertex of $G$. Such an exchange affects at most three facial walks. Either three walks transform into one, one into three, or the number of facial walks remains the same. This changes the genus by at most one. Going from a minimum genus embedding to a maximum genus embedding therefore results in all intermediate genera. $\square$

An elegant characterization of the maximum genus of graphs is due to Xuong.

THEOREM 4.5.4 (Xuong [**Xu79**]). *Let $G$ be a connected graph with $n$ vertices and $q$ edges. Then*

$$\mathbf{g}_M(G) = \tfrac{1}{2}(q - n + 1) - \tfrac{1}{2} \min_T c_{\mathrm{odd}}(G - E(T)), \qquad (4.21)$$

*where the minimum is taken over all spanning trees $T$ of $G$ and $c_{\mathrm{odd}}(G - E(T))$ denotes the number of components of $G - E(T)$ with an odd number of edges.*

PROOF. First we prove by induction on $q$ that $2\mathbf{g}_M(G) \geq q - n + 1 - c_{\text{odd}}(G - E(T))$, where $T$ is an arbitrary spanning tree of $G$, by describing an orientable embedding with at most $1 + c_{\text{odd}}(G - E(T))$ faces. If $H$ is an odd component of $G - E(T)$, then let $e$ be an edge of $H$ such that either $H - e$ is connected or it is the union of a connected graph and an isolated vertex. By the induction hypothesis, $G - e$ has an orientable embedding with at most $c_{\text{odd}}(G - E(T))$ faces. Adding $e$ increases the number of faces by at most one. We may therefore assume that $c_{\text{odd}}(G - E(T)) = 0$.

If $q = |E(T)| = n - 1$, then $G$ is a tree. Any embedding of $G$ in $\mathbb{S}_0$ has only one face, and we are done. So assume that $q \geq n$. Pick a nontrivial component of $G' = G - E(T)$. Since $c_{\text{odd}}(G') = 0$, $G'$ has a vertex $v$ and distinct edges $e_1 = vv_1$, $e_2 = vv_2$ such that no component of $G' - \{e_1, e_2\}$ has an odd number of edges. ($G'$ has a vertex $v'$ whose removal from $G'$ does not increase the number of components. If $v'$ has degree at least 2, then $v'$ can play the role of $v$. Otherwise, let $v$ be its neighbor and $e_1 = vv'$. One can easily check by parity arguments that at least one other edge $e_2 = vv_2$ at $v$ has the required property.) By induction, some orientable embedding $\Pi$ of $G - \{e_1, e_2\}$ has only one face. Suppose that $e$ is an edge of $G - \{e_1, e_2\}$ incident with $v$. If the $\Pi$-facial walk is $W = uev'\ldots f_1 v_1 f_1' \ldots f_2 v_2 f_2' \ldots$, then we insert $e_i$ in the local rotation at $v_i$ between $f_i$ and $f_i'$ $(i = 1, 2)$, and insert $e_1, e_2$ at $v$ between $e$ and $e'$ such that $\pi_v(e) = e_1$, $\pi_v(e_1) = e_2$, and $\pi_v(e_2) = e'$. It is easy to see that the resulting embedding of $G$ has only one face.

Suppose, conversely, that $G$ has an orientable embedding $\Pi$ with $f$ facial walks. We shall find a spanning tree $T$ with $c_{\text{odd}}(G - E(T)) \leq f - 1$ which will prove (4.21). We use induction on $f$. We first dispose of the case $f = 1$ by induction on $q$. If $q = n - 1$, there is nothing to prove. If $G$ contains a vertex of degree 1 we remove it and apply the induction hypothesis. So assume that $G$ has no vertices of degree 1. Consider an edge $e$ of $G$ having the property that a subwalk $W$ of the $\Pi$-facial walk from $e$ to $e$ is as short as possible. If $e$ is followed by $e'$ in $W$, then $e' \neq e$ and the minimality property of $e$ implies that the induced embedding of $G - \{e, e'\}$ has at most two facial walks. By Euler's formula, the number of facial walks is odd. Hence there is only one. (In particular, $G - \{e, e'\}$ is connected.) We apply the induction hypothesis to $G - \{e, e'\}$. This completes the proof in the case when $f = 1$.

Assume now that $f > 1$. Then some edge $e$ is contained in two distinct facial walks. Hence, $G - e$ is connected. The induced embedding of $G - e$ has $f - 1$ facial walks. By the induction hypothesis, $G - e$ has a spanning tree $T$ with $c_{\text{odd}}((G - e) - E(T)) \leq f - 2$. Clearly, $c_{\text{odd}}(G - E(T)) \leq f - 1$. The proof is complete. $\square$

Edmonds [**Ed65**], Nash-Williams [**NW61**], and Tutte [**Tu61b**] independently gave a necessary and sufficient condition for a graph to have $k$

pairwise edge-disjoint spanning trees. The following is a slight reformulation of their result for $k = 2$.

THEOREM 4.5.5 (Edmonds – Nash-Williams – Tutte). *If $G$ is a connected multigraph, then the following are equivalent:*

(a) *$G$ has no two edge-disjoint spanning trees.*
(b) *$G$ has a set $F$ of edges such that $G/F$ has $m$ vertices and at most $2m - 3$ edges.*

A graph $G$ is *$k$-edge-connected* if the removal of any $k - 1$ or fewer edges from $G$ leaves a connected graph. Theorems 4.5.4 and 4.5.5 imply:

COROLLARY 4.5.6. *If $G$ is a 4-edge-connected graph with $n$ vertices and $q$ edges, then $G$ has two edge-disjoint spanning trees and*

$$\mathbf{g}_M(G) = \left\lfloor \frac{q - n + 1}{2} \right\rfloor.$$

PROOF. $G$ cannot satisfy Theorem 4.5.5(b) since $G/F$ (and hence also $G$) would have a vertex of degree at most 3 and hence the removal of the edges incident with that vertex would leave a disconnected graph. So, by Theorem 4.5.5, $G$ has two edge-disjoint spanning trees $T_1, T_2$. Thus, $c_{\text{odd}}(G - E(T_1))$ is zero or one, and Theorem 4.5.4 applies.    □

COROLLARY 4.5.7 (Nebeský [**Ne81a**]). *If $G$ is a connected graph with $n \geq 3$ vertices and $q$ edges such that, for each vertex $v$, the neighbors of $v$ induce a connected graph, then*

$$\mathbf{g}_M(G) = \left\lfloor \frac{q - n + 1}{2} \right\rfloor.$$

PROOF. A graph $G$ satisfying the assumption of the corollary need not be 4-edge-connected and it need not contain two edge-disjoint spanning trees as shown by $K_3$ and infinitely many other graphs. However, it contains two edge-disjoint trees $T_1, T_2$ such that $|V(T_1)| = n$ and $|V(T_2)| = n - 1$. This suffices to repeat the argument in the proof of Corollary 4.5.6.

To prove the existence of $T_1$ and $T_2$, we may assume, by Theorem 4.5.5, that $G$ has a set $E \subseteq E(G)$ such that $|E| \leq 3$ and $G - E$ has distinct components $H_1, H_2$. Choose $E$ such that $H_1$ is smallest possible. If $H_1$ is a single vertex $v$, it is easy to complete the proof by induction. (For, if $v_1, v_2, v_3$ are the neighbors of $v$, then we may assume that the path $v_1v_2v_3$ is present. Then we apply the induction hypothesis to $(G - v) + v_1v_3$. If $v$ has degree at most two, we apply the induction hypothesis to $G - v$.) So assume that $|V(H_1)| \geq 2$. Then $|E| = 3$. Moreover, $E$ is the edge set of a path $xuyv$ where $xy$ is an edge in $H_1$ and $uv$ is an edge in $H_2$. Then $H_1/xy$ is 4-edge-connected by the minimality of $H_1$ and has two edge-disjoint spanning trees. Also, we can apply the induction hypothesis to $H_2$. It is now easy to complete the proof. We leave the details for the reader.    □

Another important formula for the maximum genus of a graph was derived by Nebeský [**Ne81b**]. If $G$ is a connected graph and $A \subseteq E(G)$, let $c(A)$ be the number of connected components of $G - A$, and let $b(A)$ be the number of connected components $X$ of $G - A$ such that $|E(X)| \equiv |V(X)|$ (mod 2). With this notation we have:

THEOREM 4.5.8 (Nebeský [**Ne81b**]).

$$\mathbf{g}_M(G) = \tfrac{1}{2}\big(q - n + 2 - \max_{A \subseteq E(G)} \{c(A) + b(A) - |A|\}\big). \qquad (4.22)$$

Other results on the maximum genus are surveyed in [**Ri79**]. More recent results were published by Chevalier, Jaeger, Payan, and Xuong [**CJPX83**], Mohar, Pisanski, and Škoviera [**MPŠ88**], Škoviera [**Šk92**], and Nebeský [**Ne93**]; see also their references.

Since the right hand side of (4.22) is the minimum of a positive function, Nebeský's formula (4.22) provides a *good characterization* (in the sense of Edmonds) of the maximum genus problem. More precisely, let $G$ be a connected graph and $k$ a positive integer. The question "Is $\mathbf{g}_M(G) \geq k$?" clearly belongs to **NP**. By (4.22) it also belongs to **co-NP**. This does not follow immediately from Xuong's formula. However, Furst, Gross and McGeoch [**FGM88**] observed that Xuong's formula (4.21) can be turned into a polynomially bounded algorithm using the following result of Giles [**Gi82**] (which was extended to matroids by Lovász [**Lo81**]). These results of Giles and Lovász solve the *matroid parity problem*, one of the fundamental previously unsolved problems listed by Garey and Johnson [**GJ79**].

THEOREM 4.5.9 (Giles [**Gi82**]). *There exists a polynomially bounded algorithm for the following problem: If $G$ is a graph whose edge set is partitioned into pairs, find a largest forest whose edge set is the union of some of those pairs.*

COROLLARY 4.5.10 (Furst, Gross, McGeoch [**FGM88**]). *There exists a polynomially bounded algorithm for finding the maximum genus of a connected graph.*

PROOF. For simplicity we only discuss the problem of finding an embedding of genus $(|E(G)| - |V(G)| + 1)/2$ which, by Theorem 4.5.4, amounts to the same as finding a spanning tree $T$ such that $c_{\mathrm{odd}}(T) = 0$. (The general case then follows by a straightforward modification.) For this we form a new graph $G'$ by subdividing each edge $uv$ of $G$ such that $uv$ becomes a path $P(uv)$ in $G'$ of length $\deg_G(u) + \deg_G(v) - 2$. We shall now partition $E(G')$ into pairs. Each edge in $P(uv)$ is paired with an edge in a path of the form $P(uw)$ or $P(vw)$ where $w \in V(G) \backslash \{u, v\}$. Moreover, for each path $xyz$ in $G$, there is precisely one pair with one edge in $P(xy)$ and the other in $P(yz)$.

We claim that $G'$ has a spanning tree consisting of pairs if and only if $G$ has a spanning tree $T$ such that $c_{odd}(T) = 0$. (Having proved this claim, Corollary 4.5.10 follows from Theorems 4.5.9 and 4.5.4.)

Suppose first that $G'$ has a spanning tree $T'$ consisting of pairs. Then also $G' - E(T')$ consists of pairs. Let $T$ be the spanning tree of $G$ consisting of those edges $uv$ for which $P(uv) \subseteq T'$. Then $G - E(T)$ is the union of edge-disjoint paths of length two and hence each component of $G - E(T)$ has an even number of edges.

Conversely, if $T$ is a spanning tree of $G$ such that $c_{odd}(T) = 0$, then $G - E(T)$ is the union of edge-disjoint paths of length two. For each such path $xyz$ we delete in $G'$ the unique pair in $P(xy) \cup P(yz)$ and obtain a spanning tree $T'$ of $G'$ consisting of pairs. $\qquad\Box$

In 1951 Ore [**Ore51**] raised the question of whether a given graph $G$ has a *bidirectional retracting-free double tracing*, i.e., a closed walk such that each edge is traversed once in each direction and such that an edge is never followed immediately by itself (in the opposite direction). Troy [**Tr66**] proved that a necessary condition for a cubic graph to have such a walk is that its number of vertices is congruent to 2 (mod 4). As pointed out in [**Th90c**], Troy's result has a natural explanation in terms of embeddings: For a cubic graph the walk introduced by Ore is a facial walk of an embedding whose genus is $\frac{1}{2}(q - n + 1) = \frac{1}{4}(n + 2)$, by Euler's formula. Hence $n \equiv 2$ (mod 4). This argument also shows that Theorem 4.5.4 and Corollary 4.5.10 provide a solution to Ore's question, when restricted to cubic graphs. Ore's problem for general graphs is solved by the following:

THEOREM 4.5.11 (Thomassen [**Th90c**]). *A connected graph $G$ has a bidirectional retracting-free double tracing if and only if $G$ has a spanning tree $T$ such that each component of $G - E(T)$ either has an even number of edges or contains a vertex which in $G$ has degree at least 4.*

The method of Corollary 4.5.10 also implies a polynomially bounded algorithm for testing the condition of Theorem 4.5.11, see [**Th90c**].

CHAPTER 5

# The width of embeddings

In their work on graph minors Robertson and Seymour [**RS88**] introduced the face-width (or representativity) as a measure of how dense a graph is embedded on a surface. The *face-width* of a graph embedded in $S$ is the smallest number $k$ such that $S$ contains a noncontractible closed curve that intersects the graph in $k$ points. A related concept is the *edge-width* of an embedded graph $G$ defined as the length of a shortest noncontractible cycle in $G$. Robertson and Seymour proved that any infinite sequence of graphs embedded in a fixed surface $S$ with increasing face-width can serve as a generic class of graphs on $S$ in the sense that every embedding in $S$ is a surface minor of one of these embeddings. We treat this aspect of face-width in Section 5.9. Robertson and Vitray developed the basic theory of face-width in [**RV90**]. They showed that embeddings of large face-width are minimum genus embeddings and that they share many important properties with planar embeddings. The same phenomenon was discovered independently by Thomassen [**Th90b**] under the condition that the edge-width is greater than the maximum length of a facial walk.

In this chapter we discuss embedding results involving width. In the first part we study edge-width by following [**Th90b**] (extending the results from orientable to arbitrary embeddings). Particular attention is given to the so called LEW-embeddings whose edge-width is larger than the maximum length of a facial walk. In Section 5.4 we show that for every surface and any integer $k$ there are only finitely many minimal triangulations of edge-width $k$. The rest of the chapter is devoted to face-width. In Section 5.5 the basic theory is developed. Section 5.6 treats minor minimal embeddings of a given face-width. Section 5.10 contains results about uniqueness and flexibility of embeddings of graphs. The remaining sections contain further results on embedded graphs of large face-width. For some other aspects and additional references on face-width we refer to [**RV90**] and [**Mo97c**].

## 5.1. Edge-width

Let $\Pi$ be an embedding of $G$. We define the *edge-width* $\mathbf{ew}(G, \Pi)$ of the embedding $\Pi$ as the length of a shortest $\Pi$-noncontractible cycle. If

$G$ has only contractible cycles, we put $\mathbf{ew}(G, \Pi) = \infty$. If no confusion is possible, we also write $\mathbf{ew}(G)$ for short. As noted at the end of Section 4.2, $\mathbf{ew}(G, \Pi) < \infty$ if and only if $\Pi$ has positive Euler genus. By Theorem 4.3.2, the edge-width can be computed in polynomial time.

In this section we study *large-edge-width embeddings* (abbreviated *LEW-embeddings*) that are defined as embeddings whose edge-width is larger than the maximum length of a facial walk. They were introduced by Hutchinson [**Hu84a**]. She asked if every automorphism of a 3-connected graph with an LEW-embedding takes facial walks to facial walks. This question was answered positively by Thomassen [**Th90b**] who also proved that LEW-embeddings share many other important properties with embeddings of planar graphs as presented below.

Theorem 5.1.1 below justifies the concept of LEW-embeddings.

THEOREM 5.1.1 (Thomassen [**Th90b**]). *If $\Pi$ is an LEW-embedding of the graph $G$, then $\Pi$ is a minimum Euler genus embedding. Every embedding of $G$ with the same Euler genus as $\Pi$ is an embedding in the same surface as $\Pi$.*

PROOF. The proof is by induction on $|E(G)|$. If $\mathbf{eg}(\Pi) = 0$, there is nothing to prove. So we can assume $\mathbf{eg}(\Pi) > 0$. If $G$ contains a vertex $v$ of degree 1, then the induced embedding of $G - v$ is also an LEW-embedding, and we apply induction. So we can assume that $G$ has no vertices of degree 1. Denote by $f$ the number of $\Pi$-facial walks. Let $l_1, \ldots, l_f$ be the lengths of $\Pi$-facial walks and let $m = \max\{l_i \mid 1 \leq i \leq f\} < \mathbf{ew}(G, \Pi)$. Let $\Pi_1$ be an arbitrary embedding of $G$ whose Euler genus is not larger than $\mathbf{eg}(\Pi)$.

Assume first that $G$ contains a nonfacial $\Pi$-contractible cycle $C$ whose length is at most $m$. By deleting the $\Pi$-interior of $C$ we get an LEW-embedding $\Pi'$ of a smaller graph $G'$ in the same surface (by Proposition 4.2.1). Hence we can apply induction to $\Pi'$ and the embedding $\Pi'_1$ of $G'$ induced by $\Pi_1$ to conclude that $\mathbf{eg}(\Pi'_1) \geq \mathbf{eg}(\Pi')$, and in case of equality, both embeddings are in the same surface. By Proposition 4.1.5, $\mathbf{eg}(\Pi'_1) \leq \mathbf{eg}(\Pi_1)$. This implies $\mathbf{eg}(\Pi_1) \geq \mathbf{eg}(\Pi)$. Moreover, in case of equality, $\Pi_1$ and $\Pi$ are embeddings in the same surface.

Now, assume that every $\Pi$-nonfacial cycle of $G$ is longer than $m$. Let $l'_1, \ldots, l'_{f'}$ be the lengths of $\Pi_1$-facial walks. Then

$$2|E(G)| = l'_1 + \cdots + l'_{f'} = l_1 + \cdots + l_f. \tag{5.1}$$

Since $G$ has no vertices of degree 1, each $\Pi_1$-facial walk contains a cycle. Since all cycles distinct from the $\Pi$-facial walks have length greater than $m$, (5.1) implies that $f' \leq f$, and hence $\mathbf{eg}(\Pi_1) \geq \mathbf{eg}(\Pi)$. Moreover, if $f' = f$, then $l_i = l'_i$ for $i = 1, \ldots, f$ (possibly after permuting the numbers $l'_i$). Consequently, $\Pi_1$ is equivalent to $\Pi$. This proves the theorem.  $\square$

By taking subdivisions of edges, Theorem 5.1.1 extends to graphs with weighted edges.

THEOREM 5.1.2. *Suppose that $G$ is a $\Pi$-embedded graph and that $w$ : $E(G) \to \mathbb{R}^+$ is a positive edge-weight function. If $Q$ is a walk in $G$, let $w(Q)$ be the sum of $w(e)$ over the edges of the walk $Q$. Suppose that for every $\Pi$-noncontractible cycle $C$ and for every $\Pi$-facial walk $F$ we have $w(F) < w(C)$. Then every embedding of $G$ in a surface different from the surface of $\Pi$ has larger Euler genus than $\Pi$.* In particular, $\Pi$ is a minimum Euler genus embedding.*

In Section 4.4 we introduced orientably simple graphs. Their existence is a simple corollary of Theorem 5.1.1.

COROLLARY 5.1.3. *If $G$ has an orientable LEW-embedding, then $G$ is orientably simple.*

The following results show that LEW-embeddings share many properties with planar embeddings.

PROPOSITION 5.1.4. *If $G$ is a 2-connected LEW-embedded graph, then every facial walk is a cycle.*

PROOF. Denote by $\Pi$ the LEW-embedding of $G$. Suppose (reductio ad absurdum) that $W'$ is a facial walk which is not a cycle. Let $W$ be a closed subwalk with at least one edge and with no repetition of vertices. Since $G$ is 2-connected, $G$ has no vertices of degree one, and hence $W$ is a cycle. Since $G$ is LEW-embedded, $W'$ has length less than $\mathbf{ew}(G)$ and hence $W$ is $\Pi$-contractible. Since $\mathrm{Int}(W, \Pi)$ is planar and 2-connected, every facial walk of $G$ in $\mathrm{Int}(W, \Pi)$ is a cycle by Proposition 2.1.5. Hence $W'$ is contained in $\mathrm{Ext}(W, \Pi)$. Now, $W$ is of the form $u_1 u_2 \ldots u_k u_1$ where only $u_1$ is incident with an edge in $\mathrm{ext}(W, \Pi)$. Since $W \neq W'$, there must be such an edge. Hence $u_1$ is a cutvertex, contrary to the assumption that $G$ is 2-connected. $\square$

The next result corresponds to a classical result of Tutte, Theorem 2.5.1.

THEOREM 5.1.5 (Thomassen [**Th90b**]). *If $\Pi$ is an LEW-embedding of a 3-connected graph $G$, then the $\Pi$-facial cycles are precisely those induced nonseparating cycles of $G$ whose length is smaller than $\mathbf{ew}(G, \Pi)$.*

PROOF. If $G$ is planar, then the genus of $\Pi$ is also 0, by Theorem 5.1.1, and the theorem reduces to Theorem 2.5.1. So assume that $\mathbf{eg}(\Pi) > 0$.

If $C$ is an induced nonseparating cycle of length less than $\mathbf{ew}(G)$, then $C$ is $\Pi$-contractible. Since $\mathrm{Ext}(C, \Pi) \neq C$ and $C$ is induced and nonseparating, we must have $\mathrm{Int}(C, \Pi) = C$. Hence $C$ is $\Pi$-facial.

Assume, conversely, that $C$ is a $\Pi$-facial cycle. Since $\Pi$ is an LEW-embedding, $C$ has length less than $\mathbf{ew}(G)$. If $C$ is separating or has a chord, then some $\Pi$-facial cycle $C'$ which intersects $C$ would contain a segment of the form $v_1 u v_2$ or $v_1 u' u v_2$ (where $u, u'$ are consecutive vertices of $C$) such that either the first edge $v_1u$ (or $v_1u'$) or the last edge $uv_2$ is a chord of $C$, or these two edges are both edges joining $C$ to distinct components of $G - V(C)$. Let $P$ be a shortest path of $C'$ starting with the edge $uv_2$ (say) and terminating at a vertex $z$ of $C$. Without loss of generality we can assume that $P$ has length at most $|E(C')|/2 < \mathbf{ew}(G)/2$. Hence two of the cycles in $C \cup P$ have length less than $\mathbf{ew}(G)$ and are therefore $\Pi$-contractible. By Proposition 4.3.1(a), all three cycles of $C \cup P$ are $\Pi$-contractible and one of them, say $C''$, satisfies $\text{Int}(C'', \Pi) \supseteq C \cup P$. Since $P \subseteq \text{Ext}(C, \Pi)$, we conclude that $C'' \supseteq P$. Since $P$ is a segment of a facial cycle, every path from a vertex in $\text{Int}(C'', \Pi) - C''$ to $\text{Ext}(C'', \Pi) - P$ contains $u$ or $z$. But $G$ is 3-connected and so there is no vertex in $\text{Int}(C'', \Pi) - C''$. This implies that $u$ and $z$ are consecutive on $C$ and that $v_1$ is in $\text{Ext}(C'', \Pi)$. In particular, $C' \subseteq \text{Ext}(C'', \Pi)$. Since $G$ has no multiple edges, $P$ has length at least 2. Now $G - \{u, z\}$ has no path from $v_2$ to $v_1$. This contradicts the assumption that $G$ is 3-connected. Hence $C$ is induced and nonseparating. $\qquad \square$

COROLLARY 5.1.6. *A 3-connected graph has at most one LEW-embedding (up to equivalence).*

PROOF. Let $\Pi$ and $\Pi'$ be LEW-embeddings of the 3-connected graph $G$, where $\mathbf{ew}(G, \Pi) \le \mathbf{ew}(G, \Pi')$. By Theorem 5.1.5, every $\Pi$-facial cycle is $\Pi'$-facial. Since every edge is in two $\Pi$-facial cycles and in two $\Pi'$-facial cycles, we conclude that the $\Pi$-facial cycles are the same as the $\Pi'$-facial cycles and hence $\Pi$ and $\Pi'$ are equivalent. $\qquad \square$

Theorem 5.1.7 below extends part of Tutte's characterization of planarity in terms of overlap graphs of cycles (Theorem 2.4.4).

THEOREM 5.1.7 (Thomassen [**Th90b**]). *Let $G$ be a 2-connected graph with an LEW-embedding $\Pi$ of positive Euler genus. Let $C$ be a cycle of $G$ whose length is less than $\mathbf{ew}(G, \Pi)$. Then the overlap graph $O(G, C)$ is bipartite. $G$ has precisely one $C$-bridge $H$ such that the $\Pi$-genus of $C \cup H$ is positive. Moreover, the graph $C \cup H$ is nonplanar. If $G$ is a subdivision of a 3-connected graph, then $O(G, C)$ is connected and the bipartite class containing $H$ is precisely the set of $C$-bridges in $\text{Ext}(C, \Pi)$.*

PROOF. Since the length of $C$ is less than $\mathbf{ew}(G)$, $C$ is $\Pi$-contractible and hence each $C$-bridge is either in $\text{int}(C, \Pi)$ or in $\text{ext}(C, \Pi)$. Moreover, two $C$-bridges in $\text{int}(C, \Pi)$ cannot overlap since the embedding of $\text{Int}(C, \Pi)$ is planar. We shall show that no two $C$-bridges in $\text{ext}(C, \Pi)$ overlap. Let $C'$ be a cycle of $G$ with the following properties:

    (i) $C'$ is $\Pi$-contractible and $\text{Int}(C', \Pi) \supseteq C$.

    (ii) Every vertex of $C'$ which is incident with an edge in $\text{ext}(C', \Pi)$ is in $C$.

    (iii) $\text{int}(C', \Pi)$ has as many edges as possible subject to (i) and (ii).

Since $C$ satisfies (i) and (ii), $C'$ exists. Since $\mathbf{eg}(\Pi) > 0$ and $G$ is 2-connected, $C'$ has at least two vertices incident with edges in $\text{ext}(C',\Pi)$. By (ii), $C'$ has at least two vertices in common with $C$. If $C \neq C'$, there is a path $P$ contained in $C'$ such that $P$ has precisely its two ends in common with $C$. Then $P$ is contained in a $C$-bridge $H'$. Since no intermediate vertex of $P$ is incident with an edge of $\text{ext}(C',\Pi)$, it follows that $H' \subseteq \text{Int}(C',\Pi)$. Since $\text{Int}(C',\Pi)$ is planar, it follows that $H'$ does not overlap any $C$-bridge in $\text{ext}(C,\Pi)$. More generally, no two $C$-bridges in $\text{ext}(C,\Pi) \cap \text{Int}(C',\Pi)$ overlap.

We claim that there is only one $C$-bridge $H$ in $\text{ext}(C',\Pi)$. For otherwise, $G$ would have a facial cycle $C'''$ in $\text{Ext}(C',\Pi)$ containing edges from at least two distinct $C'$-bridges. Since the length of $C'''$ is less than $\mathbf{ew}(G)$, $C'''$ has a segment $P'$ of length less than $\mathbf{ew}(G)/2$ such that $P'$ has its ends but no edge or other vertex in common with $C'$. As in the proof of Theorem 5.1.5, we conclude that the three cycles of $C \cup P'$ are contractible and that one of them, say $Q$, satisfies $Q \supseteq P'$ and $\text{Int}(Q,\Pi) \supseteq C$. We can assume that no intermediate vertex of $P'$ is incident with an edge in $\text{ext}(Q,\Pi)$. (For otherwise $C'''$ is in $\text{Int}(Q,\Pi)$ and then we consider instead of $P'$ any other segment of $C'''$ which connects two vertices of $C$ and which has no edge in common with $C'$.) Since $Q$ is contractible, each segment of $C'$ in $\text{ext}(C,\Pi)$ is either in $\text{ext}(Q,\Pi)$ or $\text{int}(Q,\Pi)$. We then obtain a contradiction to (iii) by letting $P'$ replace the segment of $C'$ in $\text{int}(Q,\Pi)$ connecting the ends of $P'$.

We have shown that there is at most one $C$-bridge $H$ in $\text{ext}(C',\Pi)$. Clearly, $H$ does not overlap any $C$-bridge in $\text{Int}(C',\Pi) \cap \text{ext}(C,\Pi)$. Hence the overlap graph $O(G,C)$ is bipartite. Since $\mathbf{eg}(\Pi) > 0$, $H$ exists. The induced embedding of $\text{Ext}(C,\Pi)$ is an LEW-embedding in the same surface as $\Pi$. By Theorem 5.1.1, the graph $\text{Ext}(C,\Pi)$ is nonplanar. It follows that also $C \cup H$ is nonplanar. If we contract all edges of $H$ not incident with $C$, then the resulting graph $G'$ is planar and $O(G',C) = O(G,C)$. If $G$ is a subdivision of a 3-connected graph, then also $G'$ is a subdivision of a 3-connected graph. Now, the last part of Theorem 5.1.7 follows from the fact that $G'$ has unique embedding of genus 0. $\qquad\square$

We now describe an algorithm for finding LEW-embeddings.

THEOREM 5.1.8 (Thomassen [**Th90b**]). *There exists a polynomial time algorithm that, given a 3-connected graph $G$, constructs an LEW-embedding of $G$ or concludes that no such embedding exists.*

PROOF. By Theorem 5.1.1 we may assume that $G$ is nonplanar. Let $k$ be a fixed integer, $3 \leq k \leq |V(G)|$. Suppose that $\Pi$ is an LEW-embedding of $G$ such that $\mathbf{ew}(G,\Pi) = k$. Let $e_1$ and $e_2$ be distinct edges of $G$ incident with the same vertex $v$. Let $C$ be a shortest cycle in $G$ through $e_1$ and $e_2$. Suppose that $e_1$ and $e_2$ are consecutive in the $\Pi$-clockwise ordering around $v$. Then they are in a $\Pi$-facial cycle $C(e_1, e_2)$, and hence the length of $C$ is less than $k$. By Theorem 5.1.7, there is unique $C$-bridge $H$ such that $C \cup H$ is nonplanar. Let $G_1$ be the union of $C$ and all $C$-bridges distinct from $H$, and let $G'$ be the graph obtained from $G_1$ by adding a new vertex $w$ adjacent to all vertices of attachment of $H$ to $C$. Then $G'$ is 3-connected and the proof of Theorem 5.1.7 shows that it is planar. Those facial cycles of its planar embedding which do not contain $w$ are also $\Pi$-facial. If one of them contains $e_1$ and $e_2$, then we have a $\Pi$-facial cycle $C(e_1, e_2)$. If not, then $e_1$ and $e_2$ are in a facial cycle through $w$. In that case, $v$ is not adjacent to $w$, and $C(e_1, e_2)$ must contain edges of $H$. So, the embedding of $G'$ determines all $\Pi$-facial cycles through $v$ except $C(e_1, e_2)$.

The algorithm is now as follows: For each triple $v, e_1, e_2$ where $e_1, e_2$ are edges incident with the vertex $v$, we construct $C$, $O(G, C)$, $H$, $G'$ as described above. Either we conclude that $G$ has no LEW-embedding, or else we find, for each vertex $v$, the clockwise or anticlockwise rotation around $v$ and all facial cycles through $v$ except possibly one for the unique LEW-embedding $\Pi$ that may exist.

If $uv \in E(G)$, we know at least one $\Pi$-facial cycle containing the edge $uv$, and this enables us to determine the signature of the edge $uv$. We now use the fundamental cycle method (Theorem 4.3.2) to check if $\Pi$ is an LEW-embedding.

By repeating the above algorithm for $k = 3, 4, \ldots, |V(G)|$, we either conclude that $G$ has no LEW-embeddings or we find one. Clearly, the worst case complexity of this algorithm is polynomial.  $\square$

If $\Pi$ is an embedding of a graph $G$, $v$ a vertex of $G$, $e$ an edge incident with $v$, and $F$ a $\Pi$-facial walk containing $e$, then we say that $(\Pi, v, e, F)$ is a *rooted embedding*. Theorem 5.1.9 below shows that almost all rooted embeddings have large edge-width and that many of them are LEW-embeddings.

THEOREM 5.1.9 (Bender, Gao, and Richmond [**BGR94**]). *For each surface $S$, there are constants $c > 0$ and $s > 0$ such that almost all rooted embeddings in $S$ of graphs with $q$ edges have edge-width at least $c \log q$. Moreover, the fraction of rooted embeddings that are LEW-embeddings and the fraction of those that are not LEW-embeddings both exceed $q^{-s}$.*

The results of this section extend to the more general embeddings $\Pi$ of a graph $G$ where there is an edge-weight function $w : E(G) \to \mathbb{R}^+$ such that $w(Q) < w(C)$ for every $\Pi$-facial walk $Q$ and every $\Pi$-noncontractible cycle $C$ (cf. Theorem 5.1.2). Let us call $w$ an *LEW-weight function*.

PROBLEM 5.1.10. *Do there exist polynomially bounded algorithms for the following problems:*

(a) *Given is a graph $G$ embedded in a surface $S$. Does $G$ have an LEW-weight function?*

(b) *Does a given graph $G$ have an embedding which admits an LEW-weight function?*

## 5.2. 2-flippings and uniqueness of LEW-embeddings

In this section we extend Whitney's 2-flipping theorem (Theorem 2.6.8) to LEW-embeddings.

Suppose that $C$ is a $\Pi$-contractible cycle of a $\Pi$-embedded 2-connected graph $G$. We may assume that the signature of $\Pi$ is positive on all edges of $\operatorname{Int}(C, \Pi)$. Suppose that only two vertices $x$ and $y$ of $C$ are incident with edges in $\operatorname{ext}(C, \Pi)$. Then we define a new embedding $\Pi'$ as follows: For each vertex $z$ in $\operatorname{ext}(C, \Pi) - C$, the $\Pi'$-clockwise ordering around $z$ is the same as the $\Pi$-clockwise ordering. For each vertex $z$ in $\operatorname{Int}(C, \Pi) - \{x, y\}$, the $\Pi'$-clockwise ordering around $z$ is the $\Pi$-anticlockwise ordering around $z$. If $e_1, e_2, \ldots, e_p$ is the $\Pi$-clockwise ordering around $x$ (or $y$) such that $e_1, \ldots, e_k$ are in $\operatorname{Int}(C, \Pi)$ and $e_{k+1}, \ldots, e_p$ are in $\operatorname{ext}(C, \Pi)$, then the $\Pi'$-clockwise ordering around $x$ (respectively, $y$) is $e_k, e_{k-1}, \ldots, e_1, e_{k+1}, \ldots, e_p$. The signature of $\Pi'$ is the same as the signature of $\Pi$. Then we say that $\Pi'$ is obtained from $\Pi$ by a *2-flipping* (of $C$), and that the edges of $\operatorname{Int}(C, \Pi)$ are *involved* in that 2-flipping. If all $\Pi$-facial walks are cycles, then all $\Pi'$-facial walks are cycles. Moreover, only two facial walks are affected by the 2-flipping. This notion of a 2-flipping coincides with 2-flippings introduced in Section 2.6 for plane graphs.

LEMMA 5.2.1. *Let $G$ be a 2-connected $\Pi$-embedded graph and let $\Pi'$ be obtained from $\Pi$ by a 2-flipping. Then $\Pi'$ is an embedding in the same surface, and an arbitrary cycle of $G$ is $\Pi$-contractible if and only if it is $\Pi'$-contractible.*

PROOF. The orientability of $\Pi'$ is the same as that of $\Pi$ since the signature is unchanged. As $\Pi$ and $\Pi'$ have the same number of facial walks, they are embeddings in the same surface.

Let $C$ be the cycle of the 2-flipping. Since $\Pi$ is also obtained by a 2-flipping of $C$ from $\Pi'$, it suffices to see that an arbitrary $\Pi$-contractible cycle $C'$ is $\Pi'$-contractible. This is clear if $C' \subseteq \operatorname{Int}(C, \Pi)$ or $C' \subseteq \operatorname{ext}(C, \Pi)$. So we may assume that $C'$ is the union of two paths $P, P'$ such that $P \subseteq \operatorname{Int}(C, \Pi)$ and $P' \subseteq \operatorname{ext}(C, \Pi)$. Using the assumption that $\operatorname{Int}(C, \Pi)$ has genus 0, it follows that $C'$ is $\Pi'$-twosided, and an easy count of facial walks shows that one of the graphs $G_r(C', \Pi') \cup C'$ or $G_l(C', \Pi') \cup C'$ has $\Pi'$-genus 0. $\qquad\square$

We shall prove later (cf. Proposition 5.10.1) that for LEW-embeddings also a converse of Lemma 5.2.1 holds.

Two embeddings of $G$ are said to be *Whitney equivalent* if one can be obtained from the other by a sequence of 2-flippings. We now generalize Whitney's 2-flipping theorem (Theorem 2.6.8) to LEW-embeddings.

THEOREM 5.2.2 (Thomassen [**Th90b**]). *If* Π *is an LEW-embedding of a 2-connected graph* G, *then every embedding* Π′ *of* G *in the same surface is Whitney equivalent with* Π.

PROOF. The proof is by induction on $q = |E(G)|$. If $G$ is a planar graph, then Π is an embedding of genus 0 by Theorem 5.1.1. Then Theorem 5.2.2 follows from Theorem 2.6.8. So assume that the Euler genus of Π is positive and hence $q > |V(G)| \geq 4$.

Suppose first that $G$ is not a subdivision of a 3-connected graph. Then $G$ is the union of two proper subgraphs $G_1, G_2$ such that $G_1 \cap G_2$ consists of two vertices $a$ and $b$ such that none of $G_1, G_2$ is just a path from $a$ to $b$. If $e_1, \ldots, e_k$ is the Π-clockwise ordering around $a$, then there are distinct indices $r, s$ such that $e_r$ and $e_{s+1}$ are in $G_1$ and $e_{r+1}, e_s$ are in $G_2$. Let $C_1$ and $C_2$ be the Π-facial walks containing consecutive edges $e_r, e_{r+1}$ and $e_s, e_{s+1}$, respectively. They are cycles by Proposition 5.1.4. The indices $r$ and $s$ can be selected such that $G_i \cap C_j$ $(i, j \in \{1, 2\})$ are four distinct paths. Let $P_2$ be a shortest path in $C_2$ such that $P_2$ has its ends but neither an edge nor an intermediate vertex in common with $C_1$. The length of $C_1$ is smaller than $\mathbf{ew}(G, \Pi)$, and $P_2$ has length smaller than $\mathbf{ew}(G, \Pi)/2$. Hence two cycles in $C_1 \cup P_2$ have length less than $\mathbf{ew}(G, \Pi)$, and by Proposition 4.3.1(a), $C_1 \cup P_2$ has a Π-contractible cycle $C'$ such that $\mathrm{Int}(C', \Pi) \supseteq C_1 \cup P_2$. Since Π has positive Euler genus, also $\mathrm{Ext}(C', \Pi)$ has positive Euler genus and therefore $C' \neq C_1$. Hence $C' \supseteq P_2$.

We claim that $G$ has a Π-contractible cycle $C$ such that only two vertices of $C$ are incident with edges in $\mathrm{ext}(C, \Pi)$. If $C_2 \subseteq \mathrm{Ext}(C', \Pi)$, then the union of $P_2$ and the segment of $C_1$ in $\mathrm{int}(C', \Pi)$ can play the role of $C$. On the other hand, if $C_2 \subseteq \mathrm{Int}(C', \Pi)$, then the union of segments of $C_1$ and $C_2$ in $\mathrm{int}(C', \Pi)$ contains a cycle (because of intersection properties of $C_1$ and $C_2$) which can play the role of $C$. This shows that $C$ exists.

We choose $C$ such that $\mathrm{Int}(C, \Pi)$ has as few edges as possible. Let $x$ and $y$ be the two vertices of $C$ incident with edges in $\mathrm{ext}(C, \Pi)$. Let $Q_1, Q_2$ be the two segments of $C$ from $x$ to $y$ such that $|E(Q_1)| \leq |E(Q_2)|$. The minimality of $C$ implies that $Q_1$ has no chord and that $H = \mathrm{Int}(C, \Pi) - Q_1$ has only one component. Now we consider the embeddings $\Pi_1$ and $\Pi_1'$ of $G - H$ induced by Π and Π′, respectively. Since $C$ is Π-contractible, the Π-embedding of $G - \mathrm{int}(C, \Pi)$ is an embedding in the same surface $S$ as Π. Then it is also easy to see that $\Pi_1$ is an embedding in $S$. Clearly, $\Pi_1$ is an LEW-embedding. By Theorem 5.1.1, it is a minimum Euler genus embedding of $G - H$. In particular, $G$ and $G - H$ have the same Euler genus. Since Π′ is an embedding in $S$, also $\Pi_1'$ is an embedding in $S$.

By the induction hypothesis, $\Pi_1'$ can be obtained from $\Pi_1$ by a sequence of 2-flippings. Since $H$ is attached only to $Q_1$ which is part of a facial cycle in both embeddings of $G - H$, we can modify the sequence of 2-flippings so that it becomes a sequence of 2-flippings of $G$. This transforms Π into an embedding $\Pi''$ of $G$ such that $\Pi''$ and Π′ agree on $G - H$. Since all $\Pi$-facial walks in $G - H$ are cycles, the same holds for the $\Pi''$-facial walks (and hence also for $\Pi'$-facial walks) of $G - H$. By the last sentence of Proposition 4.2.2, $\Pi'$ is obtained from $\Pi_1'$ by embedding $H$ into a $\Pi_1'$-face. Let $R$ be the corresponding $\Pi_1'$-facial cycle. If $R$ contains $Q_1$, then clearly $\Pi'$ is obtained from $\Pi''$ by at most one 2-flipping. (Note that the minimality of $\operatorname{Int}(C, \Pi)$ ensures that there is only one planar embedding of $\operatorname{Int}(C, \Pi)$ with $Q_1$ being a facial cycle, by Theorem 2.6.8.) On the other hand, if $R$ does not contain $Q_1$, then all intermediate vertices of $Q_1$ have degree 2 in $G$. By the minimality of $\operatorname{Int}(C, \Pi)$, all intermediate vertices of $Q_2$ have degree 2 in $G$. Since $Q_1$ has length less than $\mathbf{ew}(G, \Pi)$, $Q_1$ has the following property in the $\Pi$-embedding of $G$ (by the 3-path-property of noncontractible cycles): If $R'$ is a $\Pi$-facial cycle containing $x$ and $y$, then the three cycles of $Q_1 \cup R'$ are $\Pi$-contractible. The proof of Lemma 5.2.1 shows that $Q_1$ still has that property after a sequence of 2-flippings. In particular, the three cycles of $Q_1 \cup R$ are $\Pi''$-contractible in $G - H$ and hence also $\Pi'$-contractible in $G$. Now we can perform a sequence of 2-flippings which transforms $\Pi'$ into an embedding which is obtained from $\Pi'$ by first deleting $Q_2 - \{x, y\}$ and then adding $Q_2$ such that $Q_2$ together with $Q_1$ form a facial cycle. The resulting embedding of $G$ is the same as $\Pi''$, possibly after flipping $Q_1 \cup Q_2$.

Suppose now that $G$ is a subdivision of a 3-connected graph. We shall prove that $\Pi$ and $\Pi'$ are equivalent. Let $f$ be the number of $\Pi$-facial walks. Since $G$ has no vertices of degree 1, each $\Pi'$-facial walk $W_i$ contains a subwalk $R_i$ $(i = 1, \ldots, f)$ that is a cycle. If $\Pi$ and $\Pi'$ are not equivalent, one such cycle $R_i$, which we now call $C$, is not a $\Pi$-facial cycle and has length smaller than $\mathbf{ew}(G, \Pi)$. Clearly, $C$ is $\Pi$-contractible.

The induced embedding $\Pi$ of $Q = \operatorname{Ext}(C, \Pi)$ is an LEW-embedding. Moreover, as $G$ is a subdivision of a 3-connected graph, $Q$ is 2-connected, and any separating set of two vertices in $Q$ is either contained in $C$ or separates a path from the rest of $Q$. In particular, every 2-flipping of the $\Pi$-embedding of $Q$ must flip a cycle that contains a segment of $C$.

Now $\Pi$ is an LEW-embedding of $Q$ in the same surface as $\Pi$. By Theorem 5.1.1, it is also a minimum Euler genus embedding of $Q$. Hence $\Pi'$ is also a minimum Euler genus embedding of $Q$. By the induction hypothesis, there exists a sequence of 2-flippings transforming the $\Pi$-embedding of $Q$ into an embedding $\Pi''$ such that $\Pi'$ and $\Pi''$ agree on $Q$. Since all $\Pi$-facial walks of $Q$ are cycles, the same holds for the $\Pi''$-facial walks. By Proposition 4.2.2, the $\Pi'$-embedding of $G$ is obtained from the $\Pi''$-embedding of $Q$ by adding planar subgraphs to $\Pi''$-facial cycles.

Now consider a cycle $C'$ satisfying conditions (i), (ii), and (iii) in the proof of Theorem 5.1.7. The proof of Theorem 5.1.7 shows that $G$ has only one $C$-bridge $H$ in $\operatorname{ext}(C', \Pi)$. Moreover, $H \cup C$ is nonplanar. In particular, $H$ is not just a path and has at least three vertices on $C$.

Hence $H \cup C$ is a subdivision of a 3-connected graph, and there is no 2-flipping of $\text{Ext}(C, \Pi)$ (or any graph obtained from $\text{Ext}(C, \Pi)$ by a sequence of 2-flippings) which involves edges of $H$.

Consider any $\Pi$-facial cycle $C'''$ which contains edges both from $H$ and $C'$. Then $C'''$ cannot contain two distinct segments in $H$ connecting vertices in $C$. For if that were the case, then we would obtain a contradiction to the maximality of $C'$ precisely as in the proof of Theorem 5.1.7 (where we considered a cycle containing edges from two distinct $C$-bridges in $\text{Ext}(C', \Pi)$). So $C'''$ consists of a path $P' \subseteq C'$ and a path $R \subseteq \text{ext}(C', \Pi)$ having only the ends $u, v$ in common with $C'$. If possible, we choose $C'''$ such that $P'$ is not a segment of $C$. If this is not possible, then there is no 2-flipping of the $\Pi$-embedding of $Q$. In that case we let $C'''$ be chosen such that it is one of the $\Pi'$-facial cycles that is used when we add $C$-bridges to $Q$ in order to obtain $\Pi'$ from the $\Pi''$-embedding of $Q$. In each case $\{u, v\}$ is a separating set of $G$ which shows that $G$ is not a subdivision of a 3-connected graph. This is clear if $\Pi = \Pi''$ (on $Q$). On the other hand, if $\Pi \neq \Pi''$, then $P' \not\subseteq C$ and $\{u, v\}$ separates $Q$ into two graphs none of which is a path. Since $C'$ and $C''$ are $\Pi$-contractible, the third cycle $R$ in $C' \cup C''$ is also $\Pi$-contractible, and all 2-flippings of the $\Pi$-embedding of $Q$ involve only edges in planar subgraphs of the form $\text{Int}(R, \Pi)$. After the sequence of 2-flippings, $C$ is $\Pi''$-facial but, by definition of $C$, we cannot use $C$ when we add $C$-bridges going from the $\Pi''$-embedding of $Q$ to the $\Pi'$-embedding of $G$. Hence $\{u, v\}$ separates $G$, contradicting the assumption that $G$ is a subdivision of a 3-connected graph.      □

A graph $G$ is *uniquely embeddable* in a surface $S$ if $G$ has an embedding $\Pi$ in $S$ and every embedding of $G$ in $S$ is equivalent to $\Pi$. Theorem 5.2.2 and Lemma 5.2.1 imply the following result.

COROLLARY 5.2.3. *Let $G$ be a 2-connected graph that has an LEW-embedding in a surface $S$. Then a cycle $C$ in $G$ is contractible in every embedding of $G$ in $S$ if and only if $C$ is contractible in some embedding of $G$ in $S$. If, in addition, $G$ is a subdivision of a 3-connected graph, then $G$ is uniquely embeddable in $S$.*

The proof of Theorem 5.2.2 implies a polynomially bounded algorithm for describing minimum genus embeddings of a large class of graphs. Since the genus (and the Euler genus) of a graph is the sum of the genera of its blocks, it is sufficient to consider 2-connected graphs. If $G$ is 2-connected but not a subdivision of a 3-connected graph, then we try to find a decomposition $G = G_1 \cup G_2$ such that $G_1 \cap G_2$ consists of two vertices $x$ and $y$ and such that $G_2$ is a connected graph that can be drawn in the plane with $x$ and $y$ on the boundary of the outer face. It can be checked in polynomial time if such a decomposition exists. If it does not exist, then $G$ has no LEW-embedding (as the proof of Theorem 5.2.2 shows). If the decomposition exists, we replace $G_2$ by a shortest path in

$G_2$ from $x$ to $y$. Then we repeat the algorithm on the resulting graph which has an LEW-embedding if $G$ has an LEW-embedding and which has the same genus and the same Euler genus as $G$. Continuing like this, we either conclude that $G$ has no LEW-embedding or we end up with a subgraph $H$ of $G$ which is a subdivision of a 3-connected graph. A close inspection of the proof of Theorem 5.1.8 shows that the algorithm of that theorem also works for subdivisions of 3-connected graphs. Thus we get:

THEOREM 5.2.4 (Thomassen [**Th90b**]). *There exists a polynomially bounded algorithm with the following property: If $G$ is a given 2-connected graph, the algorithm either describes a minimum Euler genus embedding or concludes that $G$ has no LEW-embeddings.*

Note that the algorithm of Theorem 5.2.4 may produce a minimum genus embedding even if $G$ has no LEW-embeddings. On the other hand, it may also happen that $G$ has an LEW-embedding and that the algorithm does not produce one. It is not clear how to obtain an LEW-embedding in that case.

## 5.3. Triangulations

Ringel [**Ri78**] raised the question of characterizing those graphs which triangulate some surface. He pointed out that a triangulation $G$ is *locally Hamiltonian*, i.e., for each vertex $v$ of $G$, the subgraph $N(v, G)$ of $G$ induced on the neighbors of $v$ has a Hamilton cycle. It must also satisfy Euler's formula which together with the equation $2q = 3f$ (where $q = |E(G)|$ and $f$ is the number of facial 3-cycles) implies that $q - 3n + 6$ is a nonnegative integer divisible by 3. Ringel pointed out that these conditions are not sufficient for a graph to be a triangulation. The proof of Theorem 4.4.9 gives rise to many additional examples of graphs that satisfy the two necessary conditions but do not triangulate a surface. Other simple examples were described in [**Th93a**]. For example, let $G$ be a prism graph $K_2 \square C_{3n}$, where $n \geq 1$. Then no three Hamilton cycles of $G$ intersect pairwise in a matching. As pointed out in [**Th93a**], the graph $Q$ obtained from $G$ by adding three independent vertices joined to all vertices of $G$ does not triangulate any surface. On the other hand, $Q$ has $27n$ edges and is locally Hamiltonian, so $Q$ satisfies the two necessary conditions.

Lemma 2.3.3 shows that triangulations of the sphere are 3-connected, and Proposition 5.3.1 below implies that triangulations of arbitrary surfaces are 3-connected.

PROPOSITION 5.3.1 (Thomassen [**Th90b**]). *Let $G$ be a connected locally Hamiltonian graph. Then $G$ is 3-connected. Moreover, if $\{v, x, y\}$ is a separating set of three vertices in $G$, then $G$ contains the 3-cycle $vxy$, and $xy$ is not an edge of any Hamilton cycle in $N(v, G)$. The subgraph*

$G - \{v, x, y\}$ has precisely two components $H_1, H_2$ and each of the graphs $G - H_i$ $(i = 1, 2)$ is locally Hamiltonian.

PROOF. Let $S$ be a smallest separating vertex set of $G$. Pick a vertex $v$ in $S$. Since $S \backslash \{v\}$ does not separate $G$, $v$ is joined to all components of $G - S$. Since $N(v, G)$ has a Hamilton cycle $C$, we conclude that $|S| \geq 3$. If $S = \{v, x, y\}$, then $G - S$ has only two components $H_1, H_2$ (because $C - \{x, y\}$ has only two components), and $v$ is joined to both $x$ and $y$. The same argument with $x$ instead of $v$ shows that $xy \in E(G)$. Clearly, $xy$ is a chord of $C$. Since $G$ is locally Hamiltonian and $G(S)$ is a complete graph, it is easy to see that $G - H_i$ is locally Hamiltonian for $i = 1, 2$. $\square$

The next result answers Ringel's question when restricted to triangulations of edge-width at least four.

THEOREM 5.3.2 (Thomassen [**Th90b**]). *Let $G$ be a connected graph. If $G$ is a triangulation with no noncontractible 3-cycles, then:*

(1) *$G$ is locally Hamiltonian.*

(2) *Every edge of $G$ is in precisely two nonseparating 3-cycles.*

*Conversely, if $G$ satisfies (1) and (2), then $G$ triangulates some surface.*

PROOF. Statement (1) is obvious, and (2) follows from Theorem 5.1.5 and Proposition 5.3.1. To prove the converse, we first note that by (1), (2), and Proposition 5.3.1, the neighborhood graph $N(v, G)$ of an arbitrary vertex $v$ of $G$ contains a unique Hamilton cycle $H_v = v_1 v_2 \ldots v_d$ ($d = \deg(v)$) such that the nonseparating 3-cycles containing $v$ are precisely the 3-cycles $vv_i v_{i+1}$, $i = 1, \ldots, d$, where the indices are taken modulo $d$. The cycle $H_v$ defines a clockwise order $\pi_v$ around $v$. It is easy to see that an appropriately defined signature and the clockwise orderings $\{\pi_v \mid v \in V(G)\}$ define an embedding of $G$ whose facial walks are precisely the nonseparating 3-cycles of $G$. $\square$

One can show that conditions (1) and (2) are necessary and sufficient for $G$ to triangulate a surface without surface nonseparating 3-cycles.

As pointed out in [**Th90b**], we can reduce the **NP**-complete problem of deciding if a cubic graph $G$ has a Hamilton path to the decision problem about local Hamiltonicity. Just take two disjoint copies of $G$ and add three new vertices each of which is joined to all other vertices. Then the resulting graph is locally Hamiltonian if and only if $G$ has a Hamilton path. Hence it is **NP**-complete to decide whether a given graph satisfies (1). On the other hand, it is easy to check (2). Moreover, having verified (2), property (1) can be checked by a planarity algorithm as the next result shows.

PROPOSITION 5.3.3 (Thomassen [**Th90b**]). *Let G be a connected graph satisfying* (2) *in Theorem 5.3.2. Then G satisfies* (1) *if and only if every closed neighborhood graph*[1] $\overline{N}(v, G)$, $v \in V(G)$, *is planar and 3-connected.*

PROOF. If $\overline{N}(v, G)$ is planar and 3-connected, then $N(v, G)$ is 2-connected, and the facial cycle in the induced embedding corresponding to the face containing $v$ is a Hamilton cycle of $N(v, G)$.

Suppose conversely that $G$ is locally Hamiltonian. Let $v$ be any vertex of $G$ and let $C = v_1 v_2 \ldots v_d$ be a Hamilton cycle of $N(v, G)$. By Proposition 5.3.1, $G$ is 3-connected and each 3-cycle through $v$ and an edge of $C$ is nonseparating. We draw $C$ as a convex polygon in the plane, draw $v$ outside $C$, and draw all chords of $C$ in $N(v, G)$ as straight line segments. We claim that this gives a planar embedding of $\overline{N}(v, G)$. For suppose that $v_i v_j$ and $v_s v_t$ are crossing chords. Then $C - \{v_i, v_j\}$ consists of two paths which belong to the same component of $G - \{v, v_i, v_j\}$. Each vertex $z$ of $G$ is joined to $v$ by three internally disjoint paths in $G$ since $G$ is 3-connected. Hence also $z$ belongs to the same component of $G - \{v, v_i, v_j\}$ as $C - \{v_i, v_j\}$. This shows that $G - \{v, v_i, v_j\}$ is connected. But then $vv_i$ belongs to the three nonseparating 3-cycles $vv_{i-1}v_i$, $vv_iv_{i+1}$, and $vv_iv_j$, contradicting (2). □

Lavrenchenko [**La87a**] asked if every triangulation with no short noncontractible cycles on the torus is uniquely embeddable on the torus. The following result goes much further.

THEOREM 5.3.4 (Thomassen [**Th90b**]). *If a $\Pi$-embedded graph G is a triangulation, then $\Pi$ is the unique embedding of G of minimum Euler genus provided G satisfies one of the conditions* (a) *or* (b) *below.*

(a) *G has no $\Pi$-noncontractible, nonseparating 3-cycles.*

(b) *The $\Pi$-noncontractible 3-cycles and the induced $\Pi$-noncontractible 4-cycles are pairwise disjoint.*

PROOF. Suppose that $G$ satisfies (a). Let $\Pi'$ be any embedding of the same Euler genus as $\Pi$. Euler's formula implies that the $\Pi'$-embedded graph $G$ is also a triangulation. It is sufficient to prove that every $\Pi'$-facial 3-cycle $C$ is also $\Pi$-facial. It follows from Proposition 5.3.1 that $C$ is nonseparating. By (a), $C$ is $\Pi$-contractible. Since a nonseparating $\Pi$-contractible cycle is $\Pi$-facial, it follows that $\Pi = \Pi'$.

Suppose now that we have (b). We insert a new vertex of degree 2 on each edge of $G$ which belongs to a $\Pi$-noncontractible induced 4-cycle or 3-cycle. In the resulting graph $G'$ (which we may consider to be $\Pi$-embedded), every noncontractible cycle has length at least five but every facial cycle has length at most four. By Theorem 5.2.2, $G'$ has only one

---

[1]The *closed neighborhood graph* $\overline{N}(v, G)$ is the subgraph of $G$ induced by $v$ and its neighbors.

embedding of Euler genus equal to the Euler genus of $\Pi$. Hence the same
is true also for $G$.                                                □

## 5.4. Minimal triangulations of a given edge-width

Let $T$ be a triangulation of some surface. Suppose that $e$ is an edge
of $T$ such that the only 3-cycles of $T$ that contain $e$ are the two facial
triangles containing $e$. By contracting the edge $e$ and replacing the two
resulting pairs of double edges by single edges, we get a triangulation $T'$
of the same surface. We say that *e can be contracted*, that $T'$ is obtained
from $T$ by an *edge contraction*, and that $T$ is obtained from $T'$ by *vertex
splitting*. We denote $T'$ by $T/\!/e$. See Figure 5.1 for an example.

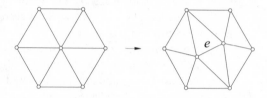

FIGURE 5.1. Vertex-splitting

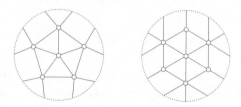

FIGURE 5.2. The minimal triangulations of the projective plane

It is easy to see that $T$ has no edges that can be contracted if and only
if each edge of $T$ is contained in a noncontractible 3-cycle. In this case we
say that $T$ is *minimal*. Since every triangulation of $S$ can be generated by
a sequence of vertex splittings from a minimal triangulation of $S$, it is of
interest to know all minimal triangulations of a surface. The only minimal
triangulation of the sphere is $K_3$ (or $K_4$ if we require that a triangulation
is 3-connected). The minimal triangulations of the projective plane have
been determined by Barnette [**Ba82a**]. They are shown in Figure 5.2.
Lawrencenko [**La87b**] described the minimal triangulations of the torus;
see Figure 5.3 (where opposite sides have to be identified). Lawrencenko
and Negami [**LN97**] determined the 25 minimal triangulations of the Klein
bottle.

FIGURE 5.3. The minimal triangulations of the torus

Barnette and Edelson [**BE88, BE89**] proved that the set of minimal triangulations is finite for every fixed surface. A simple proof of this fact was obtained independently by Gao, Richmond, and Thomassen [**GRT91**] (unpublished). Nakamoto and Ota [**NO95**] have another proof of the same result with an $O(g)$ upper bound on the size of the minimal triangulations of the orientable surface of genus $g$.

More generally, if $k \geq 3$ is an integer, a triangulation $T$ of a surface $S$ is $k$-*minimal* if the edge-width of $T$ is $k$ and each edge is contained in a noncontractible $k$-cycle. Fisk, Mohar, and Nedela [**FMN94**] determined all 4-minimal triangulations of the projective plane. They are represented in Figure 5.4.

Malnič and Mohar [**MM92**] proved that each orientable surface admits only finitely many 4-minimal triangulations. Malnič and Nedela [**MN95**] extended the proof and showed that for each surface, the class of $k$-minimal triangulations is finite for any $k$. Recently, Gao, Richter and

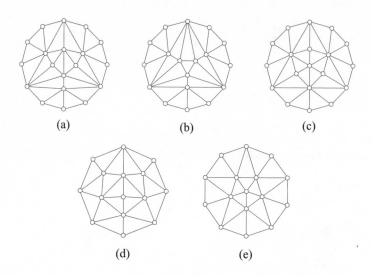

(a)               (b)               (c)

(d)               (e)

FIGURE 5.4. The 4-minimal triangulations of the projective plane

Seymour [**GRS96**] gave a different and shorter proof. We present another short proof obtained by Juvan, Malnič, and Mohar [**JMM96**].

THEOREM 5.4.1 (Malnič and Nedela [**MN95**]). *Let $S$ be a surface of Euler genus $g$ and let $k \geq 3$ be an integer. There is a constant $c_{k,g}$ such that every $k$-minimal triangulation of $S$ has at most $c_{k,g}$ edges.*

In the proof of Theorem 5.4.1 we shall use the following lemma from [**FM94a**].

LEMMA 5.4.2. *Let $k \geq 2$ and $r \geq 1$ be integers. There exists an integer $\alpha(k,r)$ such that the following holds: If a graph $H$ is the union of $\alpha(k,r)$ paths of length at most $k$ joining vertices $a$ and $b$ such that the edges of the paths incident to $a$ are all distinct, then there is a vertex $v$ of $H$ such that $a$ and $v$ are joined by $r$ internally disjoint segments of the paths.*

PROOF. We prove the lemma by induction on $k+r$ and with $\alpha(k,r) = r^{k-1}(k-1)!$. For $r = 1$, there is nothing to prove. For $k = 2$, all the paths are disjoint, so we proceed to the induction step. Now, let $H$ be a graph that is the union of $\alpha(k+1,r)$ paths of length at most $k+1$. Pick a path $P$. If there are at least $k\alpha(k,r)$ paths intersecting $P - \{a,b\}$, then some $\alpha(k,r)$ of these paths meet the same vertex of $P - \{a,b\}$, so we obtain the desired paths by induction. Otherwise, let $P_1, \ldots, P_q$ be a maximal collection of pairwise internally disjoint paths from $a$ to $b$ (taken from the $\alpha(k+1,r)$ paths). As each of the $\alpha(k+1,r)$ paths (which is not the edge $ab$) has an intermediate vertex in $P_1 \cup \cdots \cup P_q$, we have $\alpha(k+1, r) \leq qk\alpha(k, r)$, and hence $q \geq r$.     □

PROOF OF THEOREM 5.4.1. Let $T$ be a $k$-minimal triangulation of $S$ and let $v \in V(T)$ be a vertex of degree $d$. We first show that $d \leq 2\alpha$ where $\alpha = \alpha(k, 12kg)$ and the function $\alpha$ is from Lemma 5.4.2. By the $k$-minimality of $T$, each edge $e$ of $T$ is contained in a noncontractible $k$-cycle $C_e$. There is a collection $\mathcal{C}_v$ of $\lfloor d/2 \rfloor$ of these cycles such that each of them contains an edge (which we call a *distinguished edge*) incident to $v$ that is not contained in other $k$-cycles in $\mathcal{C}_v$. Let $H$ be the graph obtained from the union of the cycles in $\mathcal{C}_v$ by splitting $v$ into two vertices $v', v''$ such that $v'$ is incident with the distinguished edges, and $v''$ is incident with the other edges incident with $v$. The cycles in $\mathcal{C}_v$ become paths of length $k$ in $H$ joining $v'$ and $v''$. Suppose now that $d > 2\alpha$. By Lemma 5.4.2, there is a vertex $u$ of $H$ such that $v'$ and $u$ are joined by $12kg$ internally disjoint segments of cycles from $\mathcal{C}_v$. By Proposition 4.2.7, these segments contain a subfamily of $4k$ homotopic paths (or cycles if $u = v''$) in $T$, say $Q_1, \ldots, Q_{4k}$. We may assume that the paths $Q_1, \ldots, Q_{4k}$ are labelled such that their distinguished edges occur in clockwise order around $v$. For $1 \leq i < j \leq 4k$, let $D_{i,j} = \text{Int}(Q_i \cup Q_j)$. Now, $D_{1,2k}$ either contains all paths $Q_1, \ldots, Q_{2k}$ or $Q_{2k}, Q_{2k+1}, \ldots, Q_{4k}$. We may assume that $D_{1,2k}$ contains $Q_1, \ldots, Q_{2k}$. Consider a triangle $vxy$ in $D_{k,k+1}$ and the $k$-cycle $C_{xy}$ corresponding to the edge $xy$. By the 3-path-property of noncontractible cycles, $C_{xy}$ is an induced cycle since $\mathbf{ew}(T) = k$. Therefore, $v \notin V(C_{xy})$. Also, since $C_{xy}$ is a $k$-cycle, it cannot intersect $Q_1$ or $Q_{2k}$ in a vertex different from $u$. (For otherwise, it would intersect all paths $Q_1, \ldots, Q_k$ or all paths $Q_{k+1}, \ldots, Q_{2k}$.) Therefore $C_{xy} \subseteq D_{1,2k}$. This implies that $C_{xy}$ is contractible, a contradiction. This proves that the vertex degrees in $T$ are bounded above by $2\alpha$.

We now show that $|E(T)| \leq 3gk^2(2\alpha)^k$. Suppose therefore (reductio ad absurdum) that this is false. If $C$ is a $k$-cycle, then there are less than $k(2\alpha)^k$ edges at distance at most $\lceil k/2 \rceil$ from $C$. Hence, $T$ has edges $e_0, \ldots, e_p$, $p = 3gk$, such that the corresponding noncontractible $k$-cycles $C_{e_0}, \ldots, C_{e_p}$ are pairwise disjoint. By Proposition 4.2.6, we get a subfamily of $k+1$ pairwise homotopic cycles $Q_0, \ldots, Q_k$. By Lemma 4.2.5 we may assume that $\text{Int}(Q_0 \cup Q_j)$ contains all cycles $Q_0, \ldots, Q_j$, $j = 1, \ldots, k$. Let $xyv$ be a facial triangle such that $xy$ is an edge of $Q_t$ where $t = \lfloor k/2 \rfloor$ and $v \notin \text{Int}(Q_0 \cup Q_t)$. The $k$-cycles $C_{xv}$ and $C_{yv}$ are contained in $\text{Int}(Q_0 \cup Q_k)$. If $C_{xv}$ or $C_{yv}$ has a segment in $\text{Int}(Q_0 \cup Q_t)$, then any such maximal segment can be replaced by a segment of $Q_t$. This is proved by using the 3-path-property of noncontractible cycles and the fact that the distance between vertices on $C_{xv}$ and on $Q_t$ is equal to the distance in $G$. Similarly, we may assume that $C_{xv}$ is contained in $\text{Int}(Q_{t-1} \cup C_{yv})$. This implies that $y \in V(C_{xv})$. Hence $C_{xy}$ has a chord. This contradiction proves that $|E(T)| \le 3gk^2(2\alpha)^k$.     □

The aforementioned result of Gao, Richter, and Seymour [**GRS96**] says that any $k$-minimal triangulation contains at most $ck^2k!(4kk!)^kg^2$ vertices (where $c$ is some constant).

Seress and Szabó [**SS95**] proved that for each $\delta > 0$, there are 4-minimal triangulations $T$ of arbitrarily large Euler genus $g$ with less than $g^{1/2+\delta}$ vertices and with more than $|V(T)|^{2-\delta}$ edges. This result is best possible in the sense that $\delta$ cannot be omitted, since Clark, Entringer, McCanna, and Székely [**CEMS91**] proved that triangulations with edge-width more than three contain $o(|V(T)|^2)$ edges.

Hutchinson [**Hu88, Hu89**] showed that there is a constant $c_1$ such that a triangulation of $\mathbb{S}_g$ of edge-width $k$ has at least $c_1k^2g/\log^2 g$ vertices and she raised the question about the smallest possible triangulations of given edge-width. Przytycka and Przytycki [**PP97**] constructed triangulations of edge-width $k$ and with at most $c_2k^2g/\log g$ vertices, where $c_2$ is a constant. Similar results hold also for nonorientable surfaces.

## 5.5. Face-width

Let $\Pi$ be an embedding of a graph $G$. We define the *face-width* $\mathbf{fw}(G,\Pi)$ (or *representativity*) of $\Pi$ as the smallest number $k$ such that there exist $\Pi$-facial walks $W_1, \ldots, W_k$ whose union contains a $\Pi$-noncontractible cycle.[2] If $\Pi$ is an embedding of genus 0, then we put $\mathbf{fw}(G,\Pi) = \infty$. If no confusion is possible, we also write for short $\mathbf{fw}(G)$ or $\mathbf{fw}(\Pi)$.

PROPOSITION 5.5.1. *Let $G$ be a $\Pi$-embedded graph and let $W$ be a $\Pi$-facial walk. If $\mathbf{eg}(\Pi) > 0$ and $\mathbf{fw}(G,\Pi) \ge 2$, then $W$ contains a $\Pi$-contractible cycle $C$ such that $W \subseteq \mathrm{Int}(C,\Pi)$. The cycle $C$ is uniquely determined.*

PROOF. Since $\mathbf{fw}(G) \ge 2$, every cycle contained in $W$ is $\Pi$-contractible. If $W$ is a cycle, we put $C = W$. Otherwise, $W$ contains a subwalk $C'$ which is either a walk $vu_1v$ of length 2 or a cycle $C' = vu_1 \ldots u_mv$ such that none of the vertices $u_1, \ldots, u_m$ occur more than once in $W$. If $\mathrm{Int}(C',\Pi) \supseteq W$, then we put $C = C'$. Otherwise, let $U = V(\mathrm{Int}(C',\Pi))\backslash\{v\}$. Now we consider $G - U$ instead of $G$ and repeat the argument for $W$ minus $\{u_1, \ldots, u_m\}$.

The uniqueness of $C$ follows from the fact that for each cycle $C' \subseteq \mathrm{Int}(C,\Pi)$, we have $\mathrm{Int}(C',\Pi) \subseteq \mathrm{Int}(C,\Pi)$ if $\mathbf{eg}(\Pi) \ne 0$.     □

Proposition 5.5.1 has the following immediate consequence.

---

[2]An equivalent definition often used in the literature is that the face-width is the maximum $k$ such that every noncontractible simple closed curve in the surface intersects the graph in at least $k$ points.

PROPOSITION 5.5.2. *Let $G$ be a $\Pi$-embedded graph such that $2 \leq$ $\mathbf{fw}(G, \Pi) < \infty$. Then there is precisely one block $Q$ of $G$ that contains a $\Pi$-noncontractible cycle. The induced embedding of $Q$ is in the same surface as $\Pi$, and all facial walks of $Q$ are cycles. Each block $Q'$ of $G$ distinct from $Q$ is a planar subgraph of $G$ in the $\Pi$-interior of some facial cycle of $Q$. Finally, $\mathbf{fw}(Q, \Pi) = \mathbf{fw}(G, \Pi)$.*

Proposition 5.5.2 often enables us to consider only embeddings of 2-connected graphs.

PROPOSITION 5.5.3. *Let $G$ be a 2-connected $\Pi$-embedded graph. Then the face-width of $\Pi$ is equal to the minimum integer $k$ such that there exists a $\Pi$-noncontractible cycle which can be written as the union of $k$ segments of (distinct) $\Pi$-facial walks.*

PROOF. Let $C$ be a noncontractible cycle in $G$ such that $C \subseteq C_1 \cup \cdots \cup C_k$ where $k = \mathbf{fw}(G, \Pi)$ and $C_1, \ldots, C_k$ are $\Pi$-facial walks. If $k = 1$, then $C_1$ is not a cycle but it contains a cycle which is a segment of $C_1$. Since $G$ is 2-connected, that cycle is $\Pi$-noncontractible.

Suppose now that $k \geq 2$. Then $C_1, \ldots, C_k$ are cycles by Proposition 5.5.2. We can write $C = P_1 \cup \cdots \cup P_q$ where $P_i$ is a path in one of $C_1, \ldots, C_k$ $(i = 1, \ldots, q)$. If $q > k$, then there exist indices $i, j$ where $1 \leq i < j \leq q$ such that $P_i, P_j$ belong to the same cycle $C_l$, $1 \leq l \leq k$. We choose $C, P_1, \ldots, P_q$, and indices $i, j, l$ such that

(i) $q$ is minimum and
(ii) the number of edges in $E(C) \backslash E(C_l)$ is minimum subject to (i).

Let $P'$ be a segment of $C_l$ such that $P'$ has precisely its two ends in common with $C$. By the 3-path-condition, $C \cup P'$ has a noncontractible cycle $C'$ containing $P'$. Now, $C'$ contradicts either (i) or (ii). This shows that $k = q$ and completes the proof. $\square$

The face-width can also be expressed by using the *vertex-face multi-graph*[3] $\Gamma = \Gamma(G, \Pi)$. The multigraph $\Gamma$ is an embedded bipartite multi-graph with vertex set $V(\Gamma) = V(G) \cup V(G^*)$ where $G^*$ is the $\Pi$-dual multigraph of $G$. If $w \in V(G^*)$ corresponds to the $\Pi$-facial walk $W$, then $w$ is adjacent in $\Gamma$ to the vertices of $W$, where a repetition of vertices of $W$ corresponds to multiple edges between $w$ and the repeated vertex. The multigraph $\Gamma$ has a natural embedding in the surface of the embedding $\Pi$ such that all facial walks of $\Gamma$ have length 4. We denote this embedding by $\Pi_\Gamma$. The $\Pi_\Gamma$-facial walks are in a 1–1 correspondence with $E(G)$: If $e = uv$ is contained in $\Pi$-facial walks $w, z \in V(G^*)$ (possibly $w = z$), then the corresponding $\Pi_\Gamma$-facial walk is $uwvz$.

---

[3]This multigraph generalizes the vertex-face multigraph that we used in Section 2.8. It appears in the literature under various names, e.g., *radial graph*, *angle graph*, or *vertex-face incidence graph*.

PROPOSITION 5.5.4. *Let $G$ be a $\Pi$-embedded graph and $\Gamma$ its vertex-face multigraph. Then*

$$\mathbf{fw}(G,\Pi) = \frac{1}{2}\,\mathbf{ew}(\Gamma,\Pi_\Gamma)\,.$$

PROOF. We first observe that there is a multigraph $\tilde{G}$ that contains $G$ and $\Gamma$ as edge-disjoint subgraphs, and its embedding $\tilde{\Pi}$ extends both embeddings, $\Pi$ and $\Pi_\Gamma$. This enables us to compare noncontractible cycles of $\Gamma$ and of $G$.

Suppose first that $\mathbf{fw}(G) = 1$. Using the 3-path-condition, Proposition 4.3.1, it is not difficult to find a noncontractible cycle of length 2 in $\Gamma$. Suppose therefore that $\mathbf{fw}(G) \geq 2$. By Proposition 5.5.2, we may assume that $G$ is 2-connected and that every facial walk is a cycle.

Let $C = x_1 v_1 x_2 v_2 \ldots x_k v_k x_1$ be a shortest noncontractible cycle of $\Gamma$. Let $C_i$ be the $\Pi$-facial cycle such that the vertex $x_i \in V(G^*)$ corresponds to $C_i$, $i = 1,\ldots,k$. Note that all cycles of $\tilde{G}$ in $\mathrm{Int}(C_i,\tilde{\Pi})$ are $\tilde{\Pi}$-contractible. Also, note that any two nonconsecutive cycles among $C_1,\ldots,C_k$ are disjoint since otherwise $C$ would not be a shortest noncontractible cycle of $\Gamma$, by the 3-path-condition. Now we successively replace[4] $v_{i-1}x_iv_i$ by a $(v_{i-1},v_i)$-path in $C_i$ $(i = 1,\ldots,k)$ and obtain a $\tilde{\Pi}$-noncontractible cycle in $C_1 \cup \cdots \cup C_k$ showing that

$$\mathbf{fw}(G,\Pi) \leq k = \frac{1}{2}\mathbf{ew}(\Gamma).$$

Now, let $C$ be a noncontractible cycle in $G$ such that $C \subseteq C_1 \cup \cdots \cup C_k$ where $k = \mathbf{fw}(G,\Pi)$ and $C_1,\ldots,C_k$ are $\Pi$-facial cycles. By Proposition 5.5.3, we may assume that $C = P_1 \cup \cdots \cup P_k$ where $P_i$ is a path in $C_i$ $(i = 1,\ldots,k)$. Now we replace successively $P_i$ by a path of length 2 through the vertex of $\Gamma$ corresponding to $C_i$. We obtain a cycle of length $2k$ in $\Gamma$ which is $\tilde{\Pi}$-noncontractible because of the 3-path-condition. This shows that $\mathbf{ew}(\Gamma) \leq 2k = 2\mathbf{fw}(G,\Pi)$. $\qquad\square$

Let $G^*$ be the $\Pi$-dual multigraph of $G$ and let $\Pi^*$ be the dual embedding. Then $\Gamma(G,\Pi) = \Gamma(G^*,\Pi^*)$. Moreover, the vertex-face embedding $\Pi_\Gamma^*$ is equal to $\Pi_\Gamma$. This implies:

PROPOSITION 5.5.5. *The face-width of an embedded graph $G$ and its dual multigraph $G^*$ are the same, $\mathbf{fw}(G,\Pi) = \mathbf{fw}(G^*,\Pi^*)$.*

Proposition 5.5.4 and Theorem 4.3.2 imply:

PROPOSITION 5.5.6. *The face-width of an embedded graph can be determined in polynomial time.*

Suppose that $\mathbf{fw}(G,\Pi) \geq 2$. Let $W$ be a $\Pi$-facial walk, and let $C \subseteq W$ be the $\Pi$-contractible cycle such that $W \subseteq \mathrm{Int}(C,\Pi)$. Let us

---

[4]Special care has to be taken when $C_{i-1} \cap C_i$ has more than one vertex.

delete $int(C, \Pi)$ and contract all edges of $C$ but one, and finally delete the
remaining edge of $C$. If $W$ is a cycle, this operation geometrically shrinks
a face to a point. Therefore, we call it *face-shrinking*. We denote by $G/W$
and $\Pi/W$ the resulting graph and its induced embedding, respectively.

PROPOSITION 5.5.7. *Suppose that* $2 \leq \mathbf{fw}(G, \Pi) < \infty$. *Let $W$ be a*
$\Pi$-*facial walk. Then the embedding* $\Pi/W$ *of $G/W$ is an embedding in the*
*same surface as* $\Pi$ *and*

$$\mathbf{fw}(G, \Pi) - 1 \leq \mathbf{fw}(G/W, \Pi/W) \leq \mathbf{fw}(G, \Pi). \qquad (5.2)$$

PROOF. Let $C$ denote the cycle in Proposition 5.5.1. To obtain the
vertex-face multigraph $\Gamma/W = \Gamma(G/W, \Pi/W)$ from $\Gamma = \Gamma(G, \Pi)$ we first
delete all vertices of $G - C$ in $int(C, \Pi)$ and all vertices of $G^*$ corresponding
to the facial walks in $int(C, \Pi)$. Then we replace multiple edges between
$C$ and the vertex $w$ corresponding to $W$ by single edges. These operations
do not affect the face-width. Finally, we contract all edges incident with
$w$ and replace some of the resulting multiple edges by single edges. Now,
(5.2) follows from Proposition 5.5.4. $\qquad \square$

The dual operation of face-shrinking is the removal of a vertex. More
precisely, we delete a vertex $v$ and remove all components of $G - v$ except
the one whose $\Pi$-genus is positive. This gives rise to a dual version of
Proposition 5.5.7.

PROPOSITION 5.5.8. *Suppose that* $2 \leq \mathbf{fw}(G, \Pi) < \infty$. *Let $v$ be a*
*vertex of $G$. Then there is a unique component $G'$ of $G - v$ whose induced*
*embedding has positive Euler genus. Let $\Pi'$ be the induced embedding of*
$G'$. *Then $\Pi'$ is an embedding in the same surface as* $\Pi$ *and*

$$\mathbf{fw}(G, \Pi) - 1 \leq \mathbf{fw}(G', \Pi') \leq \mathbf{fw}(G, \Pi). \qquad (5.3)$$

COROLLARY 5.5.9. *Let $G$ be a 2-connected $\Pi$-embedded graph, and*
*let $X \subseteq V(G)$ be a separating vertex set in $G$. If $|X| < \mathbf{fw}(G, \Pi) < \infty$,*
*then there is a unique $X$-bridge $G_1$ that contains a $\Pi$-noncontractible cycle*
*disjoint from $X$. Moreover, the induced embeddings of $G_1$ and $G_1 - X$*
*are embeddings in the same surface as* $\Pi$. *The union $G_2$ of all $X$-bridges*
*distinct from $G_1$ contains only $\Pi$-contractible cycles, and its induced em-*
*bedding is of genus 0. In particular, $G_2$ is a planar graph.*

PROOF. Let $X = \{x_1, \ldots, x_k\}$. By repeated application of Proposition 5.5.8 we see that $G - X$ has precisely one component $L$ that contains
$\Pi$-noncontractible cycles. Moreover, the induced embedding of $L$ is an
embedding in the same surface as $\Pi$. In particular, each component of
$G_2 = G - L$ is embedded in a face of $L$. Hence, the induced embedding
of $G - L$ has genus 0. $\qquad \square$

If $\nu$ is a vertex of $G$ or a $\Pi$-facial walk, we define the subgraphs
$B_0(\nu), B_1(\nu), B_2(\nu), \ldots$ of $G$ recursively as follows: $B_0(\nu) = \nu$, and for

$k > 0$, $B_k(\nu)$ is the union of $B_{k-1}(\nu)$ and all $\Pi$-facial walks that have a vertex in $B_{k-1}(\nu)$. Let $\partial B_k(\nu)$ be the set of edges of $B_k(\nu)$ that are not incident with a vertex of $B_{k-1}(\nu)$.

The following extension of Proposition 5.5.1 was part of the motivation for introducing the face-width in [**RS88**].

PROPOSITION 5.5.10 (Robertson and Seymour [**RS88**]). *Let* $G$ *be a* $\Pi$-*embedded graph such that* $2 \leq \mathbf{fw}(G, \Pi) < \infty$.

(a) *Let* $\nu$ *be a vertex of* $G$ *and let* $k = \lfloor (\mathbf{fw}(G, \Pi) - 1)/2 \rfloor$. *Then there exist disjoint* $\Pi$-*contractible cycles* $C_1, \ldots, C_k$ *such that for* $i = 1, \ldots, k$, $C_i \subseteq \partial B_i(\nu)$ *and* $B_i(\nu) \subseteq \mathrm{Int}(C_i, \Pi)$.

(b) *Let* $\nu$ *be a* $\Pi$-*face and let* $k = \lfloor \mathbf{fw}(G, \Pi)/2 \rfloor - 1$. *Then there exist disjoint* $\Pi$-*contractible cycles* $C_0, \ldots, C_k$ *such that for* $i = 0, 1, \ldots, k$, $C_i \subseteq \partial B_i(\nu)$ *and* $B_i(\nu) \subseteq \mathrm{Int}(C_i, \Pi)$.

PROOF. The proof is by induction on $\mathbf{fw}(G)$. If $\mathbf{fw}(G) = 2$ and $\nu$ is a $\Pi$-face, the result follows from Proposition 5.5.1. For the inductive step we apply Propositions 5.5.7 and 5.5.8.          □

Many properties of 2-connected plane graphs generalize to embeddings of 2-connected graphs having face-width at least two.

PROPOSITION 5.5.11. *Let* $G$ *be a* $\Pi$-*embedded (multi)graph and let* $G^*$ *and* $\Pi^*$ *be its geometric dual multigraph and the dual embedding, respectively. Then the following conditions are equivalent:*

(a) *All* $\Pi$-*facial walks are cycles.*
(b) $\mathbf{fw}(G, \Pi) \geq 2$ *and* $G$ *is 2-connected.*
(c) $\Gamma(G, \Pi)$ *has no multiple edges.*
(d) $\mathbf{fw}(G^*, \Pi^*) \geq 2$ *and* $G^*$ *is 2-connected.*

PROOF. (a) implies that $\Gamma(G, \Pi)$ has no multiple edges. So, (a) implies (c). Now assume (c). Since $\Gamma$ is bipartite, $\mathbf{fw}(G, \Pi) = \mathbf{ew}(\Gamma, \Pi_\Gamma)/2 \geq 2$. If $v$ is a cutvertex of $G$, there are edges incident with $v$ that belong to distinct blocks of $G$ and that are consecutive in the $\Pi$-clockwise ordering around $v$. The $\Pi$-facial walk containing these two edges visits $v$ at least twice, a contradiction. Thus (c) implies (b). By Proposition 5.5.2, (b) implies (a). Since $\Gamma(G, \Pi) = \Gamma(G^*, \Pi^*)$, the equivalence of (c) and (d) follows by duality. This completes the proof.          □

The embeddings satisfying (a)–(d) in Proposition 5.5.11 are called *closed-2-cell embeddings* (also *strong embeddings* or *circular embeddings*).

The next result shows that embeddings of 3-connected graphs having face-width more than two share basic properties with embeddings of 3-connected graphs in the plane.

PROPOSITION 5.5.12. *Let* $G$ *be a* $\Pi$-*embedded graph. The following conditions are equivalent:*

(a) *For each vertex* $v \in V(G)$, $\partial B_1(v)$ *is a* $\Pi$-*contractible cycle which contains all neighbors of* $v$.

(b) $\mathbf{fw}(G, \Pi) \geq 3$ *and* $G$ *is* 3-*connected.*

(c) $\Gamma(G, \Pi)$ *has no multiple edges and all its* 4-*cycles are* $\Pi_\Gamma$-*facial.*

(d) *All* $\Pi$-*facial walks are cycles and any two of them are either disjoint or their intersection is just a vertex or an edge.*

(e) $\mathbf{fw}(G^*, \Pi^*) \geq 3$ *and* $G^*$ *is* 3-*connected.*

PROOF. The equivalence of (c) and (d) is easy. Let us now assume (c) (and hence also (d)), and let us prove (a). Let $v$ be a vertex of $G$. By Proposition 5.5.4, $\mathbf{fw}(G) \geq 3$. Hence Proposition 5.5.10(a) implies that $\partial B_1(v)$ contains a $\Pi$-contractible cycle $C$ such that $B_1(v) \subseteq \mathrm{Int}(C, \Pi)$. Now (d) implies that $C = \partial B_1(v)$ and that all neighbors of $v$ are on $C$. This shows that (c) implies (a).

Let us now assume (a). Since each $\Pi$-facial walk and any two $\Pi$-facial walks with a common vertex are contained in some $B_1(v)$, the union of any two facial walks contains only $\Pi$-contractible cycles. Hence $\mathbf{fw}(G) \geq 3$. If $G$ contains a cutvertex $v$ or a separating set of two vertices $v, u$, then $G - v$ has no cycle containing all neighbors of $v$, a contradiction. Hence (a) implies (b).

The implication (b) $\Rightarrow$ (c) is easy, and the equivalence of (c) and (e) follows by duality. $\square$

The embeddings satisfying (a)–(e) in Proposition 5.5.12 are called *polyhedral embeddings* as they are a natural generalization of convex 3-polytopes. See, e.g., [**Ba82b**].

Property (a) of Proposition 5.5.12 is called the *wheel neighborhood property* [**RV90**] since the neighborhood $B_1(v)$ of each vertex $v$ is a wheel (with a possibly subdivided rim). Condition (c) of Proposition 5.5.12 implies:

PROPOSITION 5.5.13. *Let* $G$ *be a* 3-*connected graph embedded with face-width at least* 3. *Then all facial walks are induced nonseparating cycles.*

The converse of Proposition 5.5.13 does not hold. Indeed, there are embeddings of face-width precisely 2 and with all faces induced and non-separating cycles.

Suppose that $G$ is a 2-connected graph and that $G = G_1 \cup G_2$ where $G_1$ and $G_2$ are connected, edge-disjoint subgraphs that intersect only in two vertices $a, b$ of $G$ and neither $G_1$ nor $G_2$ is just a path. For $i = 1, 2$, let $G_i'$ be the subgraph of $G$ which is equal to $G_i$ if $ab \in E(G_i)$, while otherwise it is obtained from $G_i$ by adding a path $P$ from $a$ to $b$ in $G_{3-i}$. This path is called a *virtual edge* of $G_i'$. If $G_1'$ and $G_2'$ can be further decomposed in that way, then we split them into smaller graphs, and repeat the process on each of the resulting graphs as long as possible.

The graphs that cannot be further decomposed are called the *3-connected blocks* of G. Each of them is a subgraph of G and is either a subdivision of a 3-connected graph, or consists of two vertices joined by two or three internally disjoint paths. The latter 3-connected blocks are called *trivial*. The nontrivial 3-connected blocks of G are uniquely determined up to the choice of the virtual edges, while the trivial 3-connected blocks may depend on the splitting process.[5]

The definition above describes the 3-connected blocks of a 2-connected graph. More generally, we define the *3-connected blocks* of an arbitrary graph G as the 3-connected blocks of the (2-connected) blocks of G.

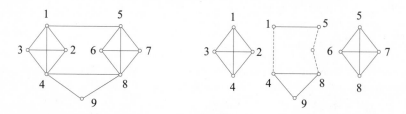

FIGURE 5.5. A 2-connected graph and its 3-connected blocks

An example of a graph and its 3-connected blocks is shown in Figure 5.5 with the virtual edges represented by broken lines. The 3-connected block B drawn in the middle is trivial. Observe that there is a different splitting process which results in two trivial blocks instead of B.

Proposition 5.5.2 reduces most embedding problems to the 2-connected case. Corollary 5.5.14 below reduces problems on embeddings of face-width at least three to the 3-connected case.

COROLLARY 5.5.14 (Robertson and Vitray [**RV90**]). *Let G be a 2-connected Π-embedded graph such that* $3 \leq \mathbf{fw}(G, \Pi) < \infty$. *Then there is a nontrivial 3-connected block Q of G such that the induced embedding* $\Pi'$ *of Q is an embedding in the same surface as* $\Pi$ *and* $\mathbf{fw}(Q, \Pi') = \mathbf{fw}(G, \Pi)$. *The 3-connected block Q is unique up to homeomorphism.*

PROOF. Suppose that $G - X$ is disconnected, where $X = \{a, b\} \subseteq V(G)$. Let $G = G_1 \cup G_2$ be the separation of Corollary 5.5.9. The induced embedding of $G_2$ is of genus 0. If P is a path from a to b in $G_1$, then the induced embedding of $G_2' = G_2 \cup P$ has face-width at most 1 by Proposition 5.5.8. Therefore, the induced embedding of any 3-connected block arising from $G_2'$ is either an embedding in the sphere or has face-width one. On the other hand, if $\Pi_1'$ is the induced embedding of $G_1'$, then

---

[5]Some authors, see e.g. [**Tu66a, HT73**], define *3-connected components* of a graph in a slightly different way. However, the difference is only in the definition of trivial 3-connected blocks.

$\Pi_1'$ is an embedding in the same surface as $\Pi$ and $\mathbf{fw}(G_1', \Pi_1') = \mathbf{fw}(G, \Pi)$. Repeating the argument for $G_1'$, we eventually find a 3-connected block $Q$ with the required properties. Since $\mathbf{fw}(Q) = \mathbf{fw}(G) \geq 3$, $Q$ is a nontrivial 3-connected block. $Q$ need not be unique. But $Q \cap G_2$ must be a path. Repeating the argument, it follows that $Q$ is unique up to homeomorphism. $\square$

Proposition 5.5.11 shows that an embedding of a 2-connected graph has face-width 2 or more if and only if all facial walks are cycles. Therefore the existence of embeddings of face-width at least 2 is related to the *Cycle Double Cover Conjecture* (Szekeres [**Sz73**], Seymour [**Se79**]) which says that every 2-connected graph contains cycles $C_1, \ldots, C_q$ such that each edge of $G$ is contained in precisely two of the cycles. Maybe even the following stronger conjectures hold.

CONJECTURE 5.5.15 (Haggard [**Ha77**]). *Every 2-connected graph has an embedding of face-width 2 or more.*

Archdeacon [**Ar84**] proved Conjecture 5.5.15 for 4-connected graphs.

CONJECTURE 5.5.16 (Jaeger [**Ja85**]). *Every 2-connected graph has an orientable embedding of face-width 2 or more.*

We refer to Jaeger [**Ja85**] and Zhang [**Zh96**] for further information on the Cycle Double Cover Conjecture.

A cubic graph is *3-edge-colorable* if $E(G)$ can be partitioned into three pairwise disjoint perfect matchings. To prove the Cycle Double Cover Conjecture, it suffices to show that every 3-connected cubic graph which is not 3-edge-colorable admits a cycle double cover (cf., e.g. [**Ja85**]). Since every cycle double cover $C_1, \ldots, C_q$ of a cubic graph $G$ determines an embedding of $G$ such that $C_1, \ldots, C_q$ are the facial cycles, Conjecture 5.5.15 restricted to cubic graphs is equivalent to the Cycle Double Cover Conjecture.

Perhaps there is a fixed upper bound on the face-width of non-3-edge-colorable cubic graphs.

PROBLEM 5.5.17 (N. Robertson, private communication). *Is there a fixed constant $k$ such that every 3-connected cubic graph with an embedding of face-width at least $k$ has a 3-edge-coloring.*

The Petersen graph in the projective plane shows that $k \geq 4$. Grünbaum [**Gr69**] conjectured that $k = 3$ may do if we consider only orientable embeddings.

Alspach, Zhang, and Goddyn [**AZ93, AGZ94**] proved that every 2-connected cubic graph with no Petersen graph minor has a cycle double cover. Tutte [**Tu66b**] conjectured that every 2-connected cubic graph with no Petersen graph minor has a 3-edge-coloring. Recently, Robertson,

Sanders, Seymour, and Thomas (private communication) proved this conjecture. These results are related to the following relaxation of Conjecture 5.5.16.

CONJECTURE 5.5.18 (Zhang, private communication). *Every 2-connected cubic graph with no Petersen graph minor has an orientable embedding of face-width 2 or more.*

PROBLEM 5.5.19. *Let $k \geq 2$ be a fixed integer. Does there exist a polynomial time algorithm for deciding if a given graph $G$ admits an embedding of face-width $k$ or more? Can we find such an embedding in polynomial time if it exists?*

Conjecture 5.5.15 suggests that the answer to Problem 5.5.19 is positive for $k = 2$. For $k = 3$, Mohar proved:

THEOREM 5.5.20 (Mohar [**Mo00**]). *It is **NP**-complete to decide if a given graph $G$ admits an embedding of face-width 3 or more.*

The problem remains **NP**-complete also for the existence of orientable embeddings of face-width at least 3 of 6-connected graphs [**Mo00**].

## 5.6. Minimal embeddings of a given face-width

Let $k \geq 2$ be an integer. An embedding $\Pi$ of a multigraph $G$ is *minimal of face-width $k$* if $\mathbf{fw}(G, \Pi) = k$ but for each edge $e$ of $G$, the face-width of $G/e$ and the face-width of $G - e$ are less than $k$. (Observe that by Proposition 5.5.2, $G$ is 2-connected.)

THEOREM 5.6.1. *For every surface $S$ and every integer $k \geq 2$ there are only finitely many minimal embeddings of face-width $k$ in $S$.*

PROOF. Let $\Pi$ be an embedding of a multigraph $G$ which is minimal of face-width $k$. We form a new graph $B = B(G, \Pi)$ embedded in the same surface, called the *barycentric subdivision* of $G$. We first subdivide each edge by replacing it by a path of length 2. If $H$ is the resulting graph, then $B = \Gamma(H, \Pi) \cup H$ is a triangulation of the surface of $\Pi$. Each triangle of $B$ corresponds to a triple $v, e, f$ where $f$ is a $\Pi$-facial walk, $e$ an edge of $G$ on $f$, and $v$ an endvertex of $e$. There are three types of edges of $B$: $ve$, $vf$, and $ef$. Since the face-width of $H$ is equal to the face-width of $G$, the edge-width of $B$ is $2k$. Let $e = uv \in E(G)$. Since the face-width of $G/e$ is less than $k$, each of the edges $ve$ and $vf$ of $B$ belongs to a noncontractible $2k$-cycle. Since $G - e$ has face-width less than $k$, also $ef \in E(B)$ is in a noncontractible $2k$-cycle, and thus $B$ is a $2k$-minimal triangulation. Now, Theorem 5.4.1 gives an upper bound on the size of $B$ and hence on $G$ as well.                                                                                  □

Let $G$ be a $\Pi$-embedded graph. A $Y\Delta$-*exchange* is an operation that replaces the edges of a facial cycle $F$ of length 3 by a vertex of degree three joined to the vertices of $F$, or conversely.

PROPOSITION 5.6.2. *Suppose that* Π *is a minimal embedding of face-width* $k$. *Then:*

(a) *Its geometric dual embedding* Π* *is also minimal of face-width* $k$.
(b) *If* Π′ *is obtained from* Π *by a sequence of* YΔ-*exchanges, then* Π′ *is also minimal of face-width* $k$.

PROOF. The first claim follows from Proposition 5.5.5 combined with the fact that edge deletion and edge contraction are dual operations. Statement (b) is easy to verify.                                              □

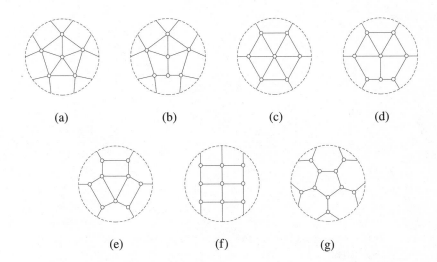

(a)               (b)               (c)               (d)

(e)               (f)               (g)

FIGURE 5.6. The minimal embeddings of face-width 3 in the projective plane

The minimal embeddings of face-width 2 in the projective plane are the embedding of $K_4$ of face-width 2 and its geometric dual [**Ba87**]. The minimal embeddings of face-width 3 in the projective plane have been determined independently by Barnette [**Ba91a**] and Vitray [**Vi92**]. They are represented in Figure 5.6. Note that in Figure 5.6(e) and (f) the same graph appears with different embeddings. The embeddings in the projective plane of face-width 3 which are minimal with respect to edge deletion only are classified in [**Ba91b**]. One can easily check that the *projective* $k \times k$ *grid* (see Figure 5.6(f) for the $3 \times 3$ grid and Figure 5.7 for $4 \times 4$ and $5 \times 5$ grids) is a minimal embedding of face-width $k$.

THEOREM 5.6.3 (Randby [**Ra97**]). *Every embedding in the projective plane which is minimal of face-width* $k$ *can be obtained from the projective* $k \times k$ *grid by a sequence of* YΔ-*exchanges. In particular, the graph of such an embedding has precisely* $2k^2 - k$ *edges.*

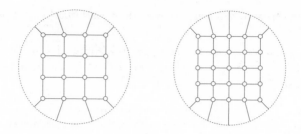

FIGURE 5.7. The projective $4 \times 4$ and $5 \times 5$ grids

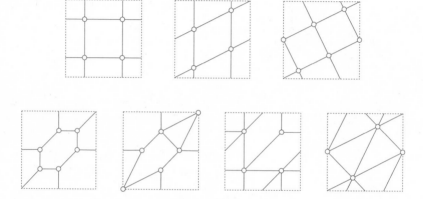

FIGURE 5.8. The minimal embeddings of face-width 2 in the torus

On the torus and the Klein bottle there are face-width minimal embeddings without facial triangles and without vertices of degree 3, see, e.g., Figure 5.8.

Two embeddings are *similar* if one can be obtained from the other by a sequence of operations each of which is a Y∆-exchange or the replacement of a graph by its geometric dual. Schrijver [**Sch94**] classified the similarity classes of minimal embeddings of face-width $k$ on the torus.

THEOREM 5.6.4 (Schrijver [**Sch94**]). *There are precisely $\frac{1}{6}(k^3 + 5k)$ (if $k$ is odd) and precisely $\frac{1}{6}(k^3 + 8k)$ (if $k$ is even) similarity classes of embeddings in the torus which are minimal of face-width $k$.*

Figure 5.8 shows the list of all embeddings in the torus which are minimal of face-width 2. There are four similarity classes of such embeddings, three of them containing only one, and one containing four distinct embeddings. They were described in [**Ba87**], except that the embedding of $K_5$ (the third embedding in Figure 5.8) was overlooked. Figures 5.9 and 5.10 show representatives of the similarity classes of the embeddings

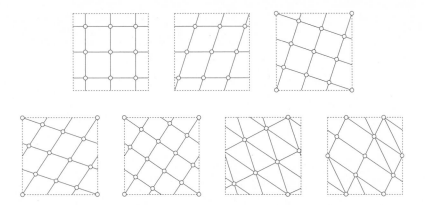

FIGURE 5.9. The similarity classes of minimal embeddings of face-width 3 in the torus

in the torus that are minimal of face-width 3 and 4, respectively. They
have been determined using the results of [**Sch94**].

Further results on the face-width of graphs embedded in the torus are
contained in [**Sch92a, Sch93, GS94**].

We conclude this section with a result of Schrijver analogous to Theorem 5.6.3. For this we need the notion of homotopy of closed walks.
Let $D$ be a closed walk in $G$. If there is an edge $e$ such that $D$ traverses $e$ and immediately after that $D$ traverses $e$ in the opposite direction,
that is $D = D_1 e e^- D_2$, then let $D' = D_1 D_2$. Similarly, if $D = D_1 P D_2$
and $W = PR$ is a $\Pi$-facial walk, we can change $D$ into the closed walk
$D' = D_1 R^- D_2$ where $R^-$ denotes the traversal of the segment $R$ in the
opposite direction. We also consider cyclic shifts that change $D = D_1 D_2$
into $D' = D_2 D_1$. These three operations of replacing $D$ by $D'$, and their
inverses, are called *elementary homotopic shifts*. Two closed walks in $G$
are $\Pi$-*homotopic* if one can be obtained from the other by a sequence of
elementary homotopic shifts.

Let $D$ be a closed walk in the vertex-face graph $\Gamma = \Gamma(G, \Pi)$. Then
we define

$$\mu(G, \Pi, D) = \frac{1}{2} \min \ell(D') \tag{5.4}$$

where $\ell(D')$ denotes the length of the walk $D'$ and where the minimum
is taken over all closed walks $D'$ in $\Gamma$ that are $\Pi_\Gamma$-homotopic to $D$. This
determines a function $\mu(G, \Pi)$ defined on $\Pi_\Gamma$-homotopy classes of closed
walks in $\Gamma$. Since $\Gamma(G, \Pi) = \Gamma(G^*, \Pi^*)$, we have $\mu(G, \Pi) = \mu(G^*, \Pi^*)$. If
$G', \Pi'$ is obtained from $G, \Pi$ by a $Y\Delta$-exchange, then there is a natural
correspondence between the closed walks in $\Gamma(G, \Pi)$ and those in $\Gamma(G', \Pi')$

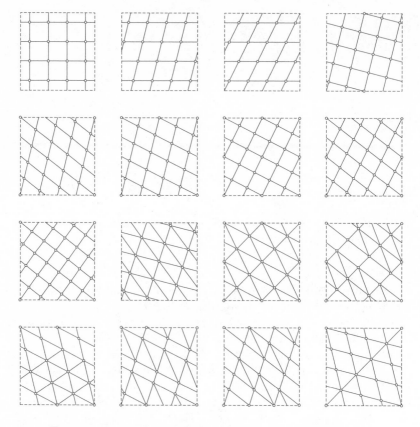

FIGURE 5.10. The similarity classes of minimal embeddings of face-width 4 in the torus

preserving homotopy. Proposition 5.6.2(b) now extends to

$$\mu(G,\Pi) = \mu(G',\Pi'). \tag{5.5}$$

Let $G$ be a $\Pi$-embedded graph and suppose that the $\Pi'$-embedded graph $G'$ is a surface minor of $G$ such that $\Pi$ and $\Pi'$ are embeddings in the same surface. Then the homotopy classes of the closed walks in $\Gamma(G',\Pi')$ are in a bijective correspondence with the homotopy classes in $\Gamma(G,\Pi)$ since each edge deletion or edge contraction in $G$ corresponds to the elimination of a face in $\Gamma$ by identifying opposite vertices of a $\Pi_\Gamma$-facial quadrangle. Consequently,

$$\mu(G',\Pi') \le \mu(G,\Pi). \tag{5.6}$$

The $\Pi$-embedded graph $G$ is called a *kernel* if $\mu(G',\Pi') \ne \mu(G,\Pi)$ for each proper surface minor $G',\Pi'$ of $G,\Pi$. Since $\mathbf{fw}(G,\Pi) \ge k$ if and only if for each $\Pi_\Gamma$-noncontractible cycle $D$ of $\Gamma$, $\mu(G,\Pi,D) \ge 2k$, kernels generalize the concept of minimal embeddings of a given face-width. Schrijver and de Graaf [**Sch92b, Gr94**] proved that kernels with the same function $\mu$ are closely related to each other.

THEOREM 5.6.5 (Schrijver [**Sch92b**], de Graaf [**Gr94**]). *If $G$, $\Pi$ and $G', \Pi'$ are kernels in the same surface, then $\mu(G, \Pi) = \mu(G', \Pi')$ if and only if $G', \Pi'$ can be obtained from $G, \Pi$ by a sequence of $Y\triangle$-exchanges and taking surface duals.*

## 5.7. Embeddings of planar graphs

Robertson and Vitray [**RV90**] proved that the face-width of a planar graph embedded in a nonplanar surface cannot be more than two. This was also observed by Thomassen [**Th90b**]. The next result is a slight extension.

THEOREM 5.7.1 (Mohar and Robertson [**MR96a**]). *Let $G$ be a 2-connected planar graph. Let $\Pi'$ be its embedding in the sphere, and let $\Pi$ be an embedding of $G$ of positive Euler genus. Suppose that the walk $W = v_0 e_1 v_1 \ldots e_k v_k$ ($k \geq 2$) is a segment of a $\Pi$-facial walk $W_1$ and that $W$ is not a segment of a $\Pi'$-facial cycle. Then there is a vertex $v_i$ ($1 \leq i < k$) such that either some $\Pi$-facial walk through $v_i$ is not a cycle, or there is a $\Pi$-facial cycle $W_2 \neq W_1$ containing $v_i$ whose intersection with $W_1$ is not just a vertex or a common segment. Consequently, $\mathbf{fw}(G, \Pi) \leq 2$.*

PROOF. Suppose that all $\Pi$-facial walks containing a vertex in $\{v_1, \ldots, v_{k-1}\}$ are cycles of $G$. Let us first assume that for some $i$, $1 \leq i < k$, $e_i$ and $e_{i+1}$ are not $\Pi'$-consecutive. Then there exist $\Pi$-consecutive edges $f, f'$ incident with $v_i$ which interlace with $e_i$ and $e_{i+1}$ in the $\Pi'$-clockwise ordering around $v_i$. Let $W_2$ be the $\Pi$-facial cycle containing $f$ and $f'$. Then $f \in \mathrm{int}(W_1, \Pi')$ and $f' \in \mathrm{ext}(W_1, \Pi')$ (say). Consequently, $W_1$ and $W_2$ have a vertex $u \neq v_i$ in common, and $W_1 \cap W_2$ is not just a vertex or a common segment. We may therefore assume that any two edges that are consecutive in $W$ are also consecutive in the $\Pi'$-clockwise or $\Pi'$-anticlockwise ordering around their common end.

We may assume that the signature of the embeddings $\Pi$ and $\Pi'$ is positive on the edges $e_2, \ldots, e_{k-1}$. Hence there are indices $i, j$, where $1 \leq i < j < k$, such that $\pi_{v_i}(e_i) = e_{i+1}$, $\pi_{v_j}(e_j) = e_{j+1}$, $\pi'_{v_i}(e_i) = e_{i+1}$, $\pi'_{v_j}(e_{j+1}) = e_j$, and all intermediate vertices $v_{i+1}, \ldots, v_{j-1}$ are of degree 2 whereas $v_i$ and $v_j$ have degree at least 3. Let $W_2$ be the $\Pi$-facial cycle distinct from $W_1$ which contains the edge $e_{i+1}$. The same arguments as above show that $W_1$ and $W_2$ have a vertex $u$ in common which is distinct from $v_i, \ldots, v_j$.

Suppose now that $\mathbf{fw}(G, \Pi) \geq 3$. By Corollary 5.5.14, there is a 3-connected block $H$ of $G$ whose face-width is equal to $\mathbf{fw}(G, \Pi)$. Since $H$

is planar and since there exists a Π-facial cycle of $H$ that is not facial in the plane embedding of $H$, the first part of Theorem 5.7.1 combined with Proposition 5.5.12 yields a contradiction. This completes the proof.   □

For a generalization of Theorem 5.7.1 and some related results we refer to [**MR96a**].

Recall that a graph is an apex graph if it contains a vertex whose removal leaves a planar graph. Robertson and Vitray [**RV90**] asked if apex graphs can have embeddings of face-width three. This question was answered in the negative for nonorientable surfaces by Robertson, Seymour, and Thomas [**RST93**] in their work about linkless embeddings of graphs in the 3-space. The proof presented below was obtained independently by Mohar in [**Mo97b**].

THEOREM 5.7.2 (Robertson, Seymour, and Thomas [**RST93**]). *If $G$ is an apex graph embedded in a nonorientable surface, then the embedding has face-width at most two.*

In the proof of Theorem 5.7.2 we will use the following lemma.

LEMMA 5.7.3. *Let $H$ be a graph which is the union of a Hamilton cycle $C$ in $H$ and pairwise disjoint cycles $C_1, \ldots, C_s$, $s \geq 1$. Suppose that the edge set $E(C_1) \cup \cdots \cup E(C_s)$ is partitioned in two classes $E_1 \cup E_2$ such that for each $i \in \{1, \ldots, s\}$ and $k \in \{1, 2\}$, no two edges of $E(C_i) \cap E_k$ overlap on $C$. If $\alpha_k$ is the number of unordered pairs of edges in $E_k$ that overlap on $C$ ($k = 1, 2$), then $\alpha_1 \equiv \alpha_2 \pmod 2$.*

PROOF. If two adjacent edges $e = yx$ and $f = xz$ of some $C_i$ are in the same class, say in $E_1$, we split $x$ into two adjacent vertices $x'$ and $x''$ and replace $yx$ and $xz$ by $yx'$ and $x''z$, respectively. We also extend $C$ by adding a path of length 2 from $x'$ to $x''$. We may assume that $yx'$ and $x''z$ do not overlap on $C$. By putting the new edge $x'x''$ of $C_i$ in $E_2$, the assumptions of the lemma remain valid and the values $\alpha_1, \alpha_2$ do not change. Therefore we may assume that for $i = 1, \ldots, s$, no two adjacent edges of $C_i$ are in the same class. We may also assume that each $C_i$ is edge-disjoint with $C$.

The proof proceeds by induction on $|E(H)|$. If $s = 1$, $\alpha_1 = \alpha_2 = 0$. If $s > 2$, we apply induction for each pair $C_i, C_j$, $1 \leq i < j \leq s$. Suppose now that $s = 2$. If $C_1$ and $C_2$ are both of length 4, the lemma is easily verified by considering all cases (up to symmetry there are 8 cases). Otherwise, let $C_2$ have consecutive edges $e_1, \ldots, e_t$, where $t > 4$. Suppose that $e_2$ is an edge of $C_2$ such that the distance between its ends on $C$ is minimum. Let $a$ and $b$ be the ends of $e_1$ and $e_3$ (respectively) that are distinct from the ends of $e_2$. Now, we add the edge $e = ab$ to $H$. Let $C_2'$ be the 4-cycle $ee_1e_2e_3$ and let $C_2''$ be the $(t-2)$-cycle $ee_4e_5 \ldots e_t$. Suppose that $e_1 \in E_1$. By our choice of $e_2$, no two edges of $C_2'$ overlap and no edge of $C_2$ that is in $E_1$ overlaps with $e$ (such an edge $f$ would overlap with either

$e_1 \in E_1$, $e_3 \in E_1$, or $e_2$; in the latter case, the edge following $f$ on $C_2$ would contradict the minimality of $e_2$). Therefore we can apply induction to $C_1, C_2'$ (with $e$ in $E_2$) and to $C_1, C_2''$ (with $e$ in $E_1$). Each overlapping of $e$ with an edge of $C_1$ is counted twice whereas each other overlapping is counted either in the first or in the second subproblem, but not both. The lemma now follows easily. $\square$

PROOF OF THEOREM 5.7.2. Suppose (reductio ad absurdum) that the theorem does not hold, and let $G, \Pi$ be a smallest counterexample. By Proposition 5.5.2 and Corollary 5.5.14, $G$ is 3-connected. Let $v$ be a vertex of $G$ such that $G_1 = G - v$ is planar. Since $G$ is 3-connected and $\mathbf{fw}(G, \Pi) \geq 3$, $G_1$ is 2-connected and the face-width of its induced embedding $\Pi_1$ is (at least) two. There is precisely one $\Pi_1$-facial walk $C$ that is not $\Pi$-facial. By Proposition 5.5.11, $C$ is a cycle.

Suppose that $e, f \in E(G_1)$. If we subdivide $e$ and $f$ by inserting vertices $v_e, v_f$ of degree 2 and then add the edge $v_e v_f$, we say that the resulting graph is obtained from $G_1$ by *sticking* $e$ and $f$. Let $\tilde{G}$ be obtained from $G_1$ by sticking as many pairs of edges of $G_1$ as possible such that $\tilde{G}$ has an embedding $\Pi'$ in the sphere and such that $\Pi_1$ can be extended to an embedding $\tilde{\Pi}$ of $\tilde{G}$ in the same surface such that none of the new edges is embedded in the (subdivided) face $C$. Then $\tilde{G} + v$ is a 3-connected apex graph and its embedding $\tilde{\Pi}_0$ extending $\tilde{\Pi}$ has face-width at least 3.

Suppose that for a vertex $u \in V(\tilde{G})$, the $\tilde{\Pi}$-clockwise and the $\Pi'$-clockwise orderings around $u$ are neither the same nor inverse to each other. Then $u$ is called a *patch vertex*. Theorem 5.7.1 implies that there are distinct $\tilde{\Pi}$-facial walks $C_1, C_2$ containing $u$ such that $C_1 \cap C_2$ is neither a vertex nor an edge. Since the $\tilde{\Pi}$-facial walks distinct from $C$ are also $\tilde{\Pi}_0$-facial, they cannot intersect in such a way by Proposition 5.5.12. Hence $C_1$ or $C_2$ is equal to $C$. In particular, all patch vertices are on $C$. Also, $\tilde{\Pi}$ agrees with the planar embedding $\Pi'$ on $\tilde{G} - V(C)$.

Let $W \neq C$ be a $\tilde{\Pi}$-facial cycle that is not $\Pi'$-facial. Then $W$ contains a vertex in $C$. By the definition of $\tilde{G}$, all vertices of $V(W) \cap V(C)$ are patch vertices. The facial segments of $W$ joining consecutive patch vertices of $W$ are called *patch segments*, and $W$ is called a *patch face*.

Let $\tilde{\lambda}$ be the signature of the embedding $\tilde{\Pi}$. We may assume that $\tilde{\lambda}$ is positive on all edges of $C$ and on all edges that are not incident with patch vertices (since the $\tilde{\Pi}$-embedding and the $\Pi'$-embedding of $\tilde{G} - V(C)$ are equivalent). If $\varepsilon$ is a patch segment, we define its signature as the product of signatures of the edges that are on $\varepsilon$. Since $\tilde{\Pi}$ is a nonorientable embedding, there is a cycle $Q$ of $\tilde{G}$ with negative signature. The cycle space of a 2-connected plane graph is generated by the facial cycles (by the Jordan Curve Theorem). Hence, $Q$ is the sum of $\Pi'$-facial cycles. Therefore, there is a $\Pi'$-facial cycle $Q'$ whose signature in $\tilde{\Pi}$ is negative. Since $Q'$ is composed of patch segments, there exists a patch segment $\varepsilon$

with negative signature. Let $\tau$ and $\nu$ be the patch vertices of $\varepsilon$. Suppose that $e_1, \ldots, e_d$ is the $\tilde{\Pi}$-clockwise ordering around $\nu$ where $e_1, e_d \in E(C)$ and $e_k, e_{k+1}$ $(1 \le k < d)$ are the edges on a patch face $C_1$ which contains $\varepsilon$. We may assume that $\varepsilon$ contains $e_{k+1}$. By changing the role of $\nu$ and $\tau$, if necessary, we may also assume that $\tilde{\lambda}(e_{k+1}) = -1$. Let $t \ge 1$ be the smallest integer such that the $\tilde{\Pi}$-facial walk $C_2$ containing $e_{k+t}$ and $e_{k+t+1}$ is a patch face. (If $C_2$ would not exist, the $\tilde{\Pi}$-facial walks containing $e_i, e_{i+1}$ $(k < i < d)$ would all be $\Pi'$-facial and hence $\tilde{\lambda}(e_{i+1})$ would be negative for $k < i < d$. This would contradict the assumption that $\tilde{\lambda}(e_d) = 1$.) Let $\delta \subseteq C_2$ be the patch segment containing $e_{k+t}$. By Proposition 5.5.12, $C_1$ and $C_2$ are cycles and hence $C_1 \ne C_2$. Moreover, $C_1$ and $C_2$ intersect only at $\nu$ (and at $\tau$ if $\delta = \varepsilon$). The segment $\delta$ is chosen so that it is in $\text{ext}(C, \Pi')$ precisely when $\varepsilon$ is in $\text{ext}(C, \Pi')$. We will show that this leads to a contradiction.

Let $H$ be the graph consisting of the cycles $C, C_1', C_2'$, where $C_i'$ is the cycle whose edges correspond to the patch segments of $C_i$ $(i = 1, 2)$. The embedding $\Pi'$ of $\tilde{G}$ determines a partition of $E(C_1') \cup E(C_2')$ into sets $E_1, E_2$ corresponding to the interior and the exterior of $C$, respectively. This partition satisfies the assumptions of Lemma 5.7.3. The only thing that we need to change in order to apply the lemma is that we split the patch vertex $\nu$ (and $\tau$ if $\delta = \varepsilon$) so that $C_1'$ and $C_2'$ become disjoint, and so that $\varepsilon$ and $\delta$ give rise to disjoint edges that overlap on $C$. Since $\tilde{G}$ is planar, the only overlapping pair of edges in $E_1$ (or $E_2$) are the edges corresponding to $\varepsilon$ and $\delta$. This contradicts Lemma 5.7.3, and we are done. □

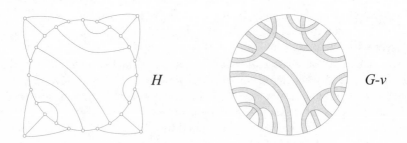

$H$           $G$-$v$

FIGURE 5.11. An apex graph embedding of face-width 3

Robertson and Vitray [**RV90**] raised the question about existence of embeddings of apex graphs of face-width 3 in orientable surfaces. Mohar [**Mo97b**] answered this question in the affirmative (in contrast to Theorem 5.7.2) by proving the existence of apex graphs with embeddings of face-width three in orientable surfaces of arbitrarily large genus. Such embeddings can be constructed as follows. Let $H$ be a planar graph and $C$ a cycle of $H$ such that the vertices of $C$ are of degree 3 and such that $H - V(C)$ consists of isolated vertices. Consider the orientable embedding of $H$ such that $C$ is a facial cycle, and such that all vertices of $H$ inside $C$ have the same clockwise rotation as in the plane and the vertices of $H$ in the exterior of $C$ have opposite rotation as in the plane. Let $G$ be a graph obtained from $H$ by subdividing $C$ and replacing all edges and vertices not in $C$ by dense planar graphs as indicated by the second graph in Figure 5.11. Now, add a vertex $v$ outside $C$ and join it to (almost) all vertices of $C$. Clearly, $G - v$ is planar. It is also easy to see that $G$ can be constructed such that it has face-width three and large genus (by having many disjoint pairs of overlapping $C$-bridges). Mohar [**Mo97b**] proved that all embeddings of 3-connected apex graphs of face-width three must have a structure similar to that of $G$ indicated in Figure 5.11.

## 5.8. The genus of a graph with a given nonorientable embedding

Proposition 4.4.1 shows that the nonorientable genus of a graph is bounded above by a linear function of the genus. On the other hand, Auslander, Brown, and Youngs [**ABY63**] proved that there are graphs embeddable in the projective plane whose genus is arbitrarily large. This phenomenon is explained completely by the following result of Fiedler, Huneke, Richter, and Robertson.

THEOREM 5.8.1 (Fiedler, Huneke, Richter, Robertson [**FHRR95**]). *Let $G$ be a graph that is $\Pi$-embedded in the projective plane. If* $\mathbf{fw}(G, \Pi) \neq 2$, *then the genus of $G$ is*

$$\mathbf{g}(G) = \left\lfloor \frac{1}{2}\mathbf{fw}(G, \Pi) \right\rfloor. \tag{5.7}$$

*If* $\mathbf{fw}(G, \Pi) = 2$, *then* $\mathbf{g}(G)$ *is either* 0 *or* 1.

In the proof of Theorem 5.8.1 we shall use the following simple result.

LEMMA 5.8.2. *Let $G$ be a graph that is $\Pi$-embedded in the projective plane. If $C$ and $C'$ are $\Pi$-noncontractible cycles, then they are both $\Pi$-onesided and $C \cap C' \neq \emptyset$. If $C$ and $C'$ are edge-disjoint, then there is a vertex $v \in V(C) \cap V(C')$ such that the edges of $C$ and $C'$ alternate in the $\Pi$-clockwise ordering around $v$.*

PROOF. Let $G'$ be the $\Pi'$-embedded graph obtained after cutting $G$ along $C$. Since $\mathbf{eg}(\Pi) = 1$ and $\mathbf{eg}(\Pi') \geq 0$, Lemma 4.2.4 implies that $C$ is $\Pi$-onesided and that $\mathbf{eg}(\Pi') = 0$. Similarly, $C'$ is $\Pi$-onesided. Since $\mathbf{eg}(\Pi') = 0$, $C'$ cannot be a cycle in $G'$. This implies the last statement of the lemma. $\square$

PROOF OF THEOREM 5.8.1. Let $w = \mathbf{fw}(G, \Pi)$. Let $\tilde{C}$ be a shortest noncontractible cycle of the vertex-face graph $\Gamma(G, \Pi)$, and let $v_1, \ldots, v_w$ and $F_1, \ldots, F_w$ be the vertices and $\Pi$-facial walks (respectively) on $\tilde{C}$. For $i = 1, \ldots, w$, we perform the following change of the embedding. Let the $\Pi$-clockwise ordering around $v_i$ be $e_i, e'_i, a_1, \ldots, a_s, f_{i+1}, f'_{i+1}, b_1, \ldots, b_r$ where $e_i, e'_i$ are consecutive edges in $F_i$ and $f_{i+1}, f'_{i+1}$ are consecutive edges in $F_{i+1}$ (where the index $i + 1$ is considered modulo $w$). Then we replace the clockwise ordering at $v_i$ by $e'_i, a_1, \ldots, a_s, f_{i+1}, e_i, b_r, \ldots, b_1, f'_{i+1}$ and we replace the signature $\lambda(e)$ by its inverse $-\lambda(e)$ for $e \in \{e_i, b_1, \ldots, b_r, f'_{i+1}\}$. Lemma 5.8.2 implies that the resulting embedding $\Pi'$ of $G$ is orientable. (We leave the details to the reader.) The $\Pi$-facial walks distinct from $F_1, \ldots, F_w$ are also $\Pi'$-facial. On the other hand, $F_1, \ldots, F_w$ are replaced by a single $\Pi'$-facial walk (if $w$ is even) or by two $\Pi'$-facial walks (if $w$ is odd). Consequently, $\chi(\Pi')$ is either $2 - w$ or $3 - w$. This proves that

$$\mathbf{g}(G) \leq \lfloor w/2 \rfloor. \tag{5.8}$$

It also proves the theorem when $\mathbf{fw}(G, \Pi) \leq 2$. We now prove by induction on $|V(G)|$ that $\mathbf{g}(G) \geq \lfloor w/2 \rfloor$ when $w \geq 3$. By Corollary 5.5.14 we may assume that $G$ is 3-connected. If $w = 3$, then $\mathbf{g}(G) \geq 1$ by Theorem 5.7.1. So assume now that $w \geq 4$. Let $\Pi'$ be an orientable embedding of $G$ of genus $\mathbf{g}(G)$. Let $C$ be a $\Pi$-facial cycle that is not $\Pi'$-facial. By Proposition 5.5.13, the cycle $C$ is nonseparating and hence $\Pi'$-surface nonseparating. Therefore the genus of the embedding of $G - C$ induced by $\Pi'$ is at most $\mathbf{g}(G) - 1$, and thus $\mathbf{g}(G - C) \leq \mathbf{g}(G) - 1$. On the other hand, the graph $H$ obtained by first shrinking the face $C$ and then removing the vertex corresponding to $C$ is a subgraph of $G - C$, and its embedding in the projective plane has face-width at least $w - 2$, by Propositions 5.5.7 and 5.5.8. If the face-width of the embedding of $H$ in the projective plane is at least three, we apply the induction hypothesis to conclude that

$$\left\lfloor \frac{w - 2}{2} \right\rfloor \leq \mathbf{g}(H) \leq \mathbf{g}(G - C) \leq \mathbf{g}(G) - 1.$$

This implies that $\mathbf{g}(G) \geq \lfloor w/2 \rfloor$ and the proof is complete.

Suppose now that the embedding of $H$ in the projective plane is of face-width 2. Since $H$ is obtained by a vertex deletion from a graph whose embedding in the projective plane has face-width three, Theorem 5.7.2 shows that $\mathbf{g}(H) \geq 1$. Now we complete the proof as in the preceding paragraph.                                                     $\square$

An analogue of Theorem 5.8.1 for the Klein bottle was obtained by Robertson and Thomas [**RT91**]. Let $\Pi$ be an embedding of $G$ in the Klein bottle. Then we denote by $\mathrm{ord}_2(G, \Pi)$ the minimum of $\lceil \ell(C)/4 \rceil$ taken over all $\Pi_\Gamma$-surface nonseparating $\Pi_\Gamma$-twosided cycles $C$ in the vertex-face graph $\Gamma(G, \Pi)$, where $\ell(C)$ denotes the length of $C$. Similarly, we denote by $\mathrm{ord}_1(G, \Pi)$ the minimum of $\lfloor \ell(C_1)/4 \rfloor + \lfloor \ell(C_2)/4 \rfloor$ taken over all pairs of $\Pi_\Gamma$-nonhomotopic $\Pi_\Gamma$-onesided cycles $C_1, C_2$ in $\Gamma(G, \Pi)$. The latter minimum restricted to all noncrossing pairs $C_1, C_2$ is denoted by $\mathrm{ord}_1'(G, \Pi)$.

THEOREM 5.8.3 (Robertson and Thomas [**RT91**]). *Let $G$ be a graph that is $\Pi$-embedded in the Klein bottle. Let*

$$g = \min\{\mathrm{ord}_1(G, \Pi), \mathrm{ord}_2(G, \Pi)\}. \tag{5.9}$$

*If $g \geq 4$, then $g = \mathbf{g}(G)$. Moreover, $g$ can be determined in polynomial time.*

We omit the proof of Theorem 5.8.3. Robertson and Thomas also proved that

$$g = \min\{\mathrm{ord}_1'(G, \Pi), \mathrm{ord}_2(G, \Pi)\}.$$

Theorems 5.8.1 and 5.8.3 imply the following:

COROLLARY 5.8.4. *The genus of graphs that can be embedded in the projective plane or the Klein bottle can be computed in polynomial time.*

The cycles $C$ used in the definition of $\mathrm{ord}_2(G, \Pi)$ and the pairs $C_1, C_2$ in the definition of $\mathrm{ord}_1(G, \Pi)$ represent the possible ways of cutting the Klein bottle so that the resulting surface is orientable. Robertson and Thomas [**RT91**] offered a natural conjecture which would extend Theorems 5.8.1 and 5.8.3 to graphs embedded in arbitrary nonorientable surfaces. That conjecture was disproved by Mohar [**Mo98**] who also proposed an alternative possibility to generalize Theorems 5.8.1 and 5.8.3.

The results of this section provide some evidence to the following

CONJECTURE 5.8.5 (Fiedler, Huneke, Richter, Robertson [**FHRR95**]). *For each fixed nonorientable surface $S$, the genus of a graph embedded in $S$ can be found in polynomial time.*

## 5.9. Face-width and surface minors

The grid graphs can serve as a generic class for planar graphs in the following sense:

PROPOSITION 5.9.1 (Robertson and Seymour [**RS86b**]). *Let $G_0$ be a plane graph. Then there is an integer $k$ such that $G_0$ is a surface minor of the $k \times k$ grid $P_k \,\square\, P_k$.*

PROOF. There is a plane graph $G_1$ with maximum degree 3 such that $G_0$ is a surface minor of $G_1$. By Theorem 2.3.4, $G_1$ has a straight line embedding in the plane. Now, every edge can be modified so that it becomes a polygonal arc whose segments are all vertical or horizontal. Then it is easy to see that, for some large $k$, the $k \times k$ grid contains a subdivision of $G_1$. This completes the proof. □

The proof of Proposition 5.9.1 does not give an explicit bound on the size of the grid. However, it is not difficult to show that the $O(n) \times O(n)$ grid suffices where $n$ is the number of vertices of $G_0$. We refer to Di Battista, Eades, Tamassia, and Tollis [**BETT94**] for further references.

Recall that a *surface minor* is defined as follows (cf. also p. 103). Let $\Pi$ be an embedding of a multigraph $G$ in a surface $S$. If $e$ is an edge of $G$ which is not a loop, then $\Pi$ can be modified to an embedding of $G/e$ in $S$. If $G - e$ is connected and $e$ is in two facial walks, then $\Pi$ induces an embedding of $G - e$ in $S$. A sequence of such contractions and deletions of edges results in a $\Pi'$-embedded graph minor $G'$ of $G$, and we say that the pair $G', \Pi'$ is a surface minor of $G, \Pi$.

Proposition 5.9.1 has the following analogue for general surfaces.

THEOREM 5.9.2 (Robertson and Seymour [**RS88**]). *Let $G_0$ be a graph that is $\Pi_0$-embedded in the surface $S \neq \mathbb{S}_0$. Then there is a constant $k$ such that for any graph $G$ which is $\Pi$-embedded in $S$ with face-width at least $k$, $G_0, \Pi_0$ is a surface minor of $G, \Pi$.*

Theorem 5.9.2 does not give explicit bounds on the face-width $k$ that guarantees the presence of $G_0, \Pi_0$ as a minor. Quantitative versions for some special cases are presented below. Graaf and Schrijver [**GS94**] proved that a graph on the torus with face-width $w \geq 5$ contains the Cartesian product $C_k \square C_k$ as the surface minor where $k = \lfloor 2w/3 \rfloor$. An analogue for general surfaces was obtained by Brunet, Mohar, and Richter [**BMR96**].

D. Barnette and X. Zha (private communication) proposed the following conjectures.

CONJECTURE 5.9.3 (Barnette, 1982). *Every triangulation of a surface of genus $g \geq 2$ contains a noncontractible surface separating cycle.*

Ellingham and Zha have informed us that they have proved Conjecture 5.9.3 for triangulations of the double torus.

CONJECTURE 5.9.4 (Zha, 1991). *Every graph embedded in a surface of genus $g \geq 2$ with face-width at least 3 contains a noncontractible surface separating cycle.*

Richter and Vitray (private communication) proved that this holds provided the face-width is at least 11. Zha and Zhao [**ZZ93**] and Brunet, Mohar, and Richter [**BMR96**] improved their result by showing that face-width 6 (even 5 for nonorientable surfaces) is sufficient. In [**BMR96**], it is also proved that a graph embedded with face-width $w$ in a surface of genus at least 2 contains $\lfloor (w-9)/8 \rfloor$ disjoint noncontractible and pairwise homotopic surface separating cycles.

If Conjecture 5.9.3 is true, also the following may hold.

CONJECTURE 5.9.5. *Let $T$ be a triangulation of an orientable surface of genus $g$, and let $h$ be an integer such that $1 \leq h < g$. Then $T$ contains a surface separating cycle $C$ such that the two surfaces separated by $C$ have genera $h$ and $g - h$, respectively.*

It is even possible that Conjecture 5.9.5 extends to all embeddings of face-width at least 3.

Barnette [**Ba88**] proved that a graph embedded in the torus with face-width at least 3 contains two disjoint noncontractible cycles. If the graph is 3-connected, then the cycles can be chosen so that one of the cylinders between them contains no vertices of the graph. Mohar and Robertson [**MR93, MR96b**] characterized the structure of embeddings that do not have two disjoint noncontractible cycles. Schrijver [**Sch93**] proved that a graph embedded in the torus with face-width $w$ contains $\lfloor 3w/4 \rfloor$ disjoint noncontractible cycles. Such cycles on the torus are necessarily homotopic to each other. For general surfaces, Brunet, Mohar, and Richter [**BMR96**] proved that a graph embedded with face-width $w$ contains at least $\lfloor (w - 1)/2 \rfloor$ disjoint noncontractible homotopic cycles. A special case of this result for $w = 5$ was obtained previously by Zha and Zhao [**ZZ93**].

Most of the results mentioned above can be viewed as special cases of Theorem 5.9.2. However, all these results give explicit bounds, some times even best possible bounds, that imply the existence of particular structures in the embedded graph $G$. Moreover, in most cases the proofs yield polynomial time algorithms to find the desired structures in $G$ whenever $G$ fulfills the required conditions.

Suppose that the embedding of the graph $G_0$ in Theorem 5.9.2 is a minimum genus embedding. If $G_0$ is a surface minor of another embedded graph $G$, then also the embedding of $G$ is a minimum genus embedding. Therefore, a consequence of Theorem 5.9.2 is that large face-width of an embedding implies that this is a minimum genus embedding. Specific bounds on the face-width that guarantee the embeddings to be minimum genus embeddings are discussed in Section 5.10 below.

Suppose now that $G_0$ is uniquely embeddable in $S$ and that its embedding has face-width at least three. By Corollary 5.1.6, such a graph $G_0$ exists. If $G$ is a 3-connected graph embedded in $S$ such that $G_0$ is a surface minor of $G$, then also the embedding of $G$ in $S$ is unique. Consequently, sufficiently large face-width of a 3-connected graph implies uniqueness of the embedding. This aspect of face-width will be treated in more detail in the next section.

## 5.10. Face-width and embedding flexibility

If there are nonequivalent embeddings of a graph $G$ in the surface $S$, then we speak of *embedding flexibility* of $G$ in $S$, and the number of nonequivalent embeddings of a graph $G$ in $S$ is called the *degree of embedding flexibility* of $G$ in $S$.

Whitney's 2-flipping theorem (Theorem 2.6.8) says that each embedding of a 2-connected graph in the sphere can be obtained from any other embedding in the sphere by a sequence of flippings. In Section 5.2 we extended Whitney's theorem to LEW-embeddings. In this section we generalize Whitney's theorem in another direction.

PROPOSITION 5.10.1 (Mohar [**Mo92**]). *Let $G$ be a 2-connected graph, and let $\Pi$ and $\Pi'$ be embeddings of $G$ such that $3 \le \mathbf{fw}(G, \Pi) < \infty$. Let $H \subseteq G$ be a nontrivial 3-connected block of $G$ containing a $\Pi$-noncontractible cycle. Then the following assertions are equivalent:*

(a) *$\Pi$ and $\Pi'$ are Whitney equivalent (and hence they are embeddings in the same surface and have equal face-width).*

(b) *A cycle $C$ of $G$ is $\Pi$-contractible if and only if it is $\Pi'$-contractible.*

(c) *An induced nonseparating cycle $C$ of $H$ is $\Pi$-contractible (in $G$) if and only if it is $\Pi'$-contractible.*

(d) *Let $\Pi_1$ and $\Pi'_1$ be the embeddings of $H$ that are induced by $\Pi$ and $\Pi'$, respectively. Then $\Pi_1$ and $\Pi'_1$ are equivalent.*

PROOF. By Corollary 5.5.14, $H$ is uniquely determined up to the choice of the virtual edges obtained in the splitting process.

By Lemma 5.2.1, (a) implies (b). Clearly, (b) implies (c). To prove that (c) implies (d), let us first recall that $\mathbf{fw}(H, \Pi_1) = \mathbf{fw}(G, \Pi) \ge 3$, by Corollary 5.5.14. Since $H$ is a subdivision of a 3-connected graph, Proposition 5.5.13 implies that the $\Pi_1$-facial walks are induced nonseparating cycles of $H$. By (c), they are $\Pi'$-contractible and hence $\Pi'_1$-facial. Therefore $\Pi'_1$ is equivalent to $\Pi_1$.

We now assume (d) and prove (a) by induction on $|E(G)|$. By Corollary 5.5.14, we have $3 \le \mathbf{fw}(H, \Pi_1) = \mathbf{fw}(H, \Pi'_1) \le \mathbf{fw}(G, \Pi')$. If $G$ is a subdivision of a 3-connected graph, then $G = H$ and (a) follows. Otherwise, $G$ is the union of two proper subgraphs $G_1, G_2$ such that $G_1 \cap G_2$ consists of two vertices $a$ and $b$ such that none of $G_1, G_2$ is just a path from $a$ to $b$. If $e_1, \ldots, e_k$ is the $\Pi$-clockwise ordering around $a$, then there are distinct indices $r, s$ such that $e_r$ and $e_{s+1}$ are in $G_1$ and $e_{r+1}, e_s$ are in $G_2$. Let $C_1$ and $C_2$ be the $\Pi$-facial walks containing the consecutive edges $e_r, e_{r+1}$ and $e_s, e_{s+1}$, respectively. $C_1$ and $C_2$ are cycles by Proposition 5.5.11 and hence $P_{ij} = G_i \cap C_j$ $(i, j \in \{1, 2\})$ are paths. The indices $r$ and $s$ can be selected such that the four paths $P_{ij}$ are distinct. Since $\mathbf{fw}(G, \Pi) \ge 3$, all cycles contained in $C_1 \cup C_2$ are $\Pi$-contractible. In particular so are $C'_1 = P_{11} \cup P_{22}$ and $C'_2 = P_{12} \cup P_{21}$. By repeated use of the 3-path-property of noncontractible cycles, we conclude that $C_1 \cup C_2$ contains a cycle whose $\Pi$-interior contains $C_1 \cup C_2$. Therefore, we may assume that $P_{12} \subseteq \mathrm{Ext}(C'_1, \Pi)$, $P_{21} \subseteq \mathrm{Int}(C'_1, \Pi)$, and also that

$G_2 = \mathrm{int}(C_1', \Pi) \cup P_{22}$, $G_1 = \mathrm{ext}(C_1', \Pi) \cup P_{11}$. Then $H \subseteq G_1 \cup P_{22}$ (assuming that we take $P_{22}$ as the virtual edge corresponding to the splitting of $G$ into $G_1'$ and $G_2'$). By the induction hypothesis applied to $G_1'$, there is a sequence of 2-flippings transforming the embedding of $G_1'$ induced by $\Pi$ into the induced embedding of $\Pi'$. Each 2-flipping in $G_1'$ determines a 2-flipping in $G$: If the virtual edge $P_{22}$ is not contained in the cycle $C$ of the flipping, then the same cycle is used for a 2-flipping in $G$; if $P_{22} \subseteq C$, then we either use $C$ or the cycle $C'$ obtained by replacing the segment $P_{22}$ of $C$ by $P_{12}$ (so that $G_2 \subseteq \mathrm{Int}(C')$). After that, we apply Theorem 2.6.8 to $G_2'$ such that the virtual edge of $G_2'$ is never flipped. The flippings transform the embedding of $G_2'$ induced by $\Pi$ into the induced embedding of $\Pi'$, and they can be viewed as 2-flippings in $G$. They complete the operation of changing $\Pi$ into $\Pi'$ by a sequence of 2-flippings, and the proof is complete. □

Whitney's Theorem 2.5.1 that a 3-connected planar graph admits a unique embedding in the sphere does not extend to other surfaces even under stronger requirements on connectivity. Lavrenchenko [**La87a**] described an infinite family of 5-connected triangulations of the torus that are not uniquely embeddable in the torus. On the other hand, Negami [**Ne83**] proved that 6-connected toroidal graphs[6] are uniquely embeddable in the torus up to the automorphisms of the graph. Similarly, if $G$ is a 5-connected projective planar graph distinct from $K_6$, and $G$ contains a subgraph homeomorphic to $K_6$, then $G$ admits a unique embedding in the projective plane [**Ne85**].

Negami [**Ne88a**] and independently Vitray [**Vi93**] described the structure of embeddings of nonplanar graphs in the projective plane that are not unique. A special case discovered independently by Barnette [**Ba89**] says that a 3-connected cubic graph embedded in the projective plane with face-width three or more has another embedding in the projective plane if and only if its dual contains a subgraph isomorphic to $K_6$ or a subgraph isomorphic to $K_4$ that is embedded with face-width 2. Lawrencenko [**La92**] considered flexibility of the triangulations of the projective plane. Mohar, Robertson, and Vitray [**MRV96**] described the structure and flexibility of embeddings of all planar graphs in the projective plane. Embedding flexibility of graphs that have embeddings of face-width four or more in the torus is described completely in a recent work of Robertson, Zha, and Zhao [**RZZ95p**].

Results about graphs that are embeddable in the torus and are not embeddable in the Klein bottle have been obtained by Riskin [**Ri94p**]. Lawrencenko and Negami [**LN99**] characterized all graphs that triangulate both the torus and the Klein bottle.

---

[6]The family of such graphs is very restricted; cf. Altschuler [**A172**].

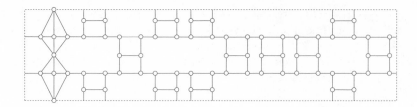

FIGURE 5.12. A 3-connected toroidal graph with many embeddings

Theorem 5.4.1 implies that for each surface $S$, there is a number $f(S)$ such that every graph that triangulates $S$ admits at most $f(S)$ nonequivalent embeddings in $S$. One can show that $f(\mathbb{S}_g) \geq 48^g$ (cf. [**MR98p**]). On the other hand, the example indicated in Figure 5.12 can be used to show that the same result does not hold for arbitrary 3-connected graphs in $S$. Increasing connectivity to 6 does not help (cf. [**MR98p**]), and it does not make sense to consider 7-connected graphs in $S$ since there are only finitely many such graphs. However, Mohar and Robertson [**MR98p**] proved that an additional constraint limits the number of embeddings.

THEOREM 5.10.2 (Mohar and Robertson [**MR98p**]). *There is a function $\xi : \mathbb{N} \to \mathbb{N}$ such that every 3-connected graph admits at most $\xi(g)$ embeddings of face-width $\geq 3$ into surfaces whose Euler genus is at most $g$.*

Flexibility of embeddings of planar graphs in general surfaces is treated by Mohar and Robertson in [**MR96a**]. To describe some of their results, we generalize flippings as follows. Suppose that the embedded graph $G$ contains vertices $v_1, v_2, v_3$ and that we can add the edges of the 3-cycle $C = v_1 v_2 v_3$ to $G$ and extend the embedding $\Pi$ of $G$ to an embedding $\tilde{\Pi}$ of $G + E(C)$ in the same surface $S$ such that $C$ is $\tilde{\Pi}$-contractible. If there is a $\tilde{\Pi}$-facial walk $W$ containing vertices $v_1, v_2$, and $v_3$, then $\tilde{\Pi}$ can be modified so that $\mathrm{int}(\tilde{\Pi}, C)$ is re-embedded in the face $W$. The resulting embedding is an embedding in $S$. If the induced embedding $\Pi'$ of $G$ is also an embedding in $S$, then we say that $\Pi'$ is obtained by a 3-*flipping* from $\Pi$. We define similarly 4-*flippings* where we use four vertices and the auxiliary contractible 4-cycle $C$. Mohar and Robertson [**MR96a**] proved that a planar graph has a bounded degree of flexibility up to Whitney's flippings, 3-flippings, and 4-flippings in the following sense:

THEOREM 5.10.3 (Mohar and Robertson [**MR96a**]). *Let $S$ be a surface. There is a constant $p = p(S)$ such that, for every embedding $\Pi$ of face-width 2 of a 2-connected planar graph $G$ in $S$, there is a sequence of 2-, 3-, and 4-flippings such that the resulting embedding $\Pi'$ of $G$ in $S$ has the following property: There is an embedding $\Pi_0$ of $G$ in the plane and a set $Q$ of at most $p$ vertices of $G$ such that the induced embeddings of $\Pi$ and $\Pi_0$ of any component of $G - Q$ are the same.*

Mohar and Robertson [**MR96a**] also proved a strengthening of Theorem 5.10.3 including embeddings of face-width 1. In addition, they proved that the $\Pi$-clockwise and the $\Pi'$-clockwise ordering around the vertices in the set $Q$ cannot differ too much. In [**Mo97c**] it is conjectured that a result similar to Theorem 5.10.3 holds also for embeddings of graphs of bounded genus.

Robertson and Vitray [**RV90**] proved that a 3-connected $\Pi$-embedded graph whose face-width is at least $\mathbf{eg}(\Pi) + 3$ is uniquely embeddable in that surface and that $\Pi$ is a minimum genus embedding. In particular, if a 3-connected graph has an embedding in the projective plane whose face-width is at least 4, then this embedding is unique. The same holds for embeddings in the torus of face-width 5 or more. Thomassen [**Th90b**] proved uniqueness and genus minimality of embeddings under a similar condition on large edge-width (Theorems 5.1.1 and 5.1.5). The result of [**RV90**] was slightly improved by Mohar [**Mo92**] who obtained a "local version" of the theorem and replaced $\mathbf{eg}(\Pi)$ in the bound on the face-width by the genus of the graph.

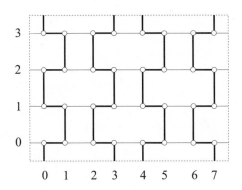

FIGURE 5.13. The graph $G_{2,2}$ embedded in the torus

N. Robertson asked if the face-width being larger than some fixed constant forces uniqueness of embeddings of 3-connected graphs. Thomassen [**Th90b**] proved that face-width four does not suffice by exhibiting toroidal embeddings of arbitrarily large face-width whose graphs admit embeddings of face-width four in some other surface. Such examples can be described as follows. We start with the quadrilateral embedding of the Cartesian product $C_{2k} \,\square\, C_{4l}$ ($k \geq 1$, $l \geq 1$) in the torus. The graph has vertices $(i,j)$, $0 \leq i < 2k$ and $0 \leq j < 4l$, and two vertices $(i,j)$, $(i',j')$ are adjacent if either $i = i'$ and $|j - j'| \in \{1, 4l - 1\}$, or $|i - i'| \in \{1, 2k - 1\}$ and $j = j'$. For $i = 1, \ldots, k$ we now remove the edges $(2i - 1, j)(2i, j)$ where $j \equiv 1, 2 \pmod 4$ and the edges $(2i - 1, j)(2i - 2, j)$ where $j \equiv 0, 3 \pmod 4$. Call the resulting graph $G_{k,l}$. The graph $G_{2,2}$ is shown in Figure 5.13. The face-width of $G_{k,l}$ is $\min\{2k, 2l\}$. Let $\Pi'$ be the embedding of $G_{k,l}$ whose facial walks are all facial 4-cycles of the embedding in the torus, all horizontal $4l$-cycles and all $4k$-cycles represented by thick edges in Figure 5.13. Clearly, the Euler genus of $\Pi'$ tends to infinity as $\min\{k, l\} \to \infty$. Proposition 5.5.12 implies that $\mathbf{fw}(G_{k,l}, \Pi') \geq 3$. Moreover, if three $\Pi'$-facial cycles mutually intersect each other, they are incident with the same vertex. This implies that $\mathbf{fw}(G_{k,l}, \Pi') \geq 4$. (In fact $\mathbf{fw}(G_{k,l}, \Pi') = 4$.)

More generally, Archdeacon proved:

THEOREM 5.10.4 (Archdeacon [**Ar92**]). *For each integer $k$ there exists a $k$-connected graph $R_k$ that has nonequivalent embeddings $\Pi$ and $\Pi'$, both of face-width more than $k$.*

Seymour and Thomas [**ST96**] improved the result of Robertson and Vitray [**RV90**] on uniqueness of embeddings:

THEOREM 5.10.5 (Seymour and Thomas [**ST96**]). *Let $G$ be a 3-connected graph and let $\Pi$ be an embedding of $G$ of face-width $k$.*

(a) *If $k \geq 100 \log \mathbf{eg}(\Pi)/\log\log \mathbf{eg}(\Pi)$ (or $k \geq 100$ if $\mathbf{eg}(\Pi) \leq 2$), then any embedding $\Pi'$ of $G$ satisfies $\mathbf{eg}(\Pi') \geq \mathbf{eg}(\Pi)$, and if $\mathbf{eg}(\Pi) = \mathbf{eg}(\Pi')$, then $\Pi$ and $\Pi'$ are equivalent (in particular, they are embeddings in the same surface).*

(b) *Let $\Pi'$ be an embedding of $G$. If $\mathbf{fw}(G, \Pi') \geq 3$ and $k \geq \min\{400, \log \mathbf{eg}(\Pi)\}$, then either $\Pi$ and $\Pi'$ are equivalent, or $\mathbf{eg}(\Pi') \geq \mathbf{eg}(\Pi) + 10^{-6}k^2$.*

Mohar [**Mo95**] obtained a related result:

THEOREM 5.10.6 (Mohar [**Mo95**]). *Let $G$ be a 2-connected graph and let $\Pi$ and $\Pi'$ be embeddings of $G$ of face-width $k$ and $k'$, respectively. Let $\kappa = \min\{k, k'\}$. Suppose that $k \geq 4$, $k' \geq 7$ and*

$$\left\lfloor \frac{k'-3}{2} \right\rfloor^{\lfloor \kappa/2 \rfloor - 2} > \mathbf{eg}(\Pi'). \tag{5.10}$$

*Then $\Pi$ and $\Pi'$ are embeddings in the same surface, and $\Pi'$ is Whitney equivalent to $\Pi$.*

Theorem 5.10.6 is weaker than Theorem 5.10.5 since it only excludes the existence of another embedding of large face-width. On the other hand, its proof is simpler and its assumption on the width is slightly weaker since the condition on $k'$ in (5.10) follows from

$$\mathbf{fw}(G, \Pi') \geq \frac{3 \log(\mathbf{eg}(\Pi'))}{\log\log(\mathbf{eg}(\Pi'))} + 5. \tag{5.11}$$

The graphs in Theorem 5.10.4 show that the bounds of Theorems 5.10.5 and 5.10.6 are close to best possible.

The rest of this section is devoted to a proof of Theorem 5.10.6. First we establish some tools for working with distinct embeddings of the same graph.

Suppose that $C$ and $C'$ are cycles of the $\Pi$-embedded graph $G$ whose intersection $x$ is either a vertex or an edge. If $x$ is a vertex, we say that $C$ and $C'$ $\Pi$-*cross* at $x$ if their edges incident to $x$ alternate in the $\Pi$-clockwise ordering around $x$. If $x$ is an edge, then $C$ and $C'$ $\Pi$-*cross* at $x$ if the same holds for the cycles and the embedding obtained after contracting $x$. Clearly, if $C$ and $C'$ $\Pi$-cross at $x$, then neither of them is $\Pi$-surface separating since $x$ is their only intersection.

LEMMA 5.10.7. *Let* $\mathcal{C}$ *be a set of cycles of* $G$. *If every cycle* $C \in \mathcal{C}$ $\Pi$-*crosses at most* $r$ *and at least one of the cycles in* $\mathcal{C}$ *and is disjoint from all other cycles in* $\mathcal{C}$, *then*

$$\mathrm{eg}(\Pi) \geq |\mathcal{C}|/(r+1).$$

PROOF. We select pairs of cycles $\mathcal{C}_i = (C_i, C_i')$ from $\mathcal{C}$, $i = 1, 2, \ldots, k$, such that:

(a) $C_1, \ldots, C_k$ are pairwise disjoint.
(b) If $1 \leq i < j \leq k$, then $C_i'$ and $C_j$ are disjoint.
(c) If $C_i$ is $\Pi$-onesided, then $C_i' = C_i$. Otherwise, $C_i'$ and $C_i$ $\Pi$-cross, $i = 1, \ldots, k$.

The pairs are obtained as follows. Suppose that we have already selected $\mathcal{C}_1, \ldots, \mathcal{C}_{i-1}$, where $i \geq 1$. Denote by $n_1$ the number of $\Pi$-onesided cycles among $C_1, \ldots, C_{i-1}$, and let $n_2$ be the number of $\Pi$-twosided cycles. If $2n_2 + n_1 < |\mathcal{C}|/(r+1)$, let $C_i$ be a cycle in $\mathcal{C}$ that is disjoint from $C_1, \ldots, C_{i-1}, C_1', \ldots, C_{i-1}'$. There are at least

$$|\mathcal{C}| - 2rn_2 - (r+1)n_1 \geq |\mathcal{C}| - (r+1)(2n_2 + n_1) > 0 \qquad (5.12)$$

candidates for $C_i$. After selecting $C_i$, let $C_i'$ be a cycle in $\mathcal{C}$ satisfying (c). It is clear that the selected pairs satisfy (a)–(c).

Let us now cut the surface along $C_k, C_{k-1}, \ldots, C_1$. After cutting along $C_i$, the graph remains connected since $C_i$ is either onesided, or $C_i$ $\Pi$-crosses $C_i'$, and $C_i'$ is disjoint from the cycles $C_{i+1}, \ldots, C_k$ used in the previous cuts. Therefore Lemma 4.2.4 implies that $\mathrm{eg}(\Pi) \geq 2n_2 + n_1$. By selecting as many pairs $\mathcal{C}_i$ as possible, (5.12) shows that $2n_2 + n_1 \geq |\mathcal{C}|/(r+1)$. $\square$

Let $\Pi$ and $\Pi'$ be embeddings of a graph $G$. A cycle of $G$ is $(\Pi, \Pi')$-*unstable* if it is $\Pi$-facial and $\Pi'$-nonfacial. Let $C$ be a $(\Pi, \Pi')$-unstable cycle, and let $D_1, \ldots, D_t$ be $(\Pi, \Pi')$-unstable cycles that $\Pi'$-cross $C$. Such cycles are called a $(\Pi, \Pi')$-*daisy* of *size* $t$ centered at $C$ if $D_i \cap D_j \cap C = \emptyset$ for $1 \leq i < j \leq t$. If $D_1, \ldots, D_t$ form a daisy, we assume that they are enumerated according to the (cyclic) order of their intersection with $C$. Figure 5.14 shows cycles of a daisy under embeddings $\Pi$ and $\Pi'$, respectively, where $\Pi'$ is an embedding in the torus.

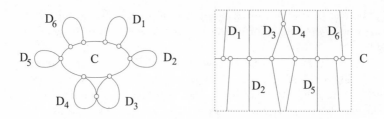

FIGURE 5.14. A daisy of size 6 centered at $C$

Suppose that $C$ is a cycle of $G$. Define $w(\Pi', C)$ as follows. If $C$ is $\Pi'$-facial, then $w(\Pi', C) = 0$. Otherwise, $w(\Pi', C)$ is the minimum number of segments of $\Pi'$-facial walks whose union is $C$.

LEMMA 5.10.8. *Let $G$ be a 3-connected graph with embeddings $\Pi$ and $\Pi'$, both of face-width at least three. Let $C$ be a $(\Pi, \Pi')$-unstable cycle and $w = w(\Pi', C)$. Then there is a $(\Pi, \Pi')$-daisy of size $w - 1$ centered at $C$. If $C'$ is a $(\Pi, \Pi')$-unstable cycle that $\Pi'$-crosses $C$, then the daisy can be chosen such that it contains $C'$.*

PROOF. Denote by $m$ the maximum size of a $(\Pi, \Pi')$-daisy centered at $C$. We claim that $m \geq w - 1$.

Let us first assume that there is a vertex $v \in V(C)$ such that the two edges of $C$ incident with $v$ are not consecutive in the $\Pi'$-clockwise ordering around $v$. Then there is a $\Pi$-facial cycle $D$ that $\Pi'$-crosses $C$ such that $D \cap C = \{v\}$. Since $C$ and $D$ $\Pi'$-cross, $D$ is $(\Pi, \Pi')$-unstable. Now, traverse $C$ starting at $v$. The (maximal) $\Pi'$-facial segments on $C$ on one or the other side of $C$ are called *left* and *right facial segments* on $C$, respectively. (Although $C$ may be $\Pi'$-onesided, the vertex $v$ enables us to speak of right and left.) Let $Q$ be the bipartite graph whose vertices are the left and the right facial segments on $C$, and two of these are adjacent if the segments share an edge. Each edge of $C$ is contained in a left and in a right facial segment and there is a bijection between $E(C)$ and $E(Q)$. Suppose that $e, f \in E(C)$ and that the corresponding edges of $Q$ do not have common endvertices. Then there are vertices $u_l, u_r \in V(C)$ between $e$ and $f$ such that $G$ has an edge incident with $u_l$ that is in a left facial segment of $C$ and an edge incident with $u_r$ that is in a right facial segment of $C$. This implies that there is a $(\Pi, \Pi')$-unstable cycle that $\Pi'$-crosses $C$ and intersects $C$ only between $e$ and $f$. It follows that every matching $M$ in $Q$ determines a set of $|M|$ $(\Pi, \Pi')$-unstable cycles that $\Pi'$-cross $C$ and whose intersections with $C$ are pairwise disjoint. This is a daisy of size $|M|$. Similarly, every vertex cover $U \subseteq V(Q)$ in $Q$ forms a set of $\Pi'$-facial segments on $C$ which cover $C$, and hence $|U| \geq w$. By the König-Egerváry Theorem (see Section 1.2), $Q$ has a matching of size $w$, and so $m \geq w$.

Since $C' \cap C$ intersects at most two cycles from the daisy, the replacement of these cycles by $C'$ gives a daisy of size $w - 1$.

By contracting an appropriately chosen edge $e$ in $C$ we may assume that $C$ has a vertex $v$ satisfying the condition at the beginning of the proof. As $w$ decreases by at most one when we contract the edge, there is a daisy of size $w - 1$. Moreover, $e$ can be chosen such that $C'$ belongs to the daisy. This completes the proof.                                        □

PROOF OF THEOREM 5.10.6. By Proposition 5.10.1 we may assume that $G$ is 3-connected. Suppose that $\Pi$ and $\Pi'$ are not equivalent. First, we claim that for each $(\Pi, \Pi')$-unstable cycle $C$, there exists a $(\Pi, \Pi')$-daisy $\mathcal{D}(C)$ centered at $C$ consisting of at least $\lambda = \lfloor (k' - 1)/2 \rfloor$ pairwise disjoint cycles. Moreover, if $C'$ is a $(\Pi, \Pi')$-unstable cycle that $\Pi'$-crosses $C$, then the daisy $\mathcal{D}(C)$ can be chosen so that it contains $C'$. To prove the claim, observe that $C$ is $\Pi'$-noncontractible because it is an induced nonseparating cycle by Proposition 5.5.13. Hence $w(\Pi', C) \geq k'$. By Lemma 5.10.8, there is a $(\Pi, \Pi')$-daisy $D_1, \ldots, D_{k'-1}$ centered at $C$ such that $D_1 = C'$. We let $\mathcal{D}(C)$ consist of $D_1, D_3, D_5, \ldots$ . It suffices to prove that any two cycles $D_i, D_j$ that are not (cyclically) consecutive cycles in the daisy $\mathcal{D}(C)$ are disjoint.

Let $y \in V(D_i) \cap V(C)$ and $z \in V(D_j) \cap V(C)$. If $D_i$ and $D_j$ intersect, then they share a vertex $x \notin V(C)$. Let us add edges $xy$, $yz$, and $zx$ in the $\Pi$-faces $D_i, C$, and $D_j$, respectively. Since $k \geq 4$, the triangle $\Delta = xyz$ is $\Pi$-contractible. One of the segments of $C$ between $y$ and $z$ is in $\text{int}(\Delta, \Pi)$, and the other is in $\text{ext}(\Delta, \Pi)$. Therefore, the $\Pi'$-noncontractible cycles $D_{i-1}$ and $D_{i+1}$ are contained in distinct $X$-bridges of $G$, where $X = \{x, y, z\}$. Since $\mathbf{fw}(\Pi') \geq 4$, we have a contradiction to Corollary 5.5.9 and we have proved the claim that $\mathcal{D}(C)$ exists.

Now we construct a family of $(\Pi, \Pi')$-unstable cycles by using the following procedure. We start by taking a $(\Pi, \Pi')$-unstable cycle $C_0$. Such a cycle exists since $\Pi$ and $\Pi'$ are not equivalent. Let $\mathcal{C}_0 = \{C_0\}$ and let $\mathcal{C}_1$ be a set of $\lambda$ pairwise disjoint cycles of a $(\Pi, \Pi')$-daisy centered at $C_0$. Having constructed $\mathcal{C}_0, \ldots, \mathcal{C}_{i-1}$ $(i \geq 2)$, we define $\mathcal{C}_i$ to be the union of daisies $\mathcal{D}(C)$, for each $C \in \mathcal{C}_{i-1} \backslash (\mathcal{C}_0 \cup \cdots \cup \mathcal{C}_{i-2})$, where $\mathcal{D}(C)$ is a $(\Pi, \Pi')$-daisy centered at $C$ that contains a cycle $C' \in \mathcal{C}_{i-2}$ and consists of $\lambda$ pairwise disjoint cycles. If $D_i, D_{i+1}$ are consecutive cycles of $\mathcal{D}(C)$, there is a $(\Pi, \Pi')$-unstable cycle intersecting $C$ between $D_i$ and $D_{i+1}$ that can be added to the daisy (but may intersect other cycles). For each $D \in \mathcal{C}_i$, there is a unique cycle, say $\varphi(D) \in \mathcal{C}_{i-1}$, whose daisy $\mathcal{D}(\varphi(D))$ contains $D$.

Suppose that $1 < 2i+1 < \kappa$, and that $C \in \mathcal{C}_i \backslash (\mathcal{C}_0 \cup \cdots \cup \mathcal{C}_{i-1})$ intersects a cycle $Q \in \mathcal{C}_q$ where $q \leq i$ and $Q \neq C, \varphi(C)$. Then the sequences $Q$, $\varphi(Q), \varphi(\varphi(Q)), \ldots, C_0$ and $C, \varphi(C), \varphi(\varphi(C)), \ldots, C_0$ contain a sequence $C = F_0, F_1, \ldots, F_r = Q$ of distinct $\Pi$-facial cycles where $2 \leq r \leq q+i \leq 2i$.

Any two consecutive cycles in this sequence $\Pi'$-cross. For $j = 0, \ldots, r-1$, let $v_j$ be a vertex in $V(F_j) \cap V(F_{j+1})$, and let $v_r$ be a vertex in $C \cap Q$. Let us add the edges $v_{j-1}v_j$ to $G$ and embed them in $F_j$ $(j = 1, \ldots, r)$, and do the same with the new edge $v_r v_0$ in $F_0$. Call the resulting cycle $\tilde{C}$. Let $\tilde{G} = G + E(\tilde{C})$ and let $\tilde{\Pi}$ be the embedding of $\tilde{G}$ extending $\Pi$. Then $\mathbf{fw}(\tilde{G}, \tilde{\Pi}) \geq \mathbf{fw}(G, \Pi) \geq \kappa \geq r + 2$. Therefore $\tilde{C}$ is $\tilde{\Pi}$-contractible. Denote by $S_j$ the segment of $F_j$ in $\mathrm{int}(\tilde{C}, \tilde{\Pi})$, $j = 1, \ldots, r - 1$. By a remark in the preceding paragraph (about a cycle $C$ between $D_i$ and $D_{i+1}$), there is a $(\Pi, \Pi')$-unstable cycle $D_j$ which $\Pi'$-crosses $F_j$ such that $D_j \cap F_j \subseteq S_j - \{v_{j-1}, v_j\}$, $j = 1, \ldots, r - 1$. In particular, $D_j \subseteq \mathrm{Int}(\tilde{C}, \tilde{\Pi})$, $j = 1, \ldots, r - 1$. An easy consideration of planar graphs shows that at least one of these cycles, say $D_s$, intersects $\tilde{C}$ in at most one vertex. Then $D_s$ is contained in a $\{v_0, \ldots, v_r\}$-bridge $B$. Similarly, there is a $(\Pi, \Pi')$-unstable cycle $D'$ in $\mathrm{Ext}(\tilde{C}, \tilde{\Pi})$. Both, $D_s$ and $D'$ are $\Pi'$-noncontractible. Since $D_s \subseteq B$ and $D'$ is in the union of other $\{v_0, \ldots, v_r\}$-bridges in $G$, we have a contradiction to Corollary 5.5.9. This proves that $C$ intersects only $\varphi(C)$ and itself in $\mathcal{C}_0 \cup \cdots \cup \mathcal{C}_i$ if $2i + 1 < \kappa$.

Let $\nu = \lfloor (\kappa - 4)/2 \rfloor$. Then $\mathcal{C}_0 \cup \cdots \cup \mathcal{C}_{\nu+1}$ consists of $(\Pi, \Pi')$-unstable cycles forming a tree-like structure by the previous claim. Their number is

$$1 + \lambda + \lambda(\lambda - 1) + \cdots + \lambda(\lambda - 1)^\nu = 1 + \lambda \frac{(\lambda - 1)^{\nu+1} - 1}{\lambda - 2}. \qquad (5.13)$$

Moreover, each of the cycles intersects exactly $\lambda$ of the other cycles in the family (if it is in $\mathcal{C}_i$ for $i \leq \nu$) or exactly one other cycle (if it is in $\mathcal{C}_{\nu+1}$). By (5.13) and Lemma 5.10.7, we have:

$$\mathbf{eg}(\Pi') \geq \left(1 + \lambda \frac{(\lambda - 1)^{\nu+1} - 1}{\lambda - 2}\right) / (\lambda + 1). \qquad (5.14)$$

Since $k' \geq 7$, we have $\lambda \geq 3$. Since $\lambda(\lambda - 1) > (\lambda - 2)(\lambda + 1)$, (5.14) yields

$$\mathbf{eg}(\Pi') \geq \frac{1}{\lambda + 1} + (\lambda - 1)^\nu - \frac{1}{(\lambda - 1)}. \qquad (5.15)$$

Since $\mathbf{eg}(\Pi')$ is an integer, we get $\mathbf{eg}(\Pi') \geq (\lambda - 1)^\nu$, a contradiction to (5.10). □

The following corollary of Theorem 5.10.5 is analogous to Corollary 5.1.3.

COROLLARY 5.10.9. *If $G$ is $\Pi$-embedded in an orientable surface of genus $g$ and $\mathbf{fw}(G, \Pi) \geq 100 \max\{1, \log(2g)/\log\log(2g)\}$, then $G$ is orientably simple.*

By Theorem 5.1.9, there is a constant $c > 0$ such that almost all (rooted) embeddings of graphs with $q$ edges in a fixed surface have face-width at least $c \log q$. Combined with the results of this section, this shows that almost all graphs on a fixed surface are minimum genus embeddings with essentially unique embeddings, and they share many properties with planar graphs.

## 5.11. Combinatorial properties of embedded graphs of large width

We conclude this chapter by reviewing results which show that conditions on the width of an embedded graph not only force the embedding to share many properties with planar embeddings, but they also imply that the embedded graphs behave like planar graphs in a more general sense. A useful tool is that of a planarizing collection of cycles.

If $C_1, \ldots, C_k$ is a collection of pairwise disjoint cycles in an embedded graph $G$, then we say that these cycles form a *planarizing collection of cycles* if the cutting along all of $C_1, \ldots, C_k$ results in a connected graph embedded in the sphere.

THEOREM 5.11.1 (Thomassen [**Th93b**]). *If $d$ and $g$ are natural numbers and $G$ is a triangulation of $\mathbb{S}_g$ of edge-width at least $8(d+1)(2^g - 1)$, then $G$ contains a planarizing collection of induced cycles $C_1, \ldots, C_g$ such that the distance between $C_i$ and $C_j$ ($1 \le i < j \le g$) is at least $d$.*

PROOF. We prove the theorem by induction on $g$. If $g = 0$ there is nothing to prove. So assume that $g \ge 1$. Denote by $\Pi$ the embedding of $G$. Let $C$ be a shortest $\Pi$-noncontractible cycle. By the 3-path-property of noncontractible cycles, $C$ is chordless. If $g = 1$, the proof is complete with $C_1 = C$. So assume that $g \ge 2$. We cut along $C$. The resulting embedded graph[7] has two copies $D_1, D_2$ of $C$ as faces. We add two new vertices $x_1, x_2$ and join $x_i$ to all vertices on $D_i$ ($i = 1, 2$) so that we get a new triangulation. Denote the resulting graph by $G'$ and let $\Pi'$ be its embedding. We claim that

$$\mathbf{ew}(G', \Pi') \ge \frac{1}{2}\mathbf{ew}(G, \Pi) \ge 4(d+1)(2^g - 1). \tag{5.16}$$

To see this, let $C'$ be a shortest $\Pi'$-noncontractible cycle. It is chordless. If $C'$ intersects $D_1$ and $D_2$, then $C'$ contains a path $P'$ from $D_1$ to $D_2$ such that $P' \cap (D_1 \cup D_2) = \{y_1, y_2\}$. Now consider $y_1$ and $y_2$ as vertices of $G$ and let $P$ be a shortest segment of $C$ from $y_1$ to $y_2$. Then $P \cup P'$ is a $\Pi$-noncontractible cycle since its intersection with $C$ is connected and $C$ comes to it from the left (say) and leaves it to the right. Hence, the length of $P \cup P'$ is at least $\mathbf{ew}(G, \Pi)$. As the length of $P$ is at most $\mathbf{ew}(G, \Pi)/2$, $P'$ and hence also $C'$ have length at least $\mathbf{ew}(G, \Pi)/2$.

We can therefore assume that $C' \cap D_2 = \emptyset$. If $C' \cap D_1$ is either empty or is a path, then it is easy to verify that $C'$ determines a $\Pi$-noncontractible cycle in $G$. Therefore its length is at least $\mathbf{ew}(G, \Pi)$. If

---

[7]If $C$ is $\Pi$-surface separating, this graph is disconnected, and all claims in the sequel in which we refer to the embedding $\Pi'$ must be interpreted as claims for each of the two embeddings.

$C'$ contains $x_1$, then $C'$ contains precisely two vertices $y, z$ of $D_1$ because $C'$ is chordless. Let $P_1$ be the shortest path in $D_1$ from $y$ to $z$. As $P_1 \cup yx_1z$ is $\Pi'$-contractible, $C'' = (C' - x_1) \cup P_1$ is $\Pi'$-noncontractible by the 3-path-property of noncontractible cycles. As above, $C''$ is also $\Pi$-noncontractible (as a cycle in $G$), and since the length of $P_1$ is at most $\mathbf{ew}(G, \Pi)/2$, we deduce that the length of $C'$ is more than $\mathbf{ew}(G, \Pi)/2$.

Assume finally that $C'$ does not contain $x_1$ and that $C' \cap D_1$ is not a path. Let $y, z$ be two vertices of $C' \cap D_1$ which are not consecutive on $D_1$. As $D_1$ is chordless, $y$ and $z$ are not consecutive on $C'$ either. One of the two cycles in $C' \cup yx_1z$ distinct from $C'$ is $\Pi'$-noncontractible by the 3-path-condition. Clearly, that cycle $C''$ is not longer than $C'$. By the minimality of $C'$, also $C''$ is minimal. Now we consider $C''$ instead of $C'$ and complete the proof as in a previous case. This proves (5.16).

We define the $q$-*canonical cycle* for $x_i$ in $G'$ as follows. The 1-canonical cycle for $x_i$ is $D_i$. Recursively, having defined the $(q - 1)$-canonical cycle $D$ for $x_i$, where $2q < \mathbf{ew}(G', \Pi') - 1$, the $q$-canonical cycle for $x_i$ is the unique chordless cycle $D'$ such that $D \subseteq \text{int}(D', \Pi')$ and all vertices of $D'$ have distance 1 from $D$.

For $i = 1, 2$, let $D_i'$ be the $(d + 2)$-canonical cycle for $x_i$ in $G'$. By (5.16), $D_1'$ and $D_2'$ exist. As every path in $G'$ from $D_1$ to $D_2$ can be extended to a $\Pi$-noncontractible cycle in $G$ by adding a segment of $C$, (5.16) implies that $\text{Int}(D_1', \Pi') \cap \text{Int}(D_2', \Pi') = \emptyset$. Now we form a new $\Pi_0$-embedded triangulation $G_0$ by adding to $\text{Ext}(D_1', \Pi') \cap \text{Ext}(D_2', \Pi')$ two new vertices $y_1, y_2$ such that $y_i$ is joined to all vertices of $D_i'$, $i = 1, 2$.

We claim that $\mathbf{ew}(G_0, \Pi_0) \geq 8(d + 1)(2^{g-1} - 1)$. Let $C_0$ be a shortest (and hence chordless) $\Pi_0$-noncontractible cycle in $G_0$. If $C_0$ contains none of $y_1, y_2$, then it is $\Pi'$-noncontractible and there is nothing to prove. So assume that $C_0$ contains a path $z_1y_1z_2$. We replace this path by a path $P$ of length at most $2(d + 2)$ in $\text{int}(D_1', \Pi')$. As the union of $P$ and a path in $D_1'$ from $z_1$ to $z_2$ is $\Pi'$-contractible, it follows by the 3-path-condition that the modified cycle is $\Pi'$-noncontractible. A similar modification is performed if $y_2 \in V(C_0)$. This proves that

$$
\begin{aligned}
\mathbf{ew}(G_0, \Pi_0) \ &\geq \ \mathbf{ew}(G', \Pi') - 4(d + 1) \\
&\geq \ 4(d + 1)(2^g - 1) - 4(d + 1) \\
&= \ 8(d + 1)(2^{g-1} - 1).
\end{aligned}
$$

We now apply the induction hypothesis to $G_0$. If $G_0$ is connected, then it contains a collection $C_1, \ldots, C_{g-1}$ of cycles satisfying the conclusion of the theorem with $G_0$ and $g-1$ instead of $G$ and $g$, respectively. In this case we put $C_g = C$. If $G_0$ has components $G_1$ and $G_2$ of $\Pi_0$-genus $g_1$ and $g_2$, respectively, then the induction hypothesis gives a planarizing collection $C_1, \ldots, C_g$ of cycles such that $G_1$ (with cycles $C_1, \ldots, C_{g_1}$) and $G_2$ (with cycles $C_{g_1+1}, \ldots, C_g$) satisfy the conclusion of the theorem. In either case, $G$ and the cycles $C_1, \ldots, C_g$ satisfy the conclusion of the theorem unless

$y_1$ or $y_2$ (or both) are in some of the cycles. Suppose, for example, that $C_1$ contains the path $z_1 y_1 z_2$. Then we modify $C_1$ into another chordless cycle by deleting $y_1$ and instead adding a path $z_1 u_1 u_2 \ldots u_r z_2$ such that $u_1 \ldots u_r$ is a segment in the $(d+1)$-canonical cycle for $x_1$. A similar modification is performed if $y_2$ belongs to one of the cycles. It is easy to verify that cutting $G$ along (the modified) collection $C_1, \ldots, C_g$ results in a connected graph. Hence the cycles form the desired collection. □

The proof of Theorem 5.11.1 also shows that in the plane graph obtained after cutting along the planarizing collection $C_1, \ldots, C_g$, not only copies of distinct cycles $C_i, C_j$ are at distance at least $d$, but also cycles corresponding to the same $C_i$ $(1 \le i \le g)$ are at distance at least $d$.

Theorem 5.11.1 extends to embedded graphs of large face-width using the vertex-face graph. It can also be extended to nonorientable surfaces.

THEOREM 5.11.2 (Yu [**Yu97**]). *Let $G$ be a graph that is $\Pi$-embedded in a surface with Euler genus $g$, and let $d$ be a positive integer. If* $\mathbf{fw}(G, \Pi)$ $\ge 8(d+1)(2^g - 1)$, *then $G$ contains a planarizing collection of induced cycles $C_1, \ldots, C_k$ $(g/2 \le k \le g)$ such that the distance between $C_i$ and $C_j$ $(1 \le i < j \le k)$ is at least $d$.*

Using Theorems 5.11.1 and 5.11.2, several combinatorial properties of planar graphs can be extended to graphs on general surfaces as described below.

Barnette [**Ba66**] proved that every 3-connected planar graph contains a spanning tree with maximum vertex degree 3. Below we present a short proof of this result following [**Th94a**] (where the result is proved only for triangulations).

THEOREM 5.11.3. *Let $G$ be a planar graph with outer cycle $C$, and let $x, y$ be vertices of $C$. Suppose that the graph obtained from $G$ by adding a new vertex adjacent to all vertices of $C$ is 3-connected. Then $G$ has a connected spanning subgraph $H$ of maximum degree at most three such that $H$ contains $C$ and such that $x$ and $y$ have degree two in $H$.*

PROOF. The proof is by induction on the number $n'$ of vertices of $G - C$. If $n' = 0$, then we take $H = C$.

Suppose now that $n' > 0$. If there are vertices $u, v \in V(C)$ such that $u, v$ are not adjacent on $C$ and such that they are contained in the same facial cycle distinct from $C$, then we add the edge $uv$ inside $C$ (if $uv$ is not already present in $G$) and we let $C_1, C_2$ be the two cycles in $C + uv$ containing $uv$. Now, apply the induction hypothesis to $G_i = \text{Int}(C_i)$ $(i = 1, 2)$. Let $H_i$ be the connected spanning subgraph of $G_i$ satisfying the conclusion of the theorem. If both $x$ and $y$ belong to $C_1$ (say), then we assume that $x$ and $y$ have degree 2 in $H_1$ and that $u$ and $v$ have degree 2 in $H_2$. If $x \in V(C_1) \backslash V(C_2)$ and $y \in V(C_2) \backslash V(C_1)$, then we assume $x, u$ have degree 2 in $H_1$ and $y, v$ have degree 2 in $H_2$. Then $(H_1 \cup H_2) - uv$

can play the role of $H$. So we may assume that if a facial cycle distinct from $C$ contains two vertices of $C$, then the two vertices are adjacent on $C$.

Let $z_1, \ldots, z_m$ be the neighbors of $x$ in that clockwise order such that $z_1, z_m \in V(C)$ and $z_1 \neq y$. By the above assumption, $m \geq 3$ and no vertex $z_2, \ldots, z_{m-1}$ is on $C$. Since $G$ plus the additional vertex adjacent to $C$ is 3-connected, the outer facial walk $C'$ of $G - x$ is a cycle. Let $H'$ be obtained by applying the induction hypothesis on $G - x$ such that $z_1$ plays the role of $x$. Now, let $H$ be obtained from $H'$ by deleting the edge of $C' - E(C)$ incident with $z_m$, and adding the two edges $z_1 x, z_m x$.   □

COROLLARY 5.11.4 (Barnette [**Ba66**]). *Every 3-connected planar graph contains a spanning tree with maximum vertex degree 3.*

Barnette [**Ba92**] proved the analogous result for the projective plane, the torus, and the Klein bottle provided the face-width is at least three. Brunet, Ellingham, Gao, Metzlar, and Richter [**BEGMR95**] extended Corollary 5.11.4 to 3-connected graphs on the torus or the Klein bottle without any assumption on the face-width. However, $K_{3,k}$ ($k \geq 8$) shows that the assumption on the face-width cannot be dropped for surfaces of negative Euler characteristic. In general we have:

THEOREM 5.11.5 (Thomassen [**Th94a**]). *If $G$ is a triangulation of an orientable surface of genus $g$ of face-width at least $8(2^{2g} + 2)(2^g - 1)$, then $G$ has a spanning tree of maximum degree at most 4.*

Archdeacon, Hartsfield, and Little [**AHL96**] constructed examples of $k$-connected graphs with an embedding of face-width more than $k$ and with no spanning tree of maximum degree less than $k$ ($k = 1, 2, 3, \ldots$). These examples show that the condition on the face-width in Theorem 5.11.5 (and in Theorems 5.11.6 and 5.11.7 below) cannot be replaced by a constant bound.

Thomassen [**Th94a**] suggested that the same method as used in his proof of Theorem 5.11.5 can be used to show that every 4-connected triangulation of an orientable surface with sufficiently large face-width contains a spanning tree with maximum degree at most three. This was verified by Ellingham and Gao [**EG94**]. A strengthening of their result was obtained by Yu [**Yu97**]. A $k$-*walk* in a graph $G$ is a walk that visits each vertex at least once and at most $k$ times. It is easy to see that the existence of a $k$-walk implies the presence of a spanning tree with maximum degree at most $k + 1$.

THEOREM 5.11.6 (Yu [**Yu97**]). *Let $G$ be a 4-connected graph embedded in a surface of Euler genus $g$. If the face-width of the embedding is at least $48(2^g - 1)$, then $G$ contains a 2-walk and hence also a spanning tree of maximum degree 3.*

By Theorem 5.11.11 below, any 4-connected graph with $n$ vertices and sufficiently large face-width has a spanning tree of maximum degree 3 such that at most $n/1000$ vertices have degree 3. At the 1995 Slovenian Conference on Graph Theory, Mohar proposed the conjecture that every 4-connected graph embedded in a surface of Euler genus $g$ with sufficiently large face-width contains a spanning tree with maximum degree 3 such that the number of vertices of degree 3 is $O(g)$.

Yu also obtained a generalization of Theorem 5.11.5.

THEOREM 5.11.7 (Yu [**Yu97**]). *Let $G$ be a 3-connected graph embedded in a surface of Euler genus $g$. If the face-width of the embedding is at least $48(2^g - 1)$, then $G$ contains a 3-walk and hence also a spanning tree of maximum degree 4.*

Further results on 2-walks in graphs on surfaces are presented in [**GR94, BEGMR95, GRY95, BMT98p**]. For a result on 4-walks cf. Gao and Wormald [**GW94**].

Barnette [**Ba94**] proved that every 3-connected planar graph contains a 2-connected spanning subgraph with maximum vertex degree at most 15, and Gao [**Ga95**] proved that 15 can be replaced by 6 (but not by 5). Combined with Theorem 5.11.2 this implies:

THEOREM 5.11.8. *Let $G$ be a 3-connected graph embedded in a surface of Euler genus $g$. If the face-width of the embedding is at least $24(2^g - 1)$, then $G$ contains a 2-connected spanning subgraph of maximum degree at most 14.*

PROOF. By Theorem 5.11.2, $G$ contains a planarizing set of induced nonadjacent cycles $Q_1, \ldots, Q_k$. Cut along these cycles to get a plane graph and add a vertex for each of the new faces joined to all vertices on that face. The resulting plane graph is 3-connected. By [**Ga95**], it has a 2-connected spanning subgraph $H$ of maximum degree at most 6. Now, $(H \cap G) \cup Q_1 \cup \cdots \cup Q_k$ is a 2-connected spanning subgraph of $G$ of maximum degree at most 14.  □

Corollary 5.11.4 is analogous to Tutte's theorem [**Tu56**] that every 4-connected planar graph contains a Hamilton cycle (which implies a spanning tree of maximum degree 2). The theorem of Tutte was extended by Thomas and Yu [**TY94**] who proved that every edge of a 4-connected projective planar graph is contained in a Hamilton cycle. They also proved that 5-connected toroidal graphs are Hamiltonian [**TY97**]. Brunet and Richter [**BR95**] proved that every 5-connected toroidal triangulation has a contractible Hamilton cycle.

These results were motivated by the following long-standing open question of Grünbaum [**Gr70**] and Nash-Williams [**NW73**].

PROBLEM 5.11.9. *Is every 4-connected graph on the torus Hamiltonian?*

For general surfaces, Yu proved the following conjecture of Thomassen [**Th94a**].

THEOREM 5.11.10 (Yu [**Yu97**]). *Let $G$ be a 5-connected triangulation of a surface with Euler genus $g$. If the face-width of $G$ is at least $96(2^g - 1)$, then $G$ has a Hamilton cycle.*

As pointed out in [**Th94a**], Theorem 5.11.10 does not extend to 4-connected triangulations. On the other hand, Böhme, Mohar, and Thomassen [**BMT98p**] proved:

THEOREM 5.11.11 (Böhme, Mohar, and Thomassen [**BMT98p**]). *For each nonnegative integer $g$ there is a constant $w_g$ such that every 4-connected graph $G$ embedded in a surface of Euler genus $g$ with face-width at least $w_g$ contains two cycles $C_1, C_2$ whose union contains all vertices of $G$ and such that the length of $C_i$ is at least $0.999|V(G)|$, $i = 1, 2$.*

Theorem 5.11.11 implies in particular the existence of a 2-walk and the existence of a 2-connected spanning subgraph of maximum degree 4 and a connected spanning subgraph of maximum degree 3 with almost all vertices of degree 2 in every 4-connected graph embedded with sufficiently large face-width, compare Theorems 5.11.6 and 5.11.8 and the remark after Theorem 5.11.6. Böhme, Mohar, and Thomassen [**BMT98p**] also proved that for every nonnegative integer $g$, there are constants $c_g > 0$ and $p_g > 0$ such that every 4-connected graph of Euler genus at most $g$ contains a cycle of length at least $c_g|V(G)|$ and contains paths $P_1, \ldots, P_{p_g}$ which cover all vertices of $G$.

PROBLEM 5.11.12. *Does Theorem 5.11.10 generalize to arbitrary 5-connected embedded graphs with sufficiently large face-width?*

CHAPTER 6

# Embedding extensions and obstructions

The results presented in Chapter 2 (Sections 2.3–2.7) demonstrate the importance of Kuratowski's theorem for the study of planar graphs. In the 1930's Erdős and König [**Kö36**] raised the question if for each surface $S$, there is a finite list **Forb**($S$) of graphs such that exclusion of these graphs and all their subdivisions as subgraphs characterizes the graphs embeddable in $S$ (not necessarily 2-cell embeddable). The first results in this direction were obtained in the 1970's when Glover and Huneke settled the problem for the projective plane [**GH78**] by showing that **Forb**($\mathbb{N}_1$) is finite. Glover, Huneke, and Wang [**GHW79**] exhibited a list of 103 graphs in **Forb**($\mathbb{N}_1$), and Archdeacon [**Ar80, Ar81**] showed that this list is complete. The finiteness of **Forb**($\mathbb{N}_1$) is proved and the list is presented in Section 6.5.

Robertson and Seymour [**RS90b, RS84a, RS85**] settled the general conjecture by proving that **Forb**($S$) is finite for each surface $S$. This result is an important step towards a proof of a more general statement, that has become known as Wagner's conjecture, which states that every family of graphs that is closed under the operation of deleting or contracting edges can be characterized by excluding a finite set of graphs as minors.[1] An independent proof of the finiteness of **Forb**($S$) for nonorientable surfaces $S$ was obtained by Archdeacon and Huneke [**ArH89**].

The original proof of Robertson and Seymour [**RS90b**] of the finiteness of **Forb**($S$) is existential whereas Seymour [**Se95p**] obtained a constructive proof by using graph minors and tree-width techniques which also gives an explicit upper bound for the number of vertices of the graphs in **Forb**($S$). An independent constructive proof (which leads to linear time embedding algorithms; cf. Theorem 6.6.3) was found by Mohar [**Mo96, Mo99**]. A much shorter proof was recently found by Thomassen [**Th97c**]. We shall present it in the next chapter. However, that proof depends on two other results of Robertson and Seymour, namely the excluded grid theorem [**RS86b**] and the result that the graphs of bounded tree-width are well-quasi-ordered with respect to the minor relation. A simple proof of the former was obtained recently by Diestel, Gorbunov,

---

[1] The reader is referred to the surveys [**RS85, RS90d**] and for a complete proof of Wagner's conjecture to [**RS83**]–[**RS92q**].

Jensen, and Thomassen [**DJGT99**] and is also included in Chapter 7. A simple proof of the latter is presented by Geelen, Gerards and Whittle [**GGW00p**].

The most natural approach (used, for example, in [**ArH89**] and in [**Mo96, Mo99**]) to study graphs in **Forb**($S$) is by induction on the (Euler) genus of $S$. The basic idea is the following. Suppose for simplicity that $S$ is orientable and that $S = \mathbb{S}_g$. Every graph $G \in$ **Forb**($S$) contains a subdivision $K$ of some graph $K_0 \in$ **Forb**($\mathbb{S}_{g-1}$). By the induction hypothesis, $K$ does not have too many vertices of degree different from 2 and hence $K$ admits only a bounded number of embeddings in $S$ (where the bound depends on $g$ only). Since $G \in$ **Forb**($S$), no embedding of $K$ in $S$ can be extended to an embedding of the whole graph. Therefore, $G$ is just the union of $K$ and (minimal) subgraphs of $K$-bridges (defined in Section 6.2) – called *obstructions* – whose presence in $G$ "obstructs" embedding extensions. If the number of $K$-bridges forming an obstruction is large, $K$ can be replaced by another subdivision of $K_0$ in $G$ such that we obtain a smaller obstruction. This idea leads to the study of the structure of obstructions for general *embedding extension problems* that ask whether a given embedding of a subgraph can be extended[2] to an embedding of the whole graph. Moreover, in such embedding extension problems we may allow only some of the possible embeddings of $K$-bridges.

In this chapter we study obstructions for embedding extensions and we demonstrate in Section 6.5 how they can be applied to show that **Forb**($\mathbb{N}_1$) is finite. In the last section we briefly discuss some results on the minimal forbidden subgraphs for other surfaces.

### 6.1. Forbidden subgraphs and forbidden minors

Let $S$ be a surface. We say that a graph $G$ is *topologically embeddable* in $S$ if there exists an embedding in $S$ as defined in Section 2.1. By the results of Chapter 3, this is equivalent to requiring that $G$ is a subgraph of a graph which has a (combinatorial) embedding in $S$. A graph $G$ is a *minimal forbidden subgraph* for $S$ if $G$ cannot be topologically embedded in $S$, $G$ has no vertices of degree 2, and every proper subgraph of $G$ can be topologically embedded in $S$. The set of all minimal forbidden subgraphs for $S$ is denoted by **Forb**($S$). The graph $G$ is a *minimal forbidden minor* (or an *excluded minor*) for $S$ if $G$ cannot be topologically embedded in $S$, but every proper minor of $G$ can be topologically embedded in $S$. Let us denote the set of all excluded minors for $S$ by **Forb**$_0$($S$). Clearly, every minimal forbidden minor for $S$ is also a minimal forbidden subgraph for $S$. A graph cannot be topologically embedded in $S$ if and only if it contains a subdivision of a minimal forbidden subgraph for $S$. Similarly, a graph

---

[2]Recall that an extension of the embedding $\Pi$ is an embedding in the same surface as $\Pi$.

cannot be embedded in $S$ if and only if it has a minor that is a minimal forbidden minor for $S$.

The main goal of this and the next chapter is to prove that for each surface $S$, $\mathbf{Forb}(S)$ is finite. Proposition 6.1.1 below shows that the finiteness of $\mathbf{Forb}_0(S)$ implies the finiteness of $\mathbf{Forb}(S)$.

PROPOSITION 6.1.1. *For each graph* $G_0$, *there is a finite collection* $G_1, \ldots, G_p$ *of graphs such that an arbitrary graph* $G$ *has* $G_0$ *as a minor if and only if* $G$ *contains a subdivision of one of* $G_1, \ldots, G_p$.

PROOF. Let $G_0'$ be the graph obtained from $G_0$ by subdividing each edge of $G_0$ twice. Consider any vertex $v$ of degree at least three in $G_0'$ and let $N(v)$ be the neighbors of $v$. Now delete $v$, partition $N(v)$ into classes and identify each class into a single vertex. Then add new vertices and edges such that these together with (the classes of) $N(v)$ form a tree where none of the new vertices has degree 1 or 2. The number of new vertices we add is less than $|N(v)|$. We perform this construction for every vertex $v$ of $G_0$ and let $G_1, \ldots, G_p$ denote all graphs which can arise in this way. Clearly, each $G_i$ has $G_0$ as a minor.

Conversely, if $G_0$ is a minor of $G$, let $T_v$ $(v \in V(G_0))$ be a tree in $G$ that is contracted to $v$ (see Proposition 1.2.2). We may assume that each vertex of degree 1 in $T_v$ has a neighbor in some other tree $T_u$. Since each $T_v$ is a subdivision of a tree without vertices of degree 2, $G$ contains a subdivision of one of the graphs $G_1, \ldots, G_p$. $\qquad\square$

If a graph $G \in \mathbf{Forb}(S)$ $(\mathbf{Forb}_0(S))$ is not 2-connected, then Theorems 4.4.2 and 4.4.3 show that the blocks of $G$ are minimal forbidden subgraphs (minors) for some surfaces of smaller genera. If $G$ has a 2-separator, the following lemma is useful.

LEMMA 6.1.2. *Let* $G$ *be a 2-connected minimal forbidden subgraph for some surface. Suppose that* $G$ *can be written as the union of two edge-disjoint subgraphs* $G_1, G_2$, *each of order at least three and with two vertices* $x, y$ *in common. For* $i = 1, 2$, *let* $G_i'$ *be the graph obtained from* $G_i$ *by adding the edge* $xy$ *if this edge is not already there. Then* $G_1'$ *and* $G_2'$ *are both nonplanar graphs.*

PROOF. Since $G$ is 2-connected, $G_2$ has a path $P$ from $x$ to $y$. Since $G_1 \cup P$ is a proper subgraph of $G$, it has a topological embedding in $S$. Hence also $G_1'$ (and similarly $G_2'$) has one. If $G_2'$, say, is planar then any topological embedding of $G_1'$ in $S$ can be extended to an embedding of $G_1' \cup G_2' \supseteq G$, a contradiction. $\qquad\square$

## 6.2. Bridges

Let $K$ be a connected graph. A vertex of degree different from 2 is a *branch vertex* of $K$. A *branch* of $K$ is any path in $K$ whose endvertices are branch vertices and such that each intermediate vertex has degree 2. The number of branches of $K$, denoted by $\mathbf{bs}(K)$, is the *branch size* of $K$. Let $G$ be a graph containing $K$ as a subgraph. A $K$-bridge in $G$ is a *local bridge* on the branch $e$ of $K$ if all its vertices of attachment are on $e$.

Let $B$ be a $K$-bridge. A path $P \subseteq B$ is *inside* $B$ if no intermediate vertex of $P$ is in $K$. Note that any two vertices of $B$ are joined by a path inside $B$.

LEMMA 6.2.1. *Let $G$ be a 3-connected graph and $K$ a connected subgraph of $G$ that is neither a path nor a cycle. Then $G$ contains a subgraph $K'$ such that:*

(a) *$K'$ is homeomorphic to $K$ and has the same branch vertices as $K$.*
(b) *For each branch $e$ of $K$, the corresponding branch $e'$ of $K'$ joins the same pair of branch vertices as $e$ and is contained in the union of $e$ and all $K$-bridges that are local on $e$.*
(c) *$K'$ has no local bridges.*

PROOF. It suffices to prove how to replace each branch $e$ of $K$ by a branch $e'$ satisfying (b) and such that $K - e + e'$ has no bridges that are local on $e'$. Denote by local$(e, K)$ the union of $e$ and all $K$-bridges that are local on $e$. We use induction on the number of vertices in local$(e, K)$. If local$(e, K) = e$, there is nothing to prove. Suppose that $B$ is a local bridge on $e$. Let $a$ and $b$ be the vertices of attachment of $B$ that are as far apart on $e$ as possible. If there is a nonlocal bridge $B'$ attached to $e$ between $a$ and $b$, then we say that $B$ *covers* an attachment of $B'$. In that case we replace the segment of $e$ from $a$ to $b$ by a path inside $B$, and let $e'$ be the new branch. Then local$(e', K - e + e') \subset$ local$(e, K)$, and we can use the induction hypothesis. So assume that no local bridge on $e$ covers an attachment of a nonlocal bridge. Then local$(e, K)$ contains a vertex $v$ such that either $v \notin e$ or $v$ lies on $e$ between attachments of a local bridge. Let $w$ be a vertex of $K$ that is not on $e$. Since $G$ is 3-connected, there are internally disjoint paths $P_1, P_2, P_3$ from $v$ to $w$. For $i = 1, 2, 3$ let $e_i$ be the first edge of $P_i$ in the direction from $v$ towards $w$ that does not belong to local$(e, K)$, and let $v_i$ be its end in $e$. If $v_1, v_2, v_3$ occur in that order on $e$, then $v_2$ belongs to a nonlocal bridge and is covered by a local bridge containing some edges of one of the other two paths, a contradiction. $\square$

We now extend Proposition 4.2.2 to the case when some facial walks are not necessarily cycles.

LEMMA 6.2.2. *Let $G$ be a connected graph. Suppose that $G = K \cup B$ where $K$ is connected and $B$ is a $K$-bridge in $G$. Let $\tilde{\Pi}$ be an extension to $G$ of an embedding $\Pi$ of $K$. Then there is exactly one $\Pi$-face that is not a $\tilde{\Pi}$-face.*

PROOF. Let $e$ and $f$ be any two feet of $B$. Let $P$ be a path inside $B$ that starts on $K$ with the edge $e$ and terminates on $K$ with $f$. If $\Pi'$ is the embedding of $K \cup P$ induced by $\tilde{\Pi}$, then $\chi(\Pi') = \chi(\Pi)$. Therefore $\Pi'$ has more faces than $\Pi$. This is possible only if all $\Pi$-facial walks except one, say $F$, are also $\Pi'$-facial. Since this holds for any choice of $e$ and $f$, $F$ is the only $\Pi$-face that is not a $\tilde{\Pi}$-face.                                                                           □

If $F$ is the $\Pi$-face in Lemma 6.2.2, then we say that $F$ *contains* the bridge $B$ or, equivalently, that $B$ is *embedded* in $F$ (in the extension $\tilde{\Pi}$). More generally, if $K$ is a $\Pi$-embedded subgraph of $G$ and $\tilde{\Pi}$ is an extension of $\Pi$ to $G$, then we extend these definitions to sets of bridges and say that a face $F$ *contains* a set $\mathcal{B}$ of $K$-bridges (or that $\mathcal{B}$ is *embedded* in $F$). We also speak about *embeddings of $K$-bridges* (in $\Pi$-faces), and this refers to extensions of the embedding $\Pi$.

If $F$ is a $\Pi$-facial walk, then some vertices in $F$ may appear more than once[3] in $F$. We speak of *appearances* of vertices in $F$. Let $\bar{x}$ be an appearance of a vertex $x$ in the $\Pi$-facial walk $F$, and let $e_1, e_2$ be the edges on $F$ which are incident with the appearance $\bar{x}$. Then $e_1, e_2$ are consecutive in the $\Pi$-clockwise ordering around $x$, say $e_1$ is followed by $e_2$. Suppose that the embedding $\Pi$ is extended to an embedding $\tilde{\Pi}$. If the $\tilde{\Pi}$-clockwise ordering around $x$ is $\ldots e_1 f_1 \ldots f_k e_2 \ldots$, then $f_1, \ldots, f_k$ are feet of $K$-bridges embedded in $F$, and we say that each of these edges *uses* the appearance $\bar{x}$ of $x$ (or that it is *attached* to that appearance of $x$).

LEMMA 6.2.3. *Let $K$ be a $\Pi$-embedded subgraph of a connected graph $G$ such that each $K$-bridge in $G$ is attached to at least two vertices of $K$. Suppose that $\tilde{\Pi}$ is an embedding extension of $\Pi$ to $G$ and that $B$ is a $K$-bridge in $G$. If $\bar{x}$ is an appearance of a vertex $x$ of $K$ on a $\Pi$-facial walk, then all feet of $B$, whose $\tilde{\Pi}$-embedding uses the appearance $\bar{x}$, are consecutive in the $\tilde{\Pi}$-clockwise ordering around $x$.*

PROOF. Let $F$ be the $\Pi$-facial walk containing the appearance $\bar{x}$, and let $e_1, e_2$ and $f_1, \ldots, f_k$ be the edges introduced in the definition before Lemma 6.2.3. Suppose that there are indices $1 \leq i < j < l \leq k$ such that $f_i$ and $f_l$ are feet of $B$ and $f_j$ is a foot of a $K$-bridge $B'$. We shall prove that $B' = B$. Since $B'$ has another vertex of attachment, there is a path $P$ inside $B'$ starting with $f_j$ and ending at a vertex $y$ on $F$ distinct from $x$. Since $\tilde{\Pi}$ is also an extension of the embedding of $K \cup P$, and in that extension $f_i$ and $f_l$ are embedded in distinct faces of $K \cup P$, $B$ is not a $(K \cup P)$-bridge in $G$. This shows that $B' = B$ and completes the proof.                                                                                    □

We use the notation introduced in the paragraph before Lemma 6.2.3. For each pair of consecutive edges in the sequence $e_1 f_1 f_2 \ldots f_k e_2$, there is a $\tilde{\Pi}$-facial walk that contains that pair of consecutive edges. We say that each such $\tilde{\Pi}$-face is *incident* with the appearance $\bar{x}$ of $x$ on $F$.

---

[3]Note that a vertex $v$ may appear in the same facial walk up to $\deg_K(v)$ times.

LEMMA 6.2.4. *Let $K$ be a $\Pi$-embedded subgraph of a connected graph $G$ such that each $K$-bridge in $G$ is attached to at least two vertices of $K$. Let $F$ be a $\Pi$-face with distinct appearances $x_1, x_2$ of the same vertex $x$ of $K$. Suppose that $\tilde{\Pi}$ is an extension of $\Pi$ to $G$ and that there is a $\tilde{\Pi}$-face $\tilde{F}$ which is incident with the appearances $x_1$ and $x_2$ of $x$. Suppose that $B$ is a $K$-bridge in $G$ such that a foot $e$ of $B$ uses the appearance $x_2$ and belongs to $\tilde{F}$. Then there exists an embedding extension $\tilde{\Pi}'$ of $\Pi$ to $G$ such that the $\tilde{\Pi}'$-embedding of all $K$-bridges in $G$ that are distinct from $B$ coincides with $\tilde{\Pi}$, and also the embeddings of $B$ are the same except that in $\tilde{\Pi}'$ the appearance $x_1$ of $x$ is used instead of $x_2$.*

PROOF. Let $f_1 \ldots f_d$ be the $\tilde{\Pi}$-clockwise ordering around $x$ where $f_1 = e$. Suppose that $e_1, e_2$ are the consecutive edges on $\tilde{F}$ incident with the appearance $x_1$, and let $S$ be the segment of $\tilde{F}$ from $x_1$ to $x_2$. We may assume that $e_1$ and $e$ are not in $S$, while $e_2$ is. By Lemma 6.2.3 we may assume that the feet of $B$ which use the appearance $x_2$ are $f_1, \ldots, f_i$ and that $\{e_1, e_2\} = \{f_j, f_{j+1}\}$ where $1 \le i < j < d$. If $e_1 = f_j$, then we change the $\tilde{\Pi}$-clockwise ordering around $x$ to $f_1 \ldots f_i f_{j+1} \ldots f_d f_{i+1} \ldots f_j$. The resulting embedding $\tilde{\Pi}'$ has same facial walks as $\tilde{\Pi}$ except that the segment $S$ has been removed from one and added to another facial walk. Therefore $\tilde{\Pi}'$ is an embedding extension of $\Pi$ with stated properties. Similarly, if $e_1 = f_{j+1}$, then the surface is nonorientable and we change the clockwise ordering around $x$ to $f_i f_{i-1} \ldots f_1 f_{j+1} \ldots f_d f_{i+1} \ldots f_j$ and also change the signature of each of $f_1, \ldots, f_i$. Then we get the same conclusion as in the previous case. This completes the proof. □

A *basic piece* of $K$ is either a branch vertex or an *open branch* of $K$ (i.e., a branch without its ends). If $F$ is a $\Pi$-facial walk, then we also speak of *appearances* of the basic pieces on $F$. If a $K$-bridge $B$ is embedded in $F$, then all basic pieces that $B$ is attached to appear on $F$. We say that $B$ *uses* an appearance or that $B$ is *attached* to an appearance $\bar{x}$ of the basic piece $x$ in $F$ if the embedding of $B$ uses an appearance of a vertex $x_0$ in $\bar{x}$. If $F$ is a $\Pi$-face, then its *branch size* is equal to the total number of appearances of branches of $K$ on $F$.

If a $K$-bridge $B$ in $G$ is attached to at least three basic pieces of $K$, then $B$ is said to be *strongly attached*. Otherwise it is *weakly attached*. The next result shows that too many strongly attached bridges obstruct embedding extensions.

LEMMA 6.2.5. *Let $K$ be a connected $\Pi$-embedded subgraph of the connected graph $G$ and suppose that there are no local $K$-bridges in $G$. Let $\tilde{\Pi}$ be an extension of $\Pi$ to $G$. If $B$ is a $K$-bridge, let $q(B)$ be the number of appearances of basic pieces on $\Pi$-faces used by the embedding of $B$. If $F$ is a $\Pi$-face of branch size $s$, and $B_1, \ldots, B_k$ are $K$-bridges embedded in*

*F, then*

$$\sum_{i=1}^{k}(q(B_i) - 2) \le 2s - 2. \qquad (6.1)$$

*Consequently, if $\mathcal{B}$ is any collection of $K$-bridges in $G$, then*

$$\sum_{B\in\mathcal{B}}(q(B) - 2) \le 4\,\mathbf{bs}(K) - 2. \qquad (6.2)$$

PROOF. The proof is by induction on $k$. Observe that $2s$ is the number of appearances of basic pieces of $K$ on $F$. We may assume that $q(B_i) \ge 3$ for $1 \le i \le k$. The case $k = 0$ is trivial. If $k \ge 1$, let $f_1, \dots, f_q$ (where $q = q(B_1)$) be feet of $B_1$ using distinct appearances of basic pieces of $K$ on $F$. Their vertices of attachment divide $F$ into $q$ segments, containing $p_1, \dots, p_q$ appearances (respectively) of basic pieces of $K$. Clearly, $p_1 + \cdots + p_q = 2s + q$. By the induction hypothesis

$$\sum_{i=1}^{k}(q(B_i) - 2) \le (p_1 - 2) + \cdots + (p_q - 2) + (q - 2) = 2s - 2.$$

This proves (6.1). The bound (6.2) follows from (6.1) by a summation over all $\Pi$-facial walks. □

## 6.3. Obstructions in a bridge

Let $G$ be a connected graph and $K$ a connected $\Pi$-embedded subgraph of $G$. A subgraph $\Omega \subseteq G - E(K)$ is an *obstruction* for extending the embedding of $K$ if $\Pi$ has no extension to $K \cup \Omega$. The obstruction $\Omega$ is *minimal* if for each edge $e \in E(\Omega)$, $\Omega - e$ is not an obstruction.

We also consider some *restricted embedding extensions* with additional requirements. If $\Omega$ is a subgraph of $G - E(K)$ such that no embedding extension of $K$ to $K \cup \Omega$ has the required property, then $\Omega$ is an *obstruction* for extensions with this property. In this restricted case we also speak of *minimal obstructions*.

In this section we consider embedding extension problems (abbreviated *EE problems* or *EEP*) where any $K$-bridge is required to be embedded so that it is attached to at most one appearance of each basic piece. Such embedding extensions are *simple*.

The *disk EEP*, is the following. Let $C$ be a cycle in a graph $G$ and let $C$ be embedded in the plane. Is there an embedding extension such that $G$ is embedded in the interior of $C$? Two types of obstructions are shown in Figure 6.1. The edges may be subdivided and $C$ is the dotted curve. The paths from $u_1, u_2, u_3$ to $C$ may have length 0. The paths $P_1$ and $P_2$ in Figure 6.1(a) are called *disjoint crossing paths* (with respect to $C$), and the obstruction in Figure 6.1(b) is called a *tripod*. Recall that a *Kuratowski subgraph* of $G$ is a subgraph homeomorphic to $K_5$ or $K_{3,3}$.

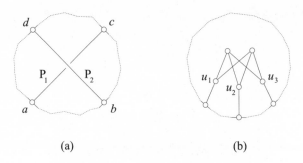

(a)                              (b)

FIGURE 6.1. Obstructions for the disk EEP

THEOREM 6.3.1. *Let $G$ and $C$ be as above. Let $\tilde{G}$ be the graph obtained from $G$ by adding a new vertex joined to all vertices of $C$. Then there is a disk embedding extension of $C$ to $G$ unless $G - E(C)$ contains an obstruction of the following type:*

(a) *a pair of disjoint crossing paths,*

(b) *a tripod, or*

(c) *a Kuratowski subgraph contained in a 3-connected block of $\tilde{G}$ distinct from the 3-connected block of $\tilde{G}$ containing $C$.*

PROOF. The proof is by induction on $|E(G)|$. Excluding case (c), it is easy to reduce the problem to the case where $\tilde{G}$ is 3-connected. If $G = C$, there is nothing to prove. If $G$ has two overlapping $C$-bridges, then $G$ either contains a pair of disjoint crossing paths (with respect to $C$) or a tripod. So assume that no two $C$-bridges in $G$ overlap. If $G$ has more than one $C$-bridge, then we use the induction hypothesis to conclude that $C$ together with any of the bridges is planar. But then $\tilde{G}$ is planar as well. So we may assume that $G$ has only one $C$-bridge. Also, we may assume that no vertex of $C$ has degree 2 in $G$. Then $G$ is 3-connected. If $G$ is planar, the proof is complete. Otherwise, let $e$ be any edge of $G$. If $G/e$ contains any of the configurations of Figure 6.1, then also $G$ does. Similarly, if $\widetilde{G/e}$ is not 3-connected and there is a Kuratowski subgraph in one of its 3-connected blocks (not containing the cycle $C$ or $C/e$), then $G$ contains a tripod or disjoint crossing paths. (We leave details for the reader.) So, we may assume that $G/e$ is planar for each edge of $G$. This implies that $G$ is a $K_5$ or $K_{3,3}$ with some additional edges. Now the proof is easy to complete.                                                          □

Mohar [**Mo94**] described a linear time algorithm that, given $G$ and $C$, either finds a disk embedding extension or one of the obstructions in Theorem 6.3.1.

We now assume that $K$ is a connected subgraph of $G$, $\Pi$ is a fixed embedding of $K$, and $B$ is a $K$-bridge in $G$. Following Mohar [**Mo94p**], we shall establish upper bounds for the branch size of minimal obstructions for embeddings of $B$.

Let $\Phi_0$ be the set of feet of $B$. If $B$ is embedded in a $\Pi$-face $F$, then the feet of $\Phi_0$ are cyclically ordered according to their attachment on $F$. Their cyclic order is well-defined even when several feet attach to the same appearance of a vertex on $F$. Let $\Omega$ be a subgraph of $B$ with feet $\Phi \subseteq \Phi_0$, and let $\Gamma$ be a cyclic ordering of $\Phi$. We define the *split auxiliary graph* $\mathrm{Aux}(\Omega, \Gamma)$ of $\Omega$ and $\Gamma$ as follows. First, split each vertex of attachment of $\Omega$ into as many new vertices as the number of feet of $\Omega$ at this vertex. Then extend the (new) endvertices of feet of $\Omega$ into the cycle (called the *auxiliary cycle*) determined by $\Gamma$. Finally, we add an additional vertex (called the *auxiliary vertex*) and join it to all vertices on the auxiliary cycle. We say that $\Omega$ is an *obstruction for the cyclic order* $\Gamma$ if $\mathrm{Aux}(\Omega, \Gamma)$ is a nonplanar graph. We also say that $\Omega$ *obstructs* $\Gamma$, or that $\Gamma$ is *obstructed* by $\Omega$. This implies that no embedding of $B$ will induce on $\Phi$ the cyclic order $\Gamma$. We can view $\Omega$ also as an obstruction for some cyclic orderings of feet distinct from $\Phi$. Thus $\Omega$ is said to *obstruct* a cyclic ordering $\Gamma'$ of feet $\Phi' \supseteq \Phi$ if $\Gamma'$ induces an obstructed cyclic ordering of $\Phi$. More generally, we say that $\Omega$ *obstructs* a cyclic ordering $\Gamma''$ of $\Phi'' \subseteq \Phi_0$ if it obstructs every cyclic ordering $\Gamma$ of $\Phi$ which induces the restriction of $\Gamma''$ to $\Phi'' \cap \Phi$. The obstruction $\Omega$ is a *total obstruction* if it obstructs every cyclic ordering of its feet (and thus also of any other set of feet of $B$).

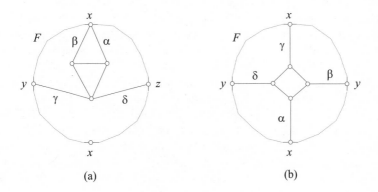

(a)                          (b)

FIGURE 6.2. Examples of obstructions

Figure 6.2 shows two examples of obstructions. The one in Figure 6.2(a) obstructs $\alpha$ and $\beta$ to be attached at different appearances of the vertex $x$ on $F$, i.e., it obstructs the cyclic ordering $(\alpha, \gamma, \beta, \delta)$. The one in Figure 6.2(b) obstructs the cyclic orderings $(\alpha, \gamma, \beta, \delta)$ and $(\alpha, \gamma, \delta, \beta)$. (Note that none of the embeddings in Figure 6.2 is simple.)[4]

To measure the size of an obstruction $\Omega$ we use the number $\mathbf{bs}_K(\Omega)$ of branches of $K \cup \Omega$ that are contained in $\Omega$.

An obstruction $\Omega \subseteq G - E(K)$ is *well connected* if $K \cup \Omega$ is connected and $\Omega$ is a $K$-bridge in $K \cup \Omega$.

LEMMA 6.3.2 (Mohar [**Mo94p**]). *Let* $\Omega \subseteq B$ *be a well connected obstruction (obstructing* $\Gamma_0$*) with* $t \geq 3$ *feet. If* $\Omega$ *is a total obstruction, then it contains a well connected total obstruction* $\Omega'$ *such that* $\mathbf{bs}_K(\Omega') \leq 12$. *Otherwise,* $\Omega$ *contains a well connected subgraph* $\Omega'$ *with the following properties:*

(a) $\Omega'$ *has the same feet as* $\Omega$.

(b) $\Omega'$ *obstructs the same cyclic orderings of feet as* $\Omega$.

(c) *For some and hence for every cyclic ordering* $\Gamma'$ *of the feet of* $\Omega'$, Aux$(\Omega', \Gamma')$ *is a subdivision of a 3-connected graph.*

(d) $\mathbf{bs}_K(\Omega') \leq 5t - 9$.

PROOF. Suppose first that $\Omega$ is a total obstruction. Let $Q$ be the graph obtained from $\Omega$ by identifying all its vertices of attachment on $K$ into a single vertex $w$. If $Q$ is planar, the local rotation at $w$ determines a cyclic ordering of feet of $\Omega$ that is not obstructed. Hence $Q$ is nonplanar. A Kuratowski subgraph of $Q$ determines a subgraph $\Omega_0$ of $\Omega$ which is a total obstruction. Since the Kuratowski graphs are 2-connected, $\Omega$ is well connected if it intersects $K$. If not, then we add a shortest path from $K$ to $\Omega$ and obtain a well connected obstruction of branch size at most 12.

Suppose now that some cyclic order, say $\Gamma$, is not obstructed by $\Omega$. Let $Q$ be the 3-connected block of Aux$(\Omega, \Gamma)$ which contains the auxiliary vertex and the auxiliary cycle. Since $\Omega$ is well connected and has at least three feet, $Q$ contains all feet of $\Omega$. Let $\Omega''$ be the subgraph of $\Omega$ corresponding to $Q$. It is easy to see that $\Omega''$ is well connected and that it satisfies (a) and (b). It also satisfies (c) since Aux$(\Omega'', \Gamma) = Q$.

Let $e_1, \ldots, e_t$ be the edges of the auxiliary cycle in Aux$(\Omega'', \Gamma)$, and let $F_i$ be the facial walk of the plane embedding of Aux$(\Omega'', \Gamma)$ which contains $e_i$ and does not contain the auxiliary vertex, $i = 1, \ldots, t$. Denote by $\Omega'$ the subgraph of $\Omega''$ consisting of all vertices and edges of $\Omega''$ that are contained in the facial walks $F_1, \ldots, F_t$. It is clear that $\Omega'$ is a well connected subgraph of $\Omega''$ and that it satisfies (a).

Let $H_i$ denote the union of the auxiliary cycle $C$, the auxiliary vertex $z$ and the walks $F_1, \ldots, F_i$, for $i = 2, 3, \ldots, t - 1$. We draw $H_i$ in the plane such that $z$ is outside $C$. We claim that $H_i$ is a subdivision of a 3-connected graph. Moreover, if $b_i$ is the number of branches of $H_i$ inside

---

[4]Recall that an embedding of $B$ is *simple* if it uses only one appearance of each basic piece that $B$ is attached to.

$C$, and if $b_i'$ is the number of branches of $H_i$ inside $C$ on the facial walk whose interior contains $F_t$, then

$$b_i + 3b_i' \leq 5i - 1. \tag{6.3}$$

We prove this claim by induction on $i$: For $i = 2$, it is trivial. Suppose it holds for $H_{i-1}$ where $i \geq 3$. We follow $F_i$ from the foot not contained in $F_{i-1}$. No subpath in $F_i$ which connects two vertices of $H_{i-1}$ and has no intermediate vertex in $H_{i-1}$ can join two vertices on the same branch of $H_{i-1}$ because $\mathrm{Aux}(\Omega'', \Gamma)$ is a subdivision of a 3-connected graph. Hence $H_i$ is a subdivision of a 3-connected graph. The inequality (6.3) on the number of branches is an easy count.

Since $b_{t-1} + 3b_{t-1}' \leq 5t - 6$, it follows easily that $\mathbf{bs}_K(\Omega') \leq 5t - 9$. This proves (c) and (d).

By Theorem 2.5.1, the facial cycles of the plane embedding of $\mathrm{Aux}(\Omega', \Gamma)$ are the induced nonseparating cycles of that graph. If $R$ is such a cycle that is disjoint from the auxiliary cycle, it is an induced nonseparating cycle (and hence facial) also in $\mathrm{Aux}(\Omega', \Gamma')$ for any $\Gamma'$. Consequently, any plane embedding of $\mathrm{Aux}(\Omega', \Gamma')$ can be extended to a plane embedding of $\mathrm{Aux}(\Omega'', \Gamma')$ by using the embedding of $\mathrm{Aux}(\Omega'', \Gamma)$. This proves (b). $\quad\square$

COROLLARY 6.3.3 (Mohar [**Mo94p**]). *Let $\Omega_1, \Omega_2$ be obstructions in the $K$-bridge $B$ such that $\Omega_1 \cup \Omega_2$ contains at least 3 feet. Then $B$ either contains a well connected total obstruction $\Omega'$ such that $\mathbf{bs}_K(\Omega') \leq 12$, or it contains a well connected subgraph $\Omega'$ with the following properties:*

(a) *$\Omega'$ has the same feet as $\Omega_1 \cup \Omega_2$.*
(b) *$\Omega'$ obstructs all the cyclic orderings of feet that are obstructed by $\Omega_1$ or by $\Omega_2$.*
(c) *For every cyclic ordering $\Gamma'$ of feet of $\Omega'$, $\mathrm{Aux}(\Omega', \Gamma')$ is a subdivision of a 3-connected graph.*
(d) *$\mathbf{bs}_K(\Omega') \leq 5t - 9$ where $t$ is the number of feet of $\Omega_1 \cup \Omega_2$.*

PROOF. By adding a path in $B - K$ between $\Omega_1$ and $\Omega_2$, we get a well connected obstruction $\Omega \supseteq \Omega_1 \cup \Omega_2$ with the same feet as $\Omega_1 \cup \Omega_2$. Now we apply Lemma 6.3.2 to the obstruction $\Omega$. $\quad\square$

To investigate obstructions for simple EE problems we shall use some special subgraphs of bridges in $G$. Let $B$ be a $K$-bridge in $G$. An E-*graph* in $B$ is defined as a minimal subgraph $H$ of $B$ such that:

(E1) $H$ is well connected.
(E2) For each branch vertex $\zeta$ in $K$ that $B$ is attached to, $H$ contains a foot incident with $\zeta$. If $\zeta$ is an open branch (with ends $x_1$ and $x_2$) in $K$ that $B$ is attached to, and $\zeta_i$ is the vertex of attachment of $B$ on $\zeta$ which is closest to $x_i$, then $H$ contains a foot incident with $\zeta_i$, $i = 1, 2$. We call $\zeta_1$ and $\zeta_2$ the *extreme attachments* of $B$ on $\zeta$.

(E3)  For every embedding $\Pi$ of $K$ and every simple extension $\tilde{\Pi}$ of $\Pi$ to $K \cup H$, there is a simple extension of $\Pi$ to $K \cup B$ which uses the same appearances of the basic pieces as $\tilde{\Pi}$.

Clearly, every bridge contains an E-graph, and the next result shows that the branch size of E-graphs is bounded.

COROLLARY 6.3.4. *Let $K$ be a connected subgraph of $G$ and let $\mathcal{B}$ be the set of $K$-bridges in $G$. There is a number $\mathbf{c}_1$ depending only on $\mathbf{bs}(K)$ such that each $B \in \mathcal{B}$ contains an E-graph $\tilde{B}$ with $\mathbf{bs}_K(\tilde{B}) \leq \mathbf{c}_1$. Suppose that $\{B_1, \dots, B_k\} \subseteq \mathcal{B}$ $(k \geq 1)$ are nonlocal $K$-bridges, and that $\tilde{B}_i$ is an E-graph in $B_i$ for $i = 1, \dots, k$. Then for every embedding $\Pi$ of $K$ and for every simple extension of $\Pi$ to $K \cup \tilde{B}_1 \cup \dots \cup \tilde{B}_k$, there is a simple extension of $\Pi$ to $K \cup B_1 \cup \dots \cup B_k$ such that each $B_i$ $(1 \leq i \leq k)$ uses the same appearances of basic pieces as $\tilde{B}_i$.*

PROOF. If $B \in \mathcal{B}$ contains a total obstruction, Lemma 6.3.2 implies that $B$ contains a well connected total obstruction $B'$ such that $\mathbf{bs}_K(B') \leq 12$. For each extreme attachment $\zeta_i$ of $B$ to the basic piece $\zeta$ of $K$, we add a foot incident with $\zeta_i$ and a path to the already constructed subgraph. This gives a subgraph of branch size at most $11 + 2(3\,\mathbf{bs}(K) + 1)$. Clearly, this subgraph contains an E-graph of bounded size.

If $B$ is not a total obstruction, then we select a minimal subgraph $B_0 \subseteq B$ satisfying (E1) and (E2). Clearly, $B_0$ is just a tree. For every embedding $\Pi$ of $K$ and every simple extension of $\Pi$ to $K \cup B_0$ we check if there is a simple extension of this embedding to $K \cup B$. If not, let $\Omega$ be an obstruction for such an extension. By Theorem 6.3.1, $\Omega$ contains at most 6 feet. We may assume that $\Omega$ is well connected. The number of simple embedding extensions of $\Pi$ to $B_0$ is bounded above by the number of cyclic orderings of feet of $B_0$. The number of feet of $B_0$ is at most $2\tau$ where $\tau$ is the number of basic pieces of $K$. By Corollary 6.3.3, the union of $B_0$ with all obstructions $\Omega$ described above contains an obstruction $B'$ with the same feet and $\mathbf{bs}_K(B') \leq 5(2\tau + 6(2\tau)!) - 9$. Clearly, $B'$ contains an E-graph $\tilde{B}$.

The last statement is an easy consequence of the fact that each $\tilde{B}_i$ $(1 \leq i \leq k)$ is an E-graph.                                                          □

If we consider a simple EEP and if there are no local $K$-bridges in $G$, then Corollary 6.3.4 shows that if we replace every $K$-bridge $B$ in $G$ by an E-graph $\tilde{B}$, then the EEP does not change. This enables us to consider only obstructions that are the union of E-graphs. Such an obstruction $\tilde{\Omega}$ is *pre-minimal* if no E-graph $\tilde{B}$ in $\tilde{\Omega}$ is redundant, i.e., $\tilde{\Omega} \backslash \tilde{B}$ is not an obstruction for the simple EEP.

If the $K$-bridge $B$ is weakly attached, then its E-graphs are particularly simple.

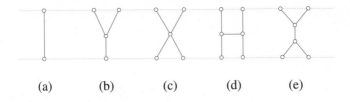

(a)     (b)     (c)     (d)     (e)

FIGURE 6.3. The E-graphs of weakly attached bridges

LEMMA 6.3.5. *Suppose that a nonlocal $K$-bridge $B$ is weakly attached. If $B$ has a simple embedding (extending some embedding of $K$), then $B$ contains an E-graph $\tilde{B}$ with $\mathbf{bs}_K(\tilde{B}) \leq 5$ which is a subdivision of one of the graphs in Figure 6.3.*

The easy proof of this lemma is left to the reader. See also [**Mo94p**].

## 6.4. 2-restricted embedding extensions

Let $K$ be a $\Pi$-embedded subgraph of a connected graph $G$. Suppose that each $K$-bridge in $G$ admits at most two embeddings in $\Pi$-faces. (Possibly both embeddings are in the same face.) Then we say that the embedding $\Pi$ of $K$ is *2-restrictive* and we speak of a *2-restricted EEP* (abbreviated *2-EEP*). For a $K$-bridge $B$ we denote by $B[1]$ and $B[2]$ its admissible embeddings. If $B$ has only one admissible embedding, we assume this is $B[1]$ and we use the symbol $B[2]$ to denote what we call the *forbidden embedding* of $B$. For $\tau \in \{1, 2\}$, we write $\neg\tau = 3 - \tau$. In particular, if $B$ has two admissible embeddings, then $B[\neg\tau]$ denotes the admissible embedding of $B$ distinct from $B[\tau]$.[5]

Suppose that $B$ and $B'$ are distinct $K$-bridges in $G$. If there is an extension of $\Pi$ to $K \cup B \cup B'$ where $B$ and $B'$ use their embeddings $B[\tau]$ and $B'[\tau']$, respectively, then we say that $B \cup B'$ has the embedding $B[\tau] \cup B'[\tau']$. If such an embedding does not exist, we say that $B[\tau]$ and $B'[\tau']$ *overlap*. If $B[\tau]$ and $B'[\tau']$ overlap, we write

$$B[\tau] \to B'[\neg\tau'] \tag{6.4}$$

and say that $B[\tau]$ *forces* the embedding $B'[\neg\tau']$ of $B'$. Clearly, (6.4) is equivalent to $B'[\tau'] \to B[\neg\tau]$.

Let $\mathcal{B}$ be the set of $K$-bridges in a pre-minimal obstruction for a 2-restricted EEP. Suppose that $\mathcal{B}$ contains a bridge $B_1$ with precisely one admissible embedding $B_1[1]$. This embedding overlaps with at least one admissible embedding of a bridge in $\mathcal{B}$ and forces an embedding of that bridge. The forced embeddings may overlap with other embeddings of

---

[5]The choice of the symbol $\neg$ stems from a close relationship between 2-EEPs and the 2-satisfiability problems (see, e.g., [**FMR79**]).

some bridges in $\mathcal{B}$, etc. As $\mathcal{B}$ is a pre-minimal obstruction, every admissible embedding of a subset $\mathcal{B}' \subset \mathcal{B}$ forces some embedding of a bridge $B \in \mathcal{B}\backslash\mathcal{B}'$. This shows that one of (a) and (b) below holds:

(a) There is a chain

$$B_1[\tau_1] \to B_2[\tau_2] \to \cdots \to B_s[\tau_s] \qquad (6.5)$$

where $\tau_1 = 1$ and $B_s[\tau_s]$ is the forbidden embedding of $B_s$.

(b) There is a chain of the form (6.5) and an index $r < s$ such that $B_s[\tau_s] \to B_r[\neg\tau_r]$.

Case (a) may be viewed as a special case of (b) with $r = s$ and hence we shall not distinguish between the two possibilities. In either case we say that $\mathcal{R} = B_1[\tau_1] \to \cdots \to B_s[\tau_s]$ together with the additional forcing $B_s[\tau_s] \to B_r[\neg\tau_r]$ ($r \le s$) is a *forcing chain*. Observe that the forcing chain shows that the bridges in the chain form an obstruction for the 2-EEP.

If no bridge in $\mathcal{B}$ has just one admissible embedding, let $B_1$ be an arbitrary bridge in $\mathcal{B}$. Fix its first embedding $B_1[1]$ and consider $B_1[2]$ as forbidden. The above proof shows that there is a forcing chain starting with $B_1[1]$. Similarly, there is a forcing chain starting with $B_1[2]$. Again, the two forcing chains show that $\mathcal{B}$ is an obstruction.

The forcing chain (6.5) is *forward minimal* if there are no forcings $B_i[\tau_i] \to B_j[\tau_j]$ for $j > i + 1$.

PROPOSITION 6.4.1. *For every pre-minimal obstruction $\mathcal{B}$ for a 2-EEP there are one or two forcing chains such that for each of these chains, say $\mathcal{R} = B_1[\tau_1] \to \cdots \to B_s[\tau_s]$, the following holds:*

(a) *All bridges in $\mathcal{R}$ are distinct.*

(b) *At most two bridges in $\mathcal{B}$ have only one admissible embedding, and if there is such a bridge $B$, then we have just one forcing chain and $B$ is either $B_1$ or $B_s$.*

(c) *Every bridge of $\mathcal{B}$ appears in at least one of the chains.*

(d) *$\mathcal{R}$ is forward minimal.*

(e) *Each of the following embeddings exists*

$$B_1[\tau_1] \cup B_2[\tau_2] \cup \cdots \cup B_{s-1}[\tau_{s-1}],$$

$$B_1[\tau_1] \cup \cdots \cup B_r[\tau_r] \cup B_s[\neg\tau_s] \cup \cdots \cup B_{r+2}[\neg\tau_{r+2}] \quad (\text{if } r < s),$$

$$B_2[\neg\tau_2] \cup B_3[\neg\tau_3] \cup \cdots \cup B_r[\neg\tau_r].$$

*where $r$, $1 \le r \le s$, is an index such that $B_s[\tau_s] \to B_r[\neg\tau_r]$.*

PROOF. Choose $\mathcal{R}$ so that $s$ is minimal. If $B_s[\tau_s] \to B_r[\neg\tau_r]$, we assume that $r$ is as small as possible. Suppose that $B_p = B_q$ where $p < q$. If $\tau_p = \tau_q$, we may delete $B_{p+1}[\tau_{p+1}], \ldots, B_q[\tau_q]$ from $\mathcal{R}$. If $\tau_p = \neg\tau_q$, then $\mathcal{R}$ could terminate at $B_{q-1}$ since $B_{q-1}[\tau_{q-1}] \to B_p[\neg\tau_p]$. This proves (a). Claim (b) is proved similarly, and (c) is clear since $\mathcal{B}$ is pre-minimal.

Suppose that $B_j[\tau_j] \to B_l[\tau_l]$ where $l > j + 1$. If $r \notin \{j+1, \ldots, l-1\}$, then $\mathcal{R}$ can be shortened by deleting the bridges $B_{j+1}, \ldots, B_{l-1}$. Suppose now that $j < r < l$ and that $j$ is minimal and, subject to that, $l$ is maximal. Now, we replace $\mathcal{R}$ by $B_1[\tau_1] \to \cdots \to B_j[\tau_j] \to B_l[\tau_l] \to \cdots \to B_s[\tau_s] \to B_r[\neg\tau_r] \to \cdots \to B_{j+1}[\neg\tau_{j+1}]$. This forcing chain is forward minimal and proves (d).

If the first two embeddings in (e) do not exist, $\mathcal{R}$ could be replaced by a shorter forcing chain of the form (6.5). If the third embedding does not exist, then there is a forcing chain $B_r[\neg\tau_r] \to \cdots \to B_j[\neg\tau_j]$ where $2 \le j \le r$. Together with the forcing chain $B_r[\tau_r] \to \cdots \to B_s[\tau_s]$, this shows that $\mathcal{B} \backslash \{B_1\}$ is an obstruction, a contradiction. This proves (e). $\square$

The structure of pre-minimal obstructions for 2-restricted EE problems was further investigated by Juvan and Mohar [**JM95p**]. They also showed how to obtain such an obstruction (or construct an embedding) by a linear time algorithm (assuming the Euler genus of the given embedding is bounded by a fixed constant).

## 6.5. The forbidden subgraphs for the projective plane

In this section we present the analogue of the Kuratowski Theorem for the projective plane. We give a relatively short proof of the fact that **Forb**($\mathbb{N}_1$) is finite. This result was proved by Glover and Huneke [**GH78**]. (Milgram [**Mi72, Mi73**] proved the weaker result that there are exactly three cubic minimal forbidden subgraphs for the projective plane, and the papers [**GH75, GH77**] contain similar results for graphs of bounded degree.) Glover, Huneke, and Wang [**GHW79**] presented a list of 103 minimal forbidden subgraphs. Finally, Archdeacon [**Ar80, Ar81**] proved that this list is complete.

THEOREM 6.5.1 (Archdeacon [**Ar80**], Glover, Huneke, Wang [**GHW79**]). *There are 103 graphs $G_1, \ldots, G_{103}$ such that a graph $G$ can be topologically embedded in the projective plane if and only if $G$ does not contain a subdivision of any of $G_1, \ldots, G_{103}$.*

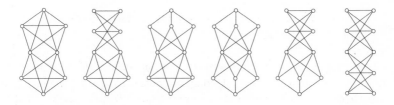

FIGURE 6.4. Two-vertex identifications of Kuratowski graphs

The list of the 103 minimal forbidden subgraphs $G_1, \ldots, G_{103}$ for the projective plane is shown in Appendix A. The following 35 graphs from that list are the minimal forbidden minors for the projective plane:

(a) The disjoint union of two Kuratowski graphs, $2K_5$, $2K_{3,3}$, and $K_5 \cup K_{3,3}$.
(b) The one-vertex identification of two Kuratowski graphs. (There are three such graphs.)
(c) The six graphs of Figure 6.4. (These graphs are obtained by identifying two vertices of two Kuratowski graphs and possibly deleting an edge.)
(d) The graphs in Figure 6.5.

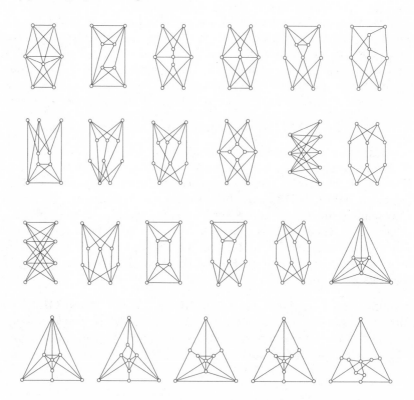

FIGURE 6.5. The 3-connected minimal forbidden minors for the projective plane

A subgraph $K$ of $G$ is called a $K_{2,3}$-*graph* in $G$ if $K$ consists of two vertices $x, y$ and three internally disjoint paths $P_1, P_2, P_3$ joining $x$ and $y$, and there is a $K$-bridge $B$ that contains an internal vertex of each of $P_1, P_2$, and $P_3$. Similarly, $K$ is a $K_4$-*graph* in $G$ if it is a subdivision of $K_4$ and there is a $K$-bridge that is attached to all four branch vertices of

$K$. A subgraph of $G$ is a *K-graph* if it is either a $K_{2,3}$-graph, or a $K_4$-graph. It is easy to see that, if $G$ is $\Pi$-embedded, then every K-graph in $G$ contains a $\Pi$-noncontractible cycle. Since a graph in the projective plane cannot contain two disjoint noncontractible cycles, we have the following [**GH75**].

LEMMA 6.5.2. *If $G$ contains two disjoint K-graphs, then $G$ cannot be embedded in the projective plane.*

We now prove the weakening of Theorem 6.5.1 that the list of minimal forbidden subgraphs for the projective plane is finite. Suppose that $G \in$ **Forb**$(\mathbb{N}_1)$. Clearly, $G$ has minimum degree at least 3. If $G$ is not 2-connected, it follows from Theorem 4.4.3 and Kuratowski's theorem that $G$ is the disjoint union or a one-vertex identification of two Kuratowski graphs. If $G$ is not 3-connected, then by Lemma 6.1.2, every splitting $G = G_1 \cup G_2$ into edge-disjoint graphs intersecting in two vertices $x, y$, where each of $G_1$ and $G_2$ has order at least three, results in two nonplanar graphs $G_1'$ and $G_2'$ where $G_i' = G_i \cup \{xy\}$ for $i = 1, 2$. It is easy to see that in this case $G$ contains one of the graphs of Figure 6.4 as a minor. On the other hand, each of these graphs contains two disjoint K-graphs and is therefore not projective planar by Lemma 6.5.2. This completes the proof when $G$ is not 3-connected.

Suppose now that $G$ is 3-connected. Let $K$ be a Kuratowski subgraph of $G$. By Lemma 2.5.5 we may assume that $K$ is a subdivision of $K_{3,3}$. By Lemma 6.2.1, we may assume that $K$ has no local bridges. Among all choices for $K$ we select one with the minimum number of bridges.

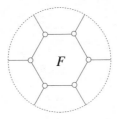

FIGURE 6.6. $K_{3,3}$ in the projective plane

Euler's formula implies that if $K_{3,3}$ is embedded in the projective plane, then there are four faces. Since the sum of the lengths of the faces is $2|E(K_{3,3})| = 18$, the facial walks have lengths $4, 4, 4, 6$. Hence, the embedding is that shown in Figure 6.6, where any cycle of length 6 in $K_{3,3}$ can play the role of $F$. This shows that there are six nonequivalent embeddings $\Pi_1, \ldots, \Pi_6$ of $K$ in the projective plane and that $G$ is the union of obstructions for extensions of $\Pi_1, \ldots, \Pi_6$. All facial walks of these embeddings are cycles, and hence Corollary 6.3.4 shows that all $K$-bridges in $G$ are E-graphs, and that their size is bounded above by a universal constant.

We complete the proof by showing that, for each of the embeddings $\Pi_k$ ($1 \leq k \leq 6$), every pre-minimal obstruction $\mathcal{B}$ contains at most 15 $K$-bridges. If a $K$-bridge in $G$ admits more than two embeddings extending the embedding $\Pi_k$, then it is attached to diametrically opposite vertices on $F$ only, and is therefore a local bridge, a contradiction. This shows that the embedding $\Pi_k$ is 2-restrictive. By Proposition 6.4.1, the pre-minimal obstruction $\mathcal{B}$ contains at most two bridges with at most one admissible embedding. All other $K$-bridges in $\mathcal{B}$ have embeddings in two $\Pi_k$-faces, and are therefore attached to two branches (possibly including their endvertices) which are opposite on $F$.

Let $B_1[\tau_1] \to \cdots \to \mathcal{B}_s[\tau_s]$ be a forcing chain of the obstruction $\mathcal{B}$ satisfying Proposition 6.4.1(a)–(e). Consider the embedding $B_2[\tau_2] \cup \cdots \cup B_r[\tau_r]$, where $r$ is defined in Proposition 6.4.1. (If $r = s$ we consider $B_2[\tau_2] \cup \cdots \cup B_{r-1}[\tau_{r-1}]$.) Then every second bridge in the chain is embedded in $F$. If the bridges embedded in $F$ contain two disjoint paths, each of which joins vertices in distinct branches on $F$, then these paths do not have simultaneous embedding in a face distinct from $F$. This contradicts Proposition 6.4.1(e) and shows that there exists a vertex $x$ in $K$ such that all bridges among $B_2, \ldots, B_r$ that are embedded in $F$, are attached to $x$ and to one or two vertices on a branch of $K$ on $F$ which does not contain $x$. A similar statement holds for the bridges that are not embedded in $F$. Denote by $y$ their common vertex of attachment. Possibly $y = x$. Since $B_{j-1}[\tau_{j-1}] \to B_j[\tau_j] \to B_{j+1}[\tau_{j+1}]$ and $B_{j+1}[\neg\tau_{j+1}] \to B_j[\neg\tau_j] \to B_{j-1}[\neg\tau_{j-1}]$, for $j = 3, \ldots, r - 1$, the admissible embeddings of $B_2, \ldots, B_r$ are all in $F$ and in another $\Pi_k$-face $F'$. Moreover, $x$ and $y$ are on the same branch. Figure 6.7 shows the two faces and some of the embedded bridges.

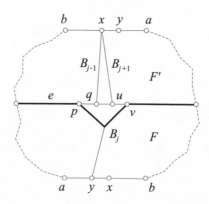

FIGURE 6.7. The $\Pi_k$-faces $F$ and $F'$

We claim that $r \leq 4$ (or that $r = s = 5$). For if this were false, there is an integer $j$ such that $2 < j < r-1$ (or $2 < j < r-2 = s-2$). Without loss of generality assume that $B_j$ is attached to $y$ as shown on Figure 6.7. Since $B_{j-1}[\tau_{j-1}] \rightarrow B_j[\tau_j]$, $B_{j-1}$ has a vertex of attachment $q$ on $e$ which is to the right[6] of an attachment $p$ of $B_j$. Since $B_{j+1}[\tau_{j+1}] \rightarrow B_{j+2}[\tau_{j+2}]$ and since $B_{j-1}[\tau_{j-1}]$ does not force $B_{j+2}[\tau_{j+2}]$, $B_{j+1}$ is embedded in $F'$ to the right of $B_{j-1}$. Since $B_j[\tau_j] \rightarrow B_{j+1}[\tau_{j+1}]$, $B_j$ has a vertex of attachment $v$ on $e$ which is to the right of $q$ and to the right of an attachment $u$ of $B_{j+1}$. By replacing the segment of $e$ between $p$ and $v$ by the path in $B_j$, we obtain a subgraph of $G$ homeomorphic to $K$ with fewer bridges, a contradiction. This proves the claim that $r \leq 4$ (or that $r = s = 5$).

Similarly we prove that $s - r \leq 4$. This shows that $\mathcal{B}$ contains at most 15 $K$-bridges and completes the proof.

We conclude this section with two applications of Theorem 6.5.1. Mohar and Robertson [**MR96b**] used it to prove that a graph contains two disjoint noncontractible cycles in every embedding if and only if it cannot be embedded in the projective plane. It suffices to verify this for the 35 minimal forbidden minors in $\mathbf{Forb_0}(\mathbb{N}_1)$. Another application is described below.

We say that a graph $G$ *covers* a graph $H$ if there exists a map $f : V(G) \rightarrow V(H)$ such that $f$ is onto and for any vertex $v$ of $G$, the restriction of $f$ to $N(v)$ is a bijection to $N(f(v))$. For example, the graphs of the icosahedron and the dodecahedron cover $K_6$ and the Petersen graph, respectively (by identifying diametrically opposite vertices). More generally, for every projective planar graph $H$ there exists a planar graph which covers $H$. (To see this, we first draw $H$ in a disc where opposite points are identified. Now take two copies of that disc, rotate one of them by $180°$, and paste their boundaries together so that we obtain a graph on the sphere.) Negami asked if the converse holds.

CONJECTURE 6.5.3 (Negami [**Ne88a**]). *If a connected graph $H$ is covered by a planar graph, then $H$ is planar or projective planar.*

If $G$ covers $H$, then every subgraph of $H$ is covered by a subgraph of $G$. So it suffices to verify Conjecture 6.5.3 for subdivisions of graphs in $\mathbf{Forb}(\mathbb{N}_1)$. Moreover, if $e$ is an edge of $H$ that is not contained in a 3-cycle, then $H/e$ is covered by the corresponding minor of $G$. Since any minor of a graph can be obtained by deleting edges and contracting edges that are not in 3-cycles, it suffices to verify the conjecture for the 35 minimal forbidden minors for the projective plane. Archdeacon (private communication) further reduced this problem to only two graphs, namely $K_7 - 3K_2 = K_{1,2,2,2}$ and $K_{4,4} - K_2$. Hliněný [**Hl98**] proved that $K_{4,4} - K_2$ has no planar cover, and therefore Conjecture 6.5.3 is equivalent to

---

[6]Right and left here refers to Figure 6.7.

CONJECTURE 6.5.4. *The graph $K_{1,2,2,2}$ is not covered by a planar graph.*

## 6.6.  The minimal forbidden subgraphs for general surfaces

Minimal obstructions for embedding extension problems were used by Mohar [**Mo99**] in a similar way as in the proof of Theorem 6.5.1 to prove the theorem of Robertson and Seymour [**RS90b**]:

> *For each surface $S$ there is a finite list of graphs $G_1, \ldots, G_N$ such that an arbitrary graph $G$ can be topologically embedded in $S$ if and only if $G$ does not contain a subdivision of any of $G_1, \ldots, G_N$.*

(See also Theorem 7.0.1 in the next chapter.) However, the complete proof of Mohar [**Mo99**] is rather complicated and technical and involves additional study of obstructions for more complicated embedding extension problems. One important case are the 2-restricted EEPs which have been considered by Juvan and Mohar [**JM95p**] in some more depth than in Section 6.4. Another special case is the EEP in which the embedded subgraph $K$ has a facial walk $F$ in which two branches $e$ and $f$ each appear twice (and all bridges should be embedded in $F$). This case which is treated by Juvan, Marinček, and Mohar [**JMM94**] is difficult and it shows up already in the torus. Perhaps this explains why the list of graphs in $\mathbf{Forb}(\mathbb{S}_1)$ is still not known.

However, several classes of minimal forbidden subgraphs for the torus are known. For example, Decker [**De78**] proved that there are exactly 259 nonisomorphic graphs in $\mathbf{Forb}(\mathbb{S}_1)$ which contain a subgraph which is the disjoint union of a subdivision of $K_5$ or $K_{3,3}$ and a K-graph (defined before Lemma 6.5.2). It is also known (private communication by Robin Thomas; also Juvan [**Ju95**]) that precisely 270 minimal forbidden minors for the torus can be embedded in the projective plane. Theorems 5.6.3 and 5.8.1 show that these are precisely the graphs which can be obtained from the projective $4 \times 4$ grid by $Y\Delta$-exchanges. Lemma 6.5.2 shows that this family is disjoint from the family determined by Decker.

Duke and Haggard [**DH72**] proved that $\mathbf{Forb}(\mathbb{S}_1)$ contains three graphs with 8 vertices, and Hlavacek [**Hl97**] showed that 48 graphs in $\mathbf{Forb}(\mathbb{S}_1)$ are of order 9. E. Neufeld (private communication) has shown, using a computer, that $\mathbf{Forb}(\mathbb{S}_1)$ contains precisely 656 2-connected graphs with 10 vertices, and 3178 2-connected graphs with 11 vertices and at most 24 edges. Among all these graphs, there are more than 2200 graphs in $\mathbf{Forb}_0(\mathbb{S}_1)$ (that is, minimal forbidden minors). Further examples of minimal forbidden graphs for the torus can be found in [**BW86**].

PROBLEM 6.6.1. *Is it true that every minimal forbidden subgraph for the torus contains less than 100 edges?*

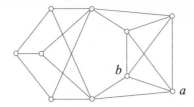

FIGURE 6.8. A minimal forbidden subgraph for the torus

Glover (cf. [**GT87**, p. 53]) asked if each edge of a minimally nonembeddable graph $G$ (in any surface) is contained in a subgraph $K$ of $G$ which is a subdivision of $K_5$ or $K_{3,3}$. This question has been answered in the affirmative if $S$ is nonorientable by Brunet, Richter, and Širáň [**BRS96**]. They also observed that the minimal forbidden subgraph $H$ for the torus represented in Figure 6.8 contains the edge $ab$ which is not part of a Kuratowski subgraph of $H$.

A more specific conjecture of Glover [**GT87**, p. 53] is still open:

CONJECTURE 6.6.2 (Glover). *Let $G$ be a minimal forbidden subgraph for the nonorientable surface $\mathbb{N}_k$. Then $G$ can be written as the union of $k + 1$ subgraphs each of which is a subdivision of $K_5$ or $K_{3,3}$.*

In Section 2.7 we observed that testing planarity [**HT74**], constructing embeddings in the sphere [**CNAO85**], or finding subgraphs that are subdivisions of Kuratowski graphs [**Wi84**] can be performed by an algorithm whose worst case running time is linear. Although the construction of minimum genus embeddings is **NP**-hard (by Theorem 4.4.11), Filotti, Miller, and Reif [**FMR79**] proved that for every fixed surface $S$, there is a polynomial time algorithm for embedding graphs in $S$. For every fixed surface $S$, Robertson and Seymour give an $O(n^3)$ algorithm for testing embeddability in $S$ using graph minors [**RS90b, RS90d**]. Robertson and Seymour recently improved their $O(n^3)$ algorithms to $O(n^2 \log n)$ [**RS95b, RS92p, RS92q**]. An embeddability testing algorithm can be extended to an algorithm which also constructs an embedding in polynomial time; see Archdeacon [**Ar90**]. Mohar [**Mo96, Mo99**] (and the papers cited therein) improved these results by showing:

THEOREM 6.6.3 (Mohar [**Mo96, Mo99**]). *Let $S$ be a fixed surface. There is a linear time algorithm that for an arbitrary graph $G$ either:*

(a) *finds an embedding of $G$ in $S$, or*
(b) *finds a subgraph $K \subseteq G$ which is a subdivision of some graph in* **Forb**$(S)$.

Simpler linear time algorithms for embedding graphs in the projective plane and the torus are described by Mohar [**Mo93**] and Juvan, Marinček, and Mohar [**JMM95**], respectively.

CHAPTER 7

# Tree-width and the excluded minor theorem

One of the highlights in the Robertson-Seymour theory on graph minors is the finiteness (for each fixed surface $S$) of the set of the minimal forbidden minors for $S$.

THEOREM 7.0.1 (Robertson and Seymour [**RS90b**]). *For each surface* $S$, $\mathbf{Forb}_0(S)$ *is finite.*

We present in Section 7.3 a short proof of Thomassen [**Th97c**] which is based on two other important results in the Robertson-Seymour theory, namely Theorem 7.2.1 (well-quasi-ordering of graphs of bounded tree-width) and Theorem 7.1.7 (the excluded grid theorem) below.

Theorem 7.2.1 was proved in the paper [**RS90a**] which is lengthy and technical as it provides general machinery for the graph minor theory. A shorter direct proof of Theorem 7.2.1 was obtained by Geelen, Gerards and Whittle [**GGW00p**]. However, to prove Theorem 7.0.1, the conclusion of Theorem 7.2.1 is needed only for graphs in $\mathbf{Forb}_0(S)$. A short proof of this weaker result was recently obtained by Mohar [**Mo01**] and is presented in Section 7.2.

Recently, Diestel, Gorbunov, Jensen, and Thomassen [**DJGT99**] obtained a short proof of Theorem 7.1.7, which we present in Section 7.1. In Section 7.3 we also summarize the few embedding results from earlier sections that are used in the short proof of Theorem 7.0.1.

## 7.1. Tree-width and the excluded grid theorem

Robertson and Seymour [**RS86a**] introduced the *tree-width* of a graph, a concept which is of importance in both discrete mathematics and in theoretical computer science, partly because many problems that are **NP**-hard for graphs in general (for example the problem of finding the chromatic number, a maximum clique, a hamiltonian cycle, etc.) can be solved by polynomial time algorithms for graphs of bounded tree-width, that is, graphs whose tree-width is bounded above by any fixed constant. (One possible exception is the genus, cf. Problem 7.3.4.)

We now define the tree-width. Let $G_1, \ldots, G_n$ be a collection of pairwise disjoint graphs. We now form a sequence of graphs $H_1, \ldots, H_n$ and a sequence of trees $T_1, \ldots, T_n$, where $T_k$ has vertex set $\{1, 2, \ldots, k\}$. Let

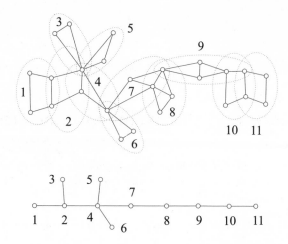

FIGURE 7.1. A graph and its tree-decomposition of width 3

$H_1 = G_1$ and let $T_1$ be the graph with a single vertex 1. Having defined $H_k$ and $T_k$ ($1 \leq k < n$), we select a vertex set $S$ in $G_{k+1}$ and a vertex set $S'$ in one of the graphs $G_1, \ldots, G_k$, say $G_j$, such that $|S'| = |S|$, and we identify $S$ with $S'$. We let $H_{k+1}$ be the graph obtained from the union of $H_k$ and $G_{k+1}$ (after identifying $S$ with $S'$) and we think of $G_k$ as a subgraph of $H_{k+1}$. We also let $T_{k+1}$ be the tree obtained from $T_k$ by adding a new vertex $k+1$ and joining it to the vertex $j$ of $T_k$. Finally, put $G = H_n$ and $T = T_n$. Then we say that $G_1, \ldots, G_n$ form a *tree-decomposition* of $G$ of *width* $\max\{|V(G_i)| - 1 \mid i = 1, \ldots, n\}$ and we denote this tree-decomposition by $(T, (G_i)_{i \in V(T)})$. (We use here $\{1, 2, \ldots, n\}$ as the vertex set of $T$ but any other index set may be used.)

The *tree-width* $\mathbf{tw}(G)$ of a graph $G$ is the smallest width of a tree-decomposition of $G$. Thus, the connected graphs of tree-width at most 1 are precisely all trees. Figure 7.1 shows a graph $G$, a tree-decomposition of width 3 and the underlying tree $T$. That graph $G$ also has a tree-decomposition of width 2.

The subgraphs $G_i$ into which a tree-decomposition decomposes a graph will be called the *parts* of that decomposition. If $C$ is a subgraph of a graph $G$, we write $N(C)$ for its set of neighbors in $G - C$, i.e., the set of vertices in $G - C$ which have a neighbor in $C$. A *separation* of $G$ is an ordered pair $(A, B)$ of subgraphs of $G$ such that $A \cup B = G$ and $E(A) \cap E(B) = \emptyset$; its *order* is the number $|V(A) \cap V(B)|$. Recall that the Cartesian product of two $n$-paths, $P_n \square P_n$, is called the $n \times n$ *grid*.

It is easy to see that a large graph of small tree-width has a separation $(A, B)$ of small order such that $|V(A) \backslash V(B)|$ and $|V(B) \backslash V(A)|$ are both big. More precisely, if the graph $G$ has a tree-decomposition of width $k$

and $|V(G)| > 4(k+1)$, then $(A, B)$ can be chosen such that $A \cap B$ is one of the parts (and hence $|V(A) \cap V(B)| \leq k+1$), and $|V(A) \backslash V(B)| \geq \frac{1}{3}(|V(G)| - k - 1)$ and $|V(B) \backslash V(A)| \geq \frac{1}{3}(|V(G)| - k - 1)$. (To see this, we pick an $i \in \{1, 2, \ldots, n\}$ such that the largest component of $G - V(G_i)$ is as small as possible. If no component of $G - V(G_i)$ has more than $\frac{2}{3}(|V(G)| - k - 1)$ vertices, we can define $A$ and $B$ such that $(A, B)$ has the desired separation properties. Otherwise, the largest component $Q$ of $G - V(G_i)$ corresponds to a component of $T - i$. Let $j$ be the vertex in that component of $T - i$ such that $G_j$ intersects $Q$ and such that the distance from $i$ to $j$ in $T$ is minimum. Then it is easy to see that the largest component of $G - V(G_j)$ has fewer vertices than $Q$, a contradiction.) If $N$ is much larger than $k$, then the $N \times N$ grid does not have the above separation property (for if we remove a vertex set $S$ from the $N \times N$ grid $G$, where $|S| < N$, there is a "row" of the grid which is disjoint from $S$ and hence at least the $N - |S|$ "columns" which are disjoint from $S$ lie in the same component of $G - S$ as that row). Hence a big grid has large tree-width. The excluded grid theorem (Theorem 7.1.7) says that, in some sense, the converse holds. For that we need some further notation.

If $Y, Z$ are subgraphs (or vertex sets) of a graph $G$, then every path with one end in $Y$ and the other end in $Z$ is called a $Y$–$Z$ *path*. We call a set $X \subseteq V(G)$ *k-connected* in $G$ if $|X| \geq k$ and for all subsets $Y, Z \subseteq X$ with $|Y| = |Z| \leq k$ there are $|Y|$ disjoint $Y$–$Z$ paths in $G$. The set $X$ is *externally k-connected* if, in addition, the required paths can be chosen without an edge or intermediate vertex in $G(X)$. For example, the vertex set of any $k$-connected subgraph of $G$ is $k$-connected in $G$ (though not necessarily externally). Also, any horizontal path of the $k \times k$ grid is $k$-connected in the grid, even externally.

PROPOSITION 7.1.1 (Diestel et al. [**DJGT99**]). *Let $G$ be a graph and $k > 0$ an integer.*

(i) *If $\mathbf{tw}(G) < k$ then $G$ contains no $(k+1)$-connected set of size $\geq 3k$.*

(ii) *Conversely, if $G$ contains no externally $(k+1)$-connected set of size $\geq 3k$, then $\mathbf{tw}(G) < 4k$.*

PROOF. (i) Choose a tree-decomposition $(T, (G_t)_{t \in V(T)})$ of $G$ of width less than $k$. We may assume that none of the parts $G_t$ is contained in another. For $t \in V(T)$, let $V_t = V(G_t)$. Then for every edge $e = rs$ of $T$, the intersection $G_r \cap G_s$ has less than $k$ vertices and separates the sets $U_r = \bigcup_{t \in V(T_r)} V_t$ and $U_s = \bigcup_{t \in V(T_s)} V_t$ in $G$; here, $T_r$ and $T_s$ denote the components of $T - e$ containing $r$ and $s$, respectively. Any separation $(A, B)$ of $G$ with vertex sets $\{V(A), V(B)\} = \{U_r, U_s\}$ (and hence $A \cap B = G_r \cap G_s$) is said to *correspond to* $e$.

Suppose that $X$ is a $(k+1)$-connected set in $G$ of size $\geq 3k$. Orient every edge $e$ of $T$ towards the component $T'$ of $T - e$ for which $|X \cap$

$\bigcup_{t \in V(T')} V_t|$ is greater, breaking ties arbitrarily. Choose a vertex $t$ of $T$ so that all the edges $e_1, \ldots, e_n$ of $T$ at $t$ point towards $t$. (For example, take the last vertex of a maximal directed path in $T$.) For every $i = 1, \ldots, n$ pick a separation $(A_i, B_i)$ corresponding to $e_i$, with $G_t \subseteq B_i$. Then $|V(A_i) \cap X| < k$: otherwise, both $A_i$ and $B_i$ would have $k$ vertices in $X$, and we could extend $V(A_i \cap B_i) \cap X$ to $k$-subsets $Y \subseteq V(A_i) \cap X$ and $Z \subseteq V(B_i) \cap X$ that cannot be linked by $k$ disjoint paths in $G$ (since $|A_i \cap B_i| < k$).

Now let $i \leq n$ be minimal such that $|V(A_1 \cup \cdots \cup A_i) \cap X| > k$, and put $A := A_1 \cup \cdots \cup A_i$ and $B := B_1 \cap \cdots \cap B_i$. By the minimality of $i$ and since $|V(A_i) \cap X| < k$, we have $|V(A) \cap X| < 2k$, so $|V(B) \cap X| > |X| - 2k \geq k$. As before, we may extend $V(A \cap B) \cap X$ to $(k + 1)$-sets $Y \subseteq V(A) \cap X$ and $Z \subseteq V(B) \cap X$. As $V_t$ separates these sets in $G$ and $|V_t| \leq k$, this contradicts our assumption that $X$ is $(k + 1)$-connected in $G$.

(ii) We prove the following more general assertion:

*If $h \geq k$ and $G$ contains no externally $k$-connected set with $h$ vertices, then $\mathbf{tw}(G) < h + k - 1$.*

Let $U \subseteq V(G)$ be maximal such that $G(U)$ has a tree-decomposition $\mathcal{D}$ of width $< h + k - 1$ such that every component $C$ of $G - U$ has at most $h$ neighbors in $U$ and these lie in one part of $\mathcal{D}$ (depending on $C$).

We claim that $U = V(G)$. Suppose not. Let $C$ be a component of $G - U$ and write $X = N(C)$. By assumption, $|X| \leq h$. In fact $|X| = h$, since otherwise for any $v \in V(C)$ we could add $X \cup \{v\}$ to $\mathcal{D}$ as a new part, contradicting the maximality of $U$. Hence by assumption, $X$ is not externally $k$-connected in $G$. Let $Y, Z \subseteq X$ be sets to witness this. By Menger's theorem, $Y$ and $Z$ are separated in $H := G(V(C) \cup Y \cup Z) - E(G(Y \cup Z))$ by a set $S$ of fewer than $|Y| = |Z| \leq k$ vertices. Let $X_Y := (X \backslash Z) \cup S$ and $X_Z := (X \backslash Y) \cup S$. Clearly, $|X \cup S| \leq h + k - 1$ and $|X_Y|, |X_Z| < |X| = h$. Moreover, any component $C' \subseteq C$ of $G - (U \cup S)$ has all its neighbors in $X \cup S$, and hence either in $X_Y$ or in $X_Z$: otherwise $H - S$ would contain a $Y$–$Z$ path through $C'$.

Extending $U$ to $U \cup S$ and adding $X \cup S$ to $\mathcal{D}$ as a new part, we obtain a contradiction to the maximality of $U$. (Note that $S \cap C \neq \emptyset$, since $|S| < |Y| = |Z|$ and $Y, Z \subseteq N(C)$.) This completes the proof of the claim which implies that $\mathbf{tw}(G) < h + k - 1$.     □

The next goal is to prove that graphs of large tree-width contain large grid minors. Roughly, we shall assume that a given graph $G$ has large tree-width, find a large highly connected set $X$ in $G$ as in Proposition 7.1.1(ii), and use its connecting paths to form a grid. Of course, this will be possible only if those paths intersect sufficiently often. If they do not, we shall try instead to partition $X$ into many sets that can be linked pairwise by mutually disjoint paths, so that contracting these sets will give us a subdivision of a large complete graph. Since we may only contract connected sets when forming a minor, our first task will thus be to strengthen Proposition 7.1.1(ii) so as to give $X$ a partition into many sets that can be made connected in a part of $G$ not used by the connecting paths.

Let us call a separation $(A, B)$ a *premesh* if all the edges of $G(V(A \cap B))$ lie in $A$ and $A$ contains a tree $T$ with the following properties:

(a) $T$ has maximum degree at most 3;
(b) every vertex of $A \cap B$ lies in $T$ and has degree at most 2 in $T$;
(c) some vertex of $A \cap B$ has degree 1 in $T$.

A premesh $(A, B)$ will be called a *k-mesh* of *order* $|V(A \cap B)|$ if $V(A \cap B)$ is externally $k$-connected in $B$, and the graph $G = A \cup B$ is said to *have* this premesh or $k$-mesh.

The lemma below is analogous to Proposition 7.1.1(ii).

LEMMA 7.1.2 (Diestel et al. [**DJGT99**]). *Let $G$ be a graph and let $h \geq k \geq 1$ be integers. If $G$ has no $k$-mesh of order $h$ then $\mathbf{tw}(G) < h + k - 1$.*

PROOF. We may assume that $G$ is connected. Let $U \subseteq V(G)$ be maximal such that $G(U)$ has a tree-decomposition $\mathcal{D}$ of width less than $h + k - 1$, with the additional property that, for every component $C$ of $G - U$, the neighbors of $C$ in $U$ lie in one part of $\mathcal{D}$ and $(G - C, \bar{C})$ is a premesh of order $\leq h$, where $\bar{C} := G(V(C) \cup N(C)) - E(G(N(C)))$. Clearly, $U \neq \emptyset$.

We claim that $U = V(G)$. Suppose not. Let $C$ be a component of $G - U$, put $X := N(C)$, and let $T$ be a tree associated with the premesh $(G - C, \bar{C})$.

By assumption, $|X| \leq h$. Let us show that equality holds here. If not, let $u \in X$ be a vertex of degree 1 in $T$ (which exists by condition (c) above) and $v$ a neighbor of $u$ in $C$. Put $U' := U \cup \{v\}$ and $X' := X \cup \{v\}$, let $T'$ be the tree obtained from $T$ by joining $v$ to $u$, and let $\mathcal{D}'$ be the tree-decomposition of $G(U')$ obtained from $\mathcal{D}$ by adding $X'$ as a new part. Clearly, the width of $\mathcal{D}'$ is less than $h + k - 1$. Consider a component $C'$ of $G - U'$. If $C' \cap C = \emptyset$ then $C'$ is also a component of $G - U$, so $N(C')$ lies inside a part of $\mathcal{D}$ (and hence of $\mathcal{D}'$), and $(G - C', \bar{C}')$ is a premesh of order $\leq h$ by assumption. If $C' \cap C \neq \emptyset$, then $C' \subseteq C$ and $N(C') \subseteq X'$. Moreover, $v \in N(C')$: otherwise $N(C') \subseteq X$ would separate $C'$ from $v$, contradicting the fact that $C'$ and $v$ lie in the same component $C$ of $G - X$. Since $v$ is a leaf of $T'$, it is straightforward to check that $(G - C', \bar{C}')$ is again a premesh of order $\leq h$, contrary to the maximality of $U$.

Thus $|X| = h$, so by assumption our premesh $(G - C, \bar{C})$ cannot be a $k$-mesh. Hence by Menger's theorem, there are sets $Y, Z \subseteq X$ of equal size $\leq k$ that are separated in $H := \bar{C}(V(C) \cup Y \cup Z)$ by a set $S$ of $k' < |Y| = |Z|$ vertices, one from each of a maximal family $(P_s \mid s \in S)$ of disjoint $Y$–$Z$ paths in $H$. Put $X' := X \cup S$ and $U' := U \cup S$, and let $\mathcal{D}'$ be the tree-decomposition of $G(U')$ obtained from $\mathcal{D}$ by adding $X'$ as a new part. Clearly, $|X'| \leq |X| + |S| \leq h + k - 1$. We show that $U'$ contradicts the maximality of $U$.

Since $Y \cup Z \subseteq N(C)$ and $|S| < |Y| = |Z|$, we have $S \cap C \neq \emptyset$. So $U'$ is larger than $U$. Let $C'$ be a component of $G - U'$. If $C' \cap C = \emptyset$, we argue as earlier. So we may assume that $C' \subseteq C$ and $N(C') \subseteq X'$. As before, at least one neighbor $v$ of $C'$ lies in $S \cap C$. By definition of $S$, $C'$ cannot have neighbors in both $Y \setminus S$ and $Z \setminus S$. Assume it has none in $Y \setminus S$. Let $T'$ be the union of $T$ and all the $Y$–$S$ subpaths of paths $P_s$ with $s \in N(C') \cap C$. Since these subpaths start in $Y \setminus S$ and have no inner vertices in $X'$, they cannot meet $C'$. Therefore $(G - C', \bar{C}')$ is a premesh with tree $T'$ and leaf $v$ (the degree conditions on $T'$ are easily checked). The order of this premesh is $|N(C')| \leq |X| - |Y| + |S| = h - |Y| + k' < h$, a contradiction to the maximality of $U$. $\qquad\square$

LEMMA 7.1.3. *Let $k \geq 2$ be an integer. Let $T$ be a tree of maximum degree $\leq 3$ and $X \subseteq V(T)$. Then $T$ has a set $E$ of edges such that every component of $T - E$ has at least $k$ and at most $2k - 1$ vertices in $X$, except that one such component may have fewer vertices in $X$.*

PROOF. Induction on $|X|$. If $|X| \leq 2k - 1$ we put $E = \emptyset$. So assume that $|X| \geq 2k$. Let $e$ be an edge of $T$ such that some component $T'$ of $T - e$ has at least $k$ vertices in $X$ and $T'$ is as small as possible. As the maximum degree of $T$ is $\leq 3$, the end of $e$ in $T'$ has degree at most two in $T'$. So the minimality of $T'$ implies that $|X \cap V(T')| \leq 2k - 1$. We finish by applying the induction hypothesis to $T - T'$. $\qquad\square$

LEMMA 7.1.4. *Let $G$ be a bipartite graph with bipartition $(A, B)$, let $a = |A|$, $b = |B|$, and let $c \leq a$ and $d \leq b$ be positive integers. Assume that $G$ has at most $(a - c)(b - d)/d$ edges. Then there exist $C \subseteq A$ and $D \subseteq B$ such that $|C| = c$ and $|D| = d$ and such that no two vertices in $C \cup D$ are adjacent in $G$.*

PROOF. As $|E(G)| \leq (a - c)(b - d)/d$, fewer than $b - d$ vertices in $B$ have more than $(a - c)/d$ neighbors in $A$. Choose $D \subseteq B$ so that $|D| = d$ and each vertex in $D$ has at most $(a - c)/d$ neighbors in $A$. Then $|N(D)| \leq a - c$, so $A$ has a subset $C$ of $c$ vertices without a neighbor in $D$. $\qquad\square$

Given a tree $T$, call an $r$-tuple $(x_1, \ldots, x_r)$ of distinct vertices of $T$ *good* if, for every $j = 1, \ldots, r - 1$, the $x_j$–$x_{j+1}$ path in $T$ contains none of the other vertices in this $r$-tuple.

LEMMA 7.1.5. *Every tree with more than $\frac{1}{2}(r - 1)(r - 2)$ edges has a good $r$-tuple of vertices.*

PROOF. Let $T$ be a tree and let $P = x_1 x_2 \ldots x_p$ be a longest path in $T$. If $p \geq r$, then $(x_1, \ldots, x_r)$ is a good $r$-tuple. Otherwise, each leaf of

$T$ is at distance at most $(p-1)/2 \leq (r-2)/2$ from $P$. This implies that $T$ has at least $2|E(T)|/(r-2)$ leaves. Since any set of $r$ leaves defines a good $r$-tuple in $T$, this proves the assertion.                                   $\square$

The next lemma from [**DJGT99**] shows how to obtain a grid from two large systems of paths that intersect in a particularly orderly way.

LEMMA 7.1.6. *Let $d, r \geq 2$ be integers such that $d \geq r^{2r+2}$. Let $G$ be a graph containing a set $\mathcal{H}$ of $r^2 - 1$ disjoint paths and a set $\mathcal{V} = \{V_1, \ldots, V_d\}$ of $d$ disjoint paths. Assume that every path in $\mathcal{V}$ intersects every path in $\mathcal{H}$, and that each path $H \in \mathcal{H}$ consists of $d$ disjoint segments (with one edge between consecutive segments) such that $V_i$ meets $H$ only in its ith segment, for every $i = 1, \ldots, d$. Then $G$ has an $r \times r$ grid minor.*

PROOF. For each $i = 1, \ldots, d$, consider the graph with vertex set $\mathcal{H}$ in which two paths are adjacent whenever $V_i$ contains a subpath between them that meets no other path in $\mathcal{H}$. Since $V_i$ meets every path in $\mathcal{H}$, this is a connected graph; let $T_i$ be a spanning tree in it. Since $|\mathcal{H}| = r^2 - 1$, Lemma 7.1.5 implies that each of these $d \geq r^2(r^2)^r$ trees $T_i$ has a good $r$-tuple of vertices. Since there are no more than $(r^2)^r$ distinct $r$-tuples on $\mathcal{H}$, some $r^2$ of the trees $T_i$ have the same good $r$-tuple $(H^1, \ldots, H^r)$. Let $I = \{i_1, \ldots, i_{r^2}\}$ be the index set of these trees (with $i_j < i_k$ for $j < k$) and put $\mathcal{H}' := \{H^1, \ldots, H^r\}$.

Here is an informal description of how we construct our $r \times r$ grid. Its 'horizontal' paths will be the paths $H^1, \ldots, H^r$. Its 'vertical' paths will be pieced together edge by edge, as follows. The $r-1$ edges of the first vertical path will come from the first $r - 1$ trees $T_i$, trees with their index $i$ among the first $r$ elements of $I$. More precisely, its 'edge' between $H^j$ and $H^{j+1}$ will be the sequence of subpaths of $V_{i_j}$ (together with some connecting horizontal bits taken from paths in $\mathcal{H} \setminus \mathcal{H}'$) induced by the edges of an $H^j$–$H^{j+1}$ path in $T_{i_j}$ that has no inner vertices in $\mathcal{H}'$. (This is why we need $(H^1, \ldots, H^r)$ to be a good $r$-tuple in every tree $T_i$.) Similarly, the $j$th edge of the second vertical path will come from an $H^j$–$H^{j+1}$ path in $T_{i_{r+j}}$, and so on. To merge these individual edges into $r$ vertical paths, we then contract in each $H^j$ the initial segment that meets the first $r$ paths $V_i$ with $i \in I$, then contract the segment that meets the following $r$ paths $V_i$ with $i \in I$, and so on.

Formally, we proceed as follows. For all $j, k \in \{1, \ldots, r\}$, consider the minimal subpath $H_k^j$ of $H^j$ that contains the $i$th segment of $H^j$ for all $i$ with $i_{(k-1)r} < i \leq i_{kr}$ (where $i_0 = 0$). Let $\hat{H}^j$ be obtained from $H^j$ by first deleting any vertices following its $i_{r^2}$th segment and then contracting every subpath $H_k^j$ to one vertex $v_k^j$. Thus, $\hat{H}^j = v_1^j \ldots v_r^j$.

Given $j \in \{1, \ldots, r-1\}$ and $k \in \{1, \ldots, r\}$, we shall define a path $V_k^j$ that will form the subdivided 'vertical edge' $v_k^j v_k^{j+1}$. This path will consist of segments of the path $V_i$ together with some otherwise unused segments of paths from $\mathcal{H} \setminus \mathcal{H}'$, for $i := i_{(k-1)r+j}$. Recall that, by definition of $\hat{H}^j$ and $\hat{H}^{j+1}$, this $V_i$ does indeed meet $H^j$ and $H^{j+1}$ precisely in vertices that were contracted into $v_k^j$ and $v_k^{j+1}$, respectively. To define $V_k^j$, consider an $H^j\!-\!H^{j+1}$ path $P = H_1 \ldots H_t$ in $T_i$ that has no inner vertices in $\mathcal{H}'$. Every edge $H_s H_{s+1}$ of $P$ corresponds to an $H_s\!-\!H_{s+1}$ subpath of $V_i$ that has no inner vertex on any path in $\mathcal{H}$. Together with (parts of) the $i$th segments of $H_2, \ldots, H_{t-1}$, these subpaths of $V_i$ form an $H^j\!-\!H^{j+1}$ path $P'$ that has no inner vertices on any of the paths $H^1, \ldots, H^r$ and meets no path from $\mathcal{H}$ outside its $i$th segment. Replacing the ends of $P'$ on $H^j$ and $H^{j+1}$ with $v_k^j$ and $v_k^{j+1}$, respectively, we obtain our desired path $V_k^j$ forming the $j$th (subdivided) edge of the $k$th 'vertical' path of the grid. Since the paths $P'$ are disjoint for different $i$ and different pairs $(j, k)$ do give rise to different $i$, the paths $V_k^j$ are disjoint except for possible common ends $v_k^j$. Moreover, they have no inner vertices on any of the paths $H^1, \ldots, H^r$, because none of these $H^j$ is an inner vertex of any of the paths $P \subseteq T_i$ used in the construction of $V_k^j$. □

We are now ready to prove the following quantitative version of the excluded grid theorem of Robertson and Seymour [**RS86b**].

THEOREM 7.1.7 (Diestel, Gorbunov, Jensen, Thomassen [**DJGT99**]). *Let $r, m > 0$ be integers, and let $G$ be a graph of tree-width at least $r^{4m^2(r+2)}$. Then $G$ contains either $K_m$ or the $r \times r$ grid as a minor.*

PROOF. Since $K_{r^2}$ contains the $r \times r$ grid as a subgraph, we may assume that $2 \le m \le r^2$ (and so also $r \ge 2$). Put $c := r^{4(r+2)}$ and $k := c^{m(m-1)}$. Then $2m + 2 \le c$, so $G$ has tree-width at least $c^{m^2} = c^m k \ge (2m+2)k$. This is enough for Lemma 7.1.2 to ensure that $G$ contains a $k$-mesh $(A, B)$ of order $m(2k-1) + (k-1)$. Let $T \subseteq A$ be a tree associated with the premesh $(A, B)$; thus, $X := V(A \cap B) \subseteq V(T)$. By Lemma 7.1.3, $T$ has $(|X| - (k-1))/(2k-1) = m$ disjoint subtrees each containing at least $k$ vertices of $X$. Let $A_1, \ldots, A_m$ be the vertex sets of these trees. By definition of a $k$-mesh, $B$ contains for all $1 \le i < j \le m$ a set $\mathcal{P}_{ij}$ of $k$ disjoint $A_i\!-\!A_j$ paths that have no inner vertices in $A$. These sets $\mathcal{P}_{ij}$ will shrink a little and be otherwise modified later in the proof, but they will always consist of several disjoint $A_i\!-\!A_j$ paths.

One case in the proof will be when we find single paths $P_{ij} \in \mathcal{P}_{ij}$ that are disjoint for different pairs $ij$ and thus link up the sets $A_i$ to form a $K_m$ minor of $G$. If this fails, we shall instead exhibit two specific sets $\mathcal{P}_{ij}$ and $\mathcal{P}_{pq}$ such that many paths of $\mathcal{P}_{ij}$ meet many paths of $\mathcal{P}_{pq}$, forming an $r \times r$ grid between them by Lemma 7.1.6.

Let us impose a linear ordering on the index pairs $ij$ by fixing an arbitrary bijection $\sigma : \{ij \mid 1 \le i < j \le m\} \to \{0, 1, \ldots, \binom{m}{2} - 1\}$. For $l = 0, 1, \ldots$ in turn, we shall consider the pair $pq$ with $\sigma(pq) = l$ and choose an $A_p\!-\!A_q$ path $P_{pq}$ that is disjoint from all previously selected such paths, i.e. from the paths $P_{st}$ with $\sigma(st) < l$. At the same time, we shall replace all the 'later' sets $\mathcal{P}_{ij}$—or what has become of them—by smaller sets containing only paths that are disjoint from $P_{pq}$. Thus for each pair $ij$, we shall define a sequence $\mathcal{P}_{ij} = \mathcal{P}_{ij}^0, \mathcal{P}_{ij}^1, \dots$ of smaller and smaller sets of paths, which eventually collapses to $\mathcal{P}_{ij}^l = \{P_{ij}\}$ when $l$ has risen to $l = \sigma(ij)$.

More formally, let $l^* \le \binom{m}{2}$ be maximal such that, for all $0 \le l < l^*$ and all $1 \le i < j \le m$, there exist sets $\mathcal{P}_{ij}^l$ satisfying the following five conditions:

(i) $\mathcal{P}_{ij}^l$ is a non-empty set of disjoint $A_i$–$A_j$ paths in $B$ that meet $A$ only in their endpoints.

(ii) If $\sigma(ij) < l$ then $\mathcal{P}_{ij}^l$ has exactly one element $P_{ij}$, and $P_{ij}$ does not meet any path belonging to a set $\mathcal{P}_{st}^l$ with $ij \ne st$.

(iii) If $\sigma(ij) = l$, then $|\mathcal{P}_{ij}^l| = k/c^{2l}$.

(iv) If $\sigma(ij) > l$, then $|\mathcal{P}_{ij}^l| = k/c^{2l+1}$.

(v) Define $H_{ij}^l := \bigcup \mathcal{P}_{ij}^l$. If $l = \sigma(pq) < \sigma(ij)$, then for every $e \in E(H_{ij}^l) \setminus E(H_{pq}^l)$ there are no $k/c^{2l+1}$ disjoint paths from $A_i$ to $A_j$ in the graph $(H_{pq}^l \cup H_{ij}^l) - e$.

Note that, since $\sigma(ij) < \binom{m}{2}$ by definition of $\sigma$, conditions (iii) and (iv) imply that $|\mathcal{P}_{ij}^l| \ge c^2$ whenever $\sigma(ij) \ge l$.

Clearly, if $l^* = \binom{m}{2}$, then by (i) and (ii) we have a (subdivided) $K_m$ minor with branch vertices corresponding to the sets $A_1, \dots, A_m$ in $G$. Suppose now that $l^* < \binom{m}{2}$. We claim that $l^* > 0$. Let $pq := \sigma^{-1}(0)$ and put $\mathcal{P}_{pq}^0 := \mathcal{P}_{pq}$. To define $\mathcal{P}_{ij}^0$ for $\sigma(ij) > 0$ put $H_{ij} := \bigcup \mathcal{P}_{ij}$, and let $F \subseteq E(H_{ij}) \setminus E(H_{pq}^0)$ be maximal such that $(H_{pq}^0 \cup H_{ij}) - F$ still contains $k/c$ disjoint paths from $A_i$ to $A_j$; then let $\mathcal{P}_{ij}^0$ be such a set of paths. As any vertex of $A$ on these paths lies in $A_i \cup A_j$ (by definition of $H_{pq}^0$ and $H_{ij}$), we may assume that they have no inner vertices in $A$. Thus our choice of $\mathcal{P}_{ij}^0$ satisfies (i)–(v).

Having shown that $l^* > 0$, let us now consider $l := l^* - 1$. Thus, conditions (i)–(v) are satisfied for $l$ but cannot be satisfied for $l+1$. Let $pq := \sigma^{-1}(l)$. Suppose that $\mathcal{P}_{pq}^l$ contains a path $P$ that avoids a set $\mathcal{Q}_{ij}$ of some $|\mathcal{P}_{ij}^l|/c$ of the paths in $\mathcal{P}_{ij}^l$ for all $ij$ with $\sigma(ij) > l$. Then we can define $\mathcal{P}_{ij}^{l+1}$ for all $ij$ as before (with a contradiction). Indeed, let $st := \sigma^{-1}(l+1)$ and put $\mathcal{P}_{st}^{l+1} := \mathcal{Q}_{st}$. For $\sigma(ij) > l+1$ write $H_{ij} := \bigcup \mathcal{Q}_{ij}$, let $F \subseteq E(H_{ij}) \setminus E(H_{st}^{l+1})$ be maximal such that $(H_{st}^{l+1} \cup H_{ij}) - F$ still contains at least $|\mathcal{P}_{ij}^l|/c^2$ disjoint paths from $A_i$ to $A_j$, and let $\mathcal{P}_{ij}^{l+1}$ be such a set of paths. Setting $\mathcal{P}_{pq}^{l+1} := \{P\}$ and $\mathcal{P}_{ij}^{l+1} := \mathcal{P}_{ij}^l = \{P_{ij}\}$ for $\sigma(ij) < l$ then gives us a family of sets $\mathcal{P}_{ij}^{l+1}$ that contradicts the maximality of $l^*$.

Thus for every path $P \in \mathcal{P}^l_{pq}$ there exists a pair $ij$ with $\sigma(ij) > l$ such that $P$ avoids fewer than $|\mathcal{P}^l_{ij}|/c$ of the paths in $\mathcal{P}^l_{ij}$. For some $\lceil |\mathcal{P}^l_{pq}|/\binom{m}{2} \rceil$ of these $P$ that pair $ij$ will be the same; let $\mathcal{P}$ denote the set of those $P$, and keep $ij$ fixed from now on. Note that $|\mathcal{P}| \geq |\mathcal{P}^l_{pq}|/\binom{m}{2} = c\,|\mathcal{P}^l_{ij}|/\binom{m}{2}$ by (iii) and (iv).

Now we use Lemma 7.1.4 to find sets $\mathcal{V} \subseteq \mathcal{P} \subseteq \mathcal{P}^l_{pq}$ and $\mathcal{H} \subseteq \mathcal{P}^l_{ij}$, where $|\mathcal{H}| = r^2$ and

$$|\mathcal{V}| \geq \tfrac{1}{2}|\mathcal{P}| \quad \left( \geq \frac{c}{m^2} |\mathcal{P}^l_{ij}| \right),$$

such that every path in $\mathcal{V}$ intersects every path in $\mathcal{H}$. Consider the bipartite graph with vertex set $\mathcal{P} \cup \mathcal{P}^l_{ij}$ in which $P \in \mathcal{P}$ is adjacent to $Q \in \mathcal{P}^l_{ij}$ whenever $P \cap Q = \emptyset$. We have to check that this graph does not have too many edges. Since every $P \in \mathcal{P}$ has fewer than $|\mathcal{P}^l_{ij}|/c$ neighbors (by definition of $\mathcal{P}$), this graph has at most

$$
\begin{aligned}
|\mathcal{P}||\mathcal{P}^l_{ij}|/c &\leq |\mathcal{P}||\mathcal{P}^l_{ij}|/6r^2 \\
&\leq \lfloor |\mathcal{P}|/2 \rfloor\, |\mathcal{P}^l_{ij}|/2r^2 \\
&\leq \lfloor |\mathcal{P}|/2 \rfloor (|\mathcal{P}^l_{ij}|/r^2 - 1) \\
&= \big(|\mathcal{P}| - \lceil |\mathcal{P}|/2 \rceil\big)\big(|\mathcal{P}^l_{ij}| - r^2\big)/r^2
\end{aligned}
$$

edges, as required. Hence, $\mathcal{V}$ and $\mathcal{H}$ exist as claimed.

Pick a path $Q \in \mathcal{H}$, and put

$$d := \lfloor \sqrt{c}/m \rfloor = \lfloor r^{2r+4}/m \rfloor \geq r^{2r+2}.$$

For $n = 1, 2, \dots, d-1$ let $e_n$ be the first edge of $Q$ (in the direction from $A_i$ to $A_j$) such that the initial component $Q_n$ of $Q - e_n$ meets at least $nd\,|\mathcal{P}^l_{ij}|$ different paths from $\mathcal{V}$, and such that $e_n$ is not an edge of $H^l_{pq}$. As any two vertices of $Q$ that lie on different paths from $\mathcal{V}$ are separated in $Q$ by an edge not in $H^l_{pq}$, each of these $Q_n$ meets exactly $nd\,|\mathcal{P}^l_{ij}|$ paths from $\mathcal{V}$. Put $Q_0 := \emptyset$ and $Q_d := Q$. Since $|\mathcal{V}| \geq d^2|\mathcal{P}^l_{ij}|$, we have thus divided $Q$ into $d$ consecutive disjoint segments $Q'_n := Q_n - Q_{n-1}$ $(n = 1, \dots, d)$ each meeting at least $d\,|\mathcal{P}^l_{ij}|$ paths from $\mathcal{V}$.

For each $n = 1, \dots, d-1$, Menger's theorem and conditions (iv) and (v) imply that $H^l_{pq} \cup H^l_{ij}$ has a set $S_n$ of $|\mathcal{P}^l_{ij}| - 1$ vertices such that $(H^l_{pq} \cup H^l_{ij}) - e_n - S_n$ contains no path from $A_i$ to $A_j$. Let $S$ denote the union of all these sets $S_n$. Then $|S| < d\,|\mathcal{P}^l_{ij}|$, so each $Q'_n$ meets at least one path $V_n \in \mathcal{V}$ that avoids $S$.

Clearly, each $S_n$ consists of a choice of exactly one vertex $x$ from every path $P \in \mathcal{P}^l_{ij} \setminus \{Q\}$. Denote the initial component of $P - x$ by $P_n$, put $P_0 := \emptyset$ and $P_d := P$, and let $P'_n := P_n - P_{n-1}$ for $n = 1, \dots, d$. The separation properties of the sets $S_n$ now imply that $V_n \cap P \subseteq P'_n$ for $n = 1, \dots, d$ (and hence in particular that $P'_n \neq \emptyset$, i.e., that $P_{n-1} \subset P_n$). Indeed, $V_n$ cannot meet $P_{n-1}$, because $P_{n-1} \cup V_n \cup (Q - Q_{n-1})$ would then contain an $A_i$–$A_j$ path in $(H^l_{pq} \cup H^l_{ij}) - e_{n-1} - S_{n-1}$, and likewise (consider $S_n$), $V_n$ cannot meet $P - P_n$. Thus for all $n = 1, \ldots, d$, the path $V_n$ meets every path $P \in \mathcal{H} \setminus \{Q\}$ precisely in its $n$th segment $P'_n$. Applying Lemma 7.1.6 to the path systems $\mathcal{H} \setminus \{Q\}$ and $\{V_1, \ldots, V_d\}$ now yields the desired grid minor. □

## 7.2. Minimal obstructions of bounded tree-width

One of the early results in the Robertson-Seymour theory on graph minors shows that graphs of bounded tree-width are well-quasi-ordered[1] with respect to the minor relation.

THEOREM 7.2.1 (Robertson and Seymour [**RS90a**]). *For every integer $w > 0$ and every infinite sequence of graphs $G_1, G_2, G_3, \ldots$ whose tree-width is at most $w$, there are indices $i$ and $j$ such that $i < j$ and $G_i$ is a minor of $G_j$.*

In this section we prove the weaker result, namely that for each surface $S$, $\mathbf{Forb}_0(S)$ contains only finitely many graphs of bounded tree-width. This is equivalent to the following:

THEOREM 7.2.2. *Let $g$ and $w$ be positive integers and let $S$ be a surface of Euler genus $g$. Then there is an integer $N$ such that every graph in $\mathbf{Forb}_0(S)$ with tree-width $< w$ has at most $N$ vertices.*

The rest of this section is devoted to the proof of Theorem 7.2.2 following Mohar [**Mo01**]. We start with some lemmas.

Suppose that $x, y$ is a separating pair of vertices of a graph $G$. An $\{x, y\}$-bridge $B$ is said to be *nonplanar* if $B + xy$ is a nonplanar graph.

LEMMA 7.2.3. *If $G \in \mathbf{Forb}_0(S)$ is 2-connected, then every $\{x, y\}$-bridge $B$ containing at least two edges is nonplanar.*

PROOF. The replacement of $B$ by the edge $xy$ results in a proper minor of $G$ which can be embedded in $S$. If $B + xy$ is planar, this embedding can be extended to an embedding of $G + xy$ in $S$, a contradiction. □

Suppose that $X$ is a separating set of vertices in a connected graph $G$ which is $\Pi$-embedded in $S$. Let $W = v_1 e_1 v_2 e_2 \ldots v_k e_k v_1$ be a $\Pi$-facial walk. Recall that the pair $\{e_{i-1}, e_i\}$ ($i \in \{1, \ldots, k\}$) is called a *mixed angle* if the edges $e_{i-1}$ and $e_i$ belong to distinct $X$-bridges in $G$. In that case $v_i \in X$, and the appearance of $v_i$ in $W$ is called *mixed*. Let $R$ be the multigraph with vertices $V(R) = X$ and edges which connect consecutive mixed appearances of vertices in $X$ in the $\Pi$-facial walks. Then every $\Pi$-facial walk with some mixed angles determines a closed walk in $R$ and

---

[1] A *well-quasi-ordering* of a set $X$ is a reflexive and transitive relation $\preceq$ such that, for every infinite sequence $x_1, x_2, x_3, \ldots$ of elements of $X$, there are indices $i$ and $j$ such that $i < j$ and $x_i \preceq x_j$.

$G \cup R$ has an embedding $\tilde{\Pi}$ in $S$ which extends the embedding $\Pi$. Let $\Pi^R$ denote the induced topological embedding of $R$ in $S$. (This embedding is not always 2-cell.)

LEMMA 7.2.4. *The faces of $\Pi^R$ in $S$ can be partitioned into two classes, $\mathcal{F}_A$ and $\mathcal{F}_B$, such that every edge of $R$ is incident with a face in $\mathcal{F}_A$ and a face in $\mathcal{F}_B$. The faces in $\mathcal{F}_A$ are 2-cells and correspond to the faces of $G$ with mixed angles and they are contained in faces of $G$. The faces in $\mathcal{F}_B$ and the $X$-bridges in $G$ which are $\tilde{\Pi}$-embedded in these faces are in bijective correspondence.*

PROOF. Let $\mathcal{F}_B$ be those faces of $R$ in $S$ that contain some $X$-bridge, and let $\mathcal{F}_A$ be the other faces. Clearly, every edge of $R$ is is incident with a face in $\mathcal{F}_A$ and a face in $\mathcal{F}_B$. It remains to prove that no face of $\mathcal{F}_B$ contains more than one $X$-bridge. Let $F \in \mathcal{F}_B$. The boundary of $F$ in $S$ consists of of one or more closed facial walks in $R$. Let $e$ be an edge in one of them, joining consecutive mixed appearances of vertices $v_i$ and $v_j$ $(i < j)$ of a $\Pi$-facial walk $W$. Since $v_{i+1}, \ldots, v_{j-1}$ are not mixed appearances in $W$, all edges $e_i, e_{i+1}, \ldots, e_{j-1}$ belong to the same $X$-bridge $B$. Consider the local clockwise rotation of $\tilde{\Pi}$ around $v_j$. We may assume that $e$ is followed by $e_{j-1}$. Then $e_{j-1}$ is followed by some other edges of $B$ (possibly none) until a mixed angle in some face is reached, in which case an edge $e'$ of $R$ would follow the edges of $B$. Clearly, $e'$ follows $e$ on the boundary of $F$. Using the same argument at $e'$, etc., we see that the edges of $G$ in the face $F$ and incident with the considered facial walk in the boundary of $F$ all belong to the same $X$-bridge $B$. If the face $F$ has another boundary component, it must be incident with the same bridge; otherwise the embedding of $G$ would not be 2-cell. Clearly, every $X$-bridge lies in a single face of $R$. This completes the proof. $\square$

LEMMA 7.2.5. *Let $G$ be a connected graph and $X \subseteq V(G)$ a separating set such that no vertex of $X$ is a cutvertex and for any two vertices $x, y \in X$, every $\{x, y\}$-bridge containing at least two edges is nonplanar. If $G$ is embedded in a surface of Euler genus $g$ and $s = |X|$, then*

$$|E(R)| \le 6g + s^2 + 5s - 12. \tag{7.1}$$

PROOF. Let $q = |E(R)|$. Since $X$ contains no cutvertices, no face of $R$ is bounded by just one edge. If a face in $\mathcal{F}_B$ is a 2-cell and is bounded by two edges, then the corresponding $X$-bridge in that face is planar, so it is just an edge joining two vertices of $X$. The number of such faces is $\le \binom{s}{2}$. By Lemma 7.2.4, the sum of the lengths of faces in $\mathcal{F}_B$ is $q$. This implies that $2\binom{s}{2} + 3(|\mathcal{F}_B| - \binom{s}{2}) \le q$, hence $3|\mathcal{F}_B| \le q + \binom{s}{2}$. Similarly, the sum of the lengths of faces in $\mathcal{F}_A$ is $q$. Therefore, $|\mathcal{F}_A| \le q/2$. Now, Euler's formula implies:

$$2 - g \le s - q + |\mathcal{F}_A| + |\mathcal{F}_B| \le s - \frac{q}{6} + \frac{1}{3}\binom{s}{2},$$

which yields (7.1). □

PROOF OF THEOREM 7.2.2. The proof is by induction on $g$ and by contradiction. We assume that there are arbitrarily large graphs of tree-width $< w$ in $\mathbf{Forb}_0(S)$. Suppose that $G \in \mathbf{Forb}_0(S)$ and that $\mathbf{tw}(G) < w$. By the genus additivity theorems (Theorems 4.4.2 and 4.4.3) and the induction hypothesis, $G$ is 2-connected. Let $(T, (Y_t)_{t \in V(T)})$ be a tree decomposition of $G$ of width $< w$. We shall view the parts $Y_t$ of the tree decomposition as being subgraphs in $G$. We may assume that $Y_t \neq Y_{t'}$ if $t \neq t'$, and that for every $t_0 \in V(T)$ and every component $B$ of $T - t_0$, the union $\cup_{t \in V(B)} Y_t \setminus Y_{t_0}$ is nonempty. This implies that $Y_{t_0}$ is a separating set in $G$ if $t_0$ is not an endvertex of $T$.

Let $X = Y_t$ be a vertex separating set in $G$. By contracting an edge in one of the $X$-bridges, we obtain a graph embeddable in $S$. Lemmas 7.2.3–7.2.5 and the upper bound on $|\mathcal{F}_B|$ in the proof of Lemma 7.2.5 imply that there are at most $d := 2g + 2w + \binom{w}{2} - 4$ $X$-bridges in $G$. By the above assumptions on the tree decomposition, every edge of $T$ incident with $t$ determines at least one $X$-bridge. Therefore, every vertex of $T$ has degree $\leq d$. Clearly, $|V(T)| \geq \frac{|V(G)|}{w}$. So, assuming $G$ may be arbitrarily large, $T$ contains a path which is arbitrarily large. Applying Menger's Theorem and the pigeonhole principle on the longest path in $T$ and its subpaths one or more (but at most $w$) times, one can conclude that there exists an integer $s \leq w$ and there exist separating sets $X_0, \ldots, X_r$ (where $r$ is arbitrarily large) such that the following holds (cf., e.g., [BMM99p] for more details):

(i) $|X_i| = s$, $i = 0, \ldots, r$.

(ii) There exist disjoint paths $P_1, \ldots, P_s$ from $X_0$ to $X_r$ which intersect $X_0, X_1, \ldots, X_r$ in that order. Denote by $u_l^i$ the vertex of $P_l \cap X_i$, $l = 1, \ldots, s$, $i = 0, \ldots, r$.

(iii) The path $P_1$ is *everywhere nontrivial*, i.e., $P_1$ has an edge $e_i$ between its intersection with $X_{i-1}$ and $X_i$ and not incident with $X_{i-1} \cup X_i$, $i = 1, \ldots, r$.

For $i = 1, \ldots, r$, let $G_i$ be the graph obtained from $G$ by contracting the edge $e_i$ of $P_1$. Let $\Pi_i$ be an embedding of $G_i$ in $S$, and let $R_i$ be the corresponding multigraph with vertex set $V(R_i) = X_i$ whose edges correspond to consecutive mixed appearances of vertices of $X_i$ in $\Pi_i$-facial walks (with respect to the separator $X_i$ of $G_i$).

For $i = 1, \ldots, r - 1$, let $B^{(i)}$ be the $X_i$-bridge in $G_i$ which contains the segment of $P_1$ from $X_0$ to $X_i$ with $e_i$ contracted, and let $B_0^{(i)}$ be the $X_i$-bridge in $G$ containing the same segment of $P_1$ (where $e_i$ is not contracted).

Let $\Pi_i^R$ be the topological embedding of $R_i$ in $S$. We say that $(R_i, \Pi_i^R)$ is *strongly homeomorphic* to $(R_j, \Pi_j^R)$ if there is a homeomorphism $S \to S$

whose restriction to $R_i$ induces an isomorphism of the $\Pi_i^R$-embedded graph $R_i$ onto the $\Pi_j^R$-embedded graph $R_j$ such that $u_l^i \mapsto u_l^j$, $l = 1, \ldots, s$, and such that the face of $R_i$ corresponding to the bridge $B^{(i)}$ (cf. Lemma 7.2.4) is mapped onto the face of $R_j$ corresponding to $B^{(j)}$.

Lemma 7.2.5 combined with the surface classification theorem, Theorem 3.1.3, implies that the number of strong homeomorphism types of pairs $(R_i, \Pi_i^R)$ is bounded in terms of $g$ and $w$. As $r$ can be arbitrarily large, there are indices $i$ and $j > i$ such that $(R_i, \Pi_i^R)$ and $(R_j, \Pi_j^R)$ are strongly homeomorphic.

Take the embedding $\Pi_i$ and delete the $X_i$-bridge $B^{(i)}$. Let $F$ denote the resulting face in $S$. Since $(R_i, \Pi_i^R)$ and $(R_j, \Pi_j^R)$ are strongly homeomorphic, the $X_j$-bridge $B^{(j)}$ can be embedded in $F$ so that any vertex $u_l^j$ of $B^{(j)}$ is identified with $u_l^i$ ($l = 1, \ldots, s$) on the boundary of $F$. This gives rise to an embedding in $S$ of the graph $G'$ obtained from $G_i \setminus B^{(i)}$ by adding a disjoint copy of $B^{(j)}$ and identifying each $u_l^i \in V(G_i \setminus B^{(i)})$ with the vertex $u_l^j \in V(B^{(j)})$, $l = 1, \ldots, s$. Although $B^{(j)}$ is a bridge in $G_j$ but not a bridge in $G$, it contains as a minor a copy of the $X_i$-bridge $B_0^{(i)}$ of $G$. In order to get $B_0^{(i)}$ as a minor, we contract all edges of the paths $P_l$ ($l = 1, \ldots, s$) between $X_i$ and $X_j$ in the copy of $B^{(j)}$ in $G'$. Now it is clear that the graph $G'$ contains $G$ as a minor. Since $G'$ is embedded in $S$, also its minor $G$ admits an embedding in $S$. This contradiction completes the proof.     □

## 7.3. The excluded minor theorem for any fixed surface

Theorem 7.3.3 below, combined with Theorems 7.2.2 and 7.1.7, easily implies the excluded minor theorem (Theorem 7.0.1) for any fixed surface. For simplicity we restrict ourselves to orientable surfaces and orientable embeddings.

As pointed out in [**Th97c**], the statements (1)–(6) below are the only embedding results needed for the excluded minor theorem.

(1) If $G'$ is a connected subgraph of the connected graph $G$ and $\Pi$ is an embedding of $G$, then

$$0 \leq \mathbf{g}(G', \Pi) \leq \mathbf{g}(G, \Pi).$$

Recall that an edge $e$ of the connected graph $G$ is a *cutedge* if $G - e$ has two components $G_1$ and $G_2$. In that case a simple count shows that

(2)  $\mathbf{g}(G, \Pi) = \mathbf{g}(G_1, \Pi) + \mathbf{g}(G_2, \Pi)$.

From the genus additivity theorem (Theorem 4.4.2) it follows that:

(3) If $v$ is a cutvertex of $G$ and $G = G_1 \cup G_2$ where $G_1 \cap G_2 = \{v\}$ and $G_2$ is nonplanar, then $\mathbf{g}(G) > \mathbf{g}(G_1)$.

(4) Let $G_1, G_2$ be disjoint connected graphs and let $e = xy$ be an edge of $G_2$. Let $G$ be obtained from $G_1 \cup G_2$ by first deleting $e$ and then

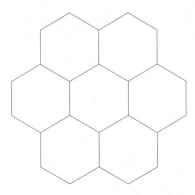

FIGURE 7.2. The hexagonal grid $J_2$

adding an edge from $x$ to $G_1$ and an edge from $y$ to $G_1$. If $G_2$ is nonplanar, then $\mathbf{g}(G) > \mathbf{g}(G_1)$.

The conclusion of (4) can be proved by a simple count of facial walks in a hypothetical embedding of $G$ whose genus is equal to $\mathbf{g}(G_1)$.

By Theorem 2.5.1 we have

(5) If $G$ is a subdivision of a 3-connected planar graph, then $G$ has precisely one embedding of genus 0 (up to equivalence).

This implies:

(6) If $G$ is as in (5) and $x, y \in V(G)$, then $G + xy$ is planar if and only if $G$ has a facial cycle containing $x$ and $y$.

We define the *hexagonal grid* $J_k$ as a finite subgraph of the hexagonal tiling of the Euclidean plane. We let $J_1$ be a cycle of length 6. For each $k \geq 2$, we define $J_k$ as the union of $J_{k-1}$ and all those 6-cycles in the hexagonal tiling which intersect $J_{k-1}$ (cf. Figure 7.2). Clearly, $J_k$ is a subdivision of a 3-connected graph when $k > 1$. It is easy to see that $J_k$ is a subgraph of the $4k \times 2k$ grid and that $J_k$ contains the $k \times k$ grid as a minor.

Suppose now that $H$ is a subgraph of the graph $G$ such that $H$ is a subdivision of a hexagonal grid. Let $C$ be the outer cycle of $H$. We say that $H$ is *good* in $G$ if the union of $H$ and all those $H$-bridges which have a vertex of attachment in $H - V(C)$ is planar.

PROPOSITION 7.3.1 (Thomassen [**Th97c**]). *Let $G$ be a connected graph of genus $g$ and let $H'$ be a subgraph of $G$ such that $H'$ is a subdivision of the hexagonal grid $J_m$. Let $k$ be a natural number. If $m > 100k\sqrt{g}$, then $H'$ contains a subdivision $H$ of the hexagonal grid $J_k$ such that $H$ is good in $G$.*

PROOF. Let $Q_1, Q_2, \ldots, Q_{2g+2}$ be pairwise disjoint subgraphs in $H'$ each of which is a subdivision of $J_k$, and such that, for any $i, j$ where

$1 \leq i < j \leq 2g + 2$, the following holds: If $x_i$ (respectively $x_j$) is on the outer cycle of $Q_i$ (respectively $Q_j$) and has a neighbor $y_i$ (respectively $y_j$) in $H'$ but not in $Q_i$ (respectively $Q_j$) then $H'$ has a path from $x_i$ to $x_j$ having only $x_i$ and $x_j$ in common with $Q_1 \cup Q_2 \cup \cdots \cup Q_{2g+2}$. We claim that at least one of $Q_1, \ldots, Q_{2g+2}$ is good in $G$. Suppose (reductio ad absurdum) that this is not the case. Then we construct an increasing sequence $M_1, \ldots, M_{g+2}$ of connected subgraphs in $G$ such that

  (i) $M_1 = Q_1$,
  (ii) $M_i$ intersects at most $2i - 1$ of the graphs $Q_1, \ldots, Q_{2g+2}$, and $M_i$
       includes each $Q_j$ that it intersects for $i = 1, 2, \ldots, g + 1$ and $j =$
       $1, 2, \ldots, 2g + 2$.
 (iii) $\mathbf{g}(M_i) \geq i - 1$ for $i = 1, 2, \ldots, g + 2$.

Note that (iii) gives a contradiction for $i = g + 2$.

Suppose we have constructed $M_i$ ($1 \leq i \leq g + 1$) such that (i), (ii), and (iii) are satisfied. We shall then construct $M_{i+1}$.

By (ii), there is a $j \in \{1, 2, \ldots, 2g + 2\}$ such that $M_i \cap Q_j = \emptyset$.

Let $R$ be any $Q_j$-bridge which has a vertex of attachment $x$ such that $x$ is not on the outer cycle of $Q_j$. Let $y$ be a vertex which is on the outer cycle of $Q_j$ such that $y$ is not on the same facial cycle as $x$, but $y$ is adjacent to a vertex of $H'$ outside the outer cycle of $Q_j$. Let $P_1$ be a path in $H'$ from $y$ to $M_i$ such that $P_1$ has only its ends in common with

$$M_i \cup Q_1 \cup Q_2 \cup \cdots \cup Q_{2g+2}.$$

Assume first that $R$ intersects $M_i$. Let $P_2$ be a path in $R$ from $x$ to $M_i$ such that $P_2$ has only its ends in common with $M_i \cup Q_j$. As $Q_j + xy$ is nonplanar, by (6), it follows by (3) or (4) that $M_i \cup P_1 \cup P_2 \cup Q_j$ can play the role of $M_{i+1}$. We may therefore assume that $R \cap M_i = \emptyset$.

Assume next that $R$ contains a vertex $z$ of $H'$ not in $Q_j$. Then we let $P_2$ consist of a path in $R$ from $x$ to $z$ followed by a path in $H'$ from $z$ to $M_i \cup P_1 \cup Q_j$ such that $P_2$ has only its ends in common with $M_i \cup Q_j \cup P_1$ and intersects at most one of those grids $Q_1, Q_2, \ldots, Q_{2g+2}$, say $Q_r$, which are not contained in $M_i \cup Q_j$. Then we let $M_{i+1} = M_i \cup Q_j \cup P_1 \cup P_2$ or $M_i \cup Q_j \cup P_1 \cup P_2 \cup Q_r$ (if $Q_r$ exists). By (3) or (4), $\mathbf{g}(M_{i+1}) > \mathbf{g}(M_i)$, so (ii), (iii) are satisfied with $i + 1$ instead of $i$. So we may assume that $R$ contains no vertex of $H'$ not in $Q_j$.

Since $Q_j$ is not good in $G$, the union of $Q_j$ and all those $Q_j$-bridges $R$ which intersect $Q_j$ minus its outer cycle forms a nonplanar graph $Q_j'$. Now $Q_j' \cup P_1$ can play the role of $M_{i+1}$.                                    □

Proposition 7.3.2 below is a special case of a result of Mohar [**Mo92**, Theorem 5.1]. The proof is from [**Th97c**].

PROPOSITION 7.3.2. *Let $G$ be a connected graph of genus $g$, and let $H$ be a subgraph of $G$ such that $H$ is a subdivision of the hexagonal grid $J_k$ and such that $H$ is good in $G$. Let $\Pi$ be an embedding of $G$ such that* *the* $\Pi$-*genus is* $g$. *If* $k \geq 4g + 6$, *then the* $\Pi$-*genus of the subdivision of* $J_{k-4g-4}$ *in* $J_k$ *is zero. In other words,* $\Pi$ *induces a planar embedding of* $J_{k-4g-4}$.

PROOF. For each natural number $j \leq k$, $\Pi$ induces an embedding of $J_j$ which we also denote by $\Pi$. We first claim that there exists a natural number $i$, $1 \leq i \leq g + 1$, such that all those 6-cycles of $J_{k-4i+3}$ which intersect the outer cycle of $J_{k-4i+3}$ are $\Pi$-facial cycles in $J_k$. Suppose (reductio ad absurdum) that this is false. Then pick such a nonfacial 6-cycle, $R_i'$ say, in $J_{k-4i+3}$, and let $R_i$ be the union of $R_i'$ and those six 6-cycles in $J_k$ which intersect $R_i'$. By (5), the $\Pi$-embedding of $R_i$ has genus greater than 0. We extend $R_1 \cup \cdots \cup R_{g+1}$ to a connected graph $R$ in $G$ such that only the edges of $R_1 \cup \cdots \cup R_{g+1}$ are contained in cycles of $R$. By (2), the $\Pi$-genus of $R$ is greater than $g$. This contradiction proves the claim.

Consider an $H$-bridge $Q$ which has not all its vertices of attachment on the outer cycle of $H$. Since $H$ is good in $G$, $H \cup Q$ is planar. In particular, there is a 6-cycle $C$ in $J_k$, such that all vertices of attachment of $Q$ are in the cycle in $H$ corresponding to $C$. If there are two possible choices for $C$ we say that $Q$ is *weakly attached* to $C$. Otherwise, $Q$ is *strongly attached* to $C$. If $Q$ is strongly attached to $C$, then we say that in the planar embedding of $H \cup Q$, $Q$ is embedded *inside* $C$. Now, if $Q$ is strongly attached to $C$ where $C$ is a 6-cycle in $J_{k-4i+3}$ intersecting the outer cycle of $J_{k-4i+3}$, then we may assume not only that $C$ is a $\Pi$-facial cycle in $H$ (as proved in the claim) but also that $Q$ is $\Pi$-embedded inside $C$ and that $\Pi$ induces a planar embedding of $C \cup Q$. If $Q$ is weakly attached to $C$, we may assume that $\Pi$ induces a planar embedding of $Q \cup C \cup C'$ where $C'$ is another 6-cycle to which $Q$ is attached. We simply repeat the proof of the claim.

Now consider the subgraph $M$ of $G$ obtained by deleting the subgraph of $H$ inside the outer cycle of $J_{k-4i+2}$, and all $H$-bridges attached to some vertex of $J_{k-4i+2}$, except those $H$-bridges which are also attached to a vertex in $J_{k-4i+3}$ but not in $J_{k-4i+2}$. Clearly, $M$ is connected, and the $\Pi$-genus of $M$ is at most $g$.

The above claim implies that the outer cycle of $J_{k-4i+2}$ is a $\Pi$-facial walk of $M$. Therefore the $\Pi$-embedding of $M$ can be extended to an embedding of $G$ with the same genus as the $\Pi$-genus of $M$. So, the $\Pi$-genus of $M$ equals $g$.

Now, every 6-cycle $R$ in $J_{k-4i}$ must be $\Pi$-facial. For otherwise, the union of $R$, the six 6-cycles in $J_k$ which intersect $R$, and a path from $R$ to $M$ would have $\Pi$-genus greater than $g$, by (3). This contradiction proves that the $\Pi$-embedding of $J_{k-4g-4}$ is of genus 0.    $\square$

Theorem 7.3.3 below is closely related to a result of Seymour [**Se95p**, (3.3)]. We consider only the orientable case.

THEOREM 7.3.3 (Thomassen [**Th97c**]). *Let $G$ be a 2-edge-connected graph such that $\mathbf{g}(G) = g$ and $\mathbf{g}(G - e) < g$ for every edge $e$ in $G$. Then $G$ contains no subdivision of the hexagonal grid $J_m$, $m = \lceil 1100g^{3/2} \rceil$.*

PROOF. Suppose (reductio ad absurdum) that $G$ contains a subdivision of $J_m$. By Proposition 7.3.1, $G$ contains a subdivision $H$ of $J_k$ ($k \geq 4g + 6$) which is good in $G$. Let $e$ be any edge in the subdivision of $J_1$ in $H$. Let $\Pi$ be an embedding of $G - e$ of genus $g - 1$. By (the proof of) Proposition 7.3.2, the induced embedding of $J_3$ is of genus zero. But then $\Pi$ can be modified to an embedding of $G$ of genus $g - 1$, a contradiction. □

Fellows and Langston [**FL89**] showed that an explicit upper bound on the tree-width of the graphs in $\mathbf{Forb}_0(S)$ yields an algorithm for finding them. Such an upper bound was found by Seymour [**Se95p**, (3.3)]. It also follows from Theorem 7.3.3 combined with the excluded grid theorem 7.1.7. A stronger result of Seymour [**Se95p**] is an explicit upper bound on the number of vertices of graphs in $\mathbf{Forb}_0(S)$.

PROBLEM 7.3.4. *Let $w_0$ be any fixed integer. Does there exist a poly-nomially bounded algorithm to find the (orientable) genus of any graph of tree-width less than $w_0$?*

CHAPTER 8

# Colorings of graphs on surfaces

Let $G$ be a graph and $k$ a natural number. A $k$-coloring of $G$ is a map $c$ that maps the vertices of $G$ into the set $\{1, 2, \ldots, k\}$ (whose elements are called colors) such that no two adjacent vertices are mapped to the same color. A graph is $k$-colorable if it admits a $k$-coloring. The chromatic number $\chi(G)^1$ of $G$ is equal to the smallest $k$ such that $G$ is $k$-colorable. Clearly, $\chi(G) \leq 2$ if and only if $G$ is bipartite. By König's theorem 1.2.3, $\chi(G) \leq 2$ if and only if $G$ has no odd cycle. Since odd cycles are easy to find, one can easily recognize graphs with $\chi(G) \leq 2$. In contrast to this, it is **NP**-complete to decide if $\chi(G) \leq 3$, even for planar graphs, cf. Garey and Johnson [**GJ79**].

Graph coloring is one of the central subjects in discrete mathematics. The recent book by Jensen and Toft [**JT95**] is an excellent reference. In this chapter we focus on coloring graphs on surfaces. In Section 8.2 we describe briefly some of the ideas in the proof of the Four Color Theorem. In Section 8.3 we consider its generalization to general surfaces, known as the Heawood Problem: What is the largest chromatic number of a graph that can be embedded in a given surface. A solution was conjectured by Heawood [**He890**] and proved by Ringel and Youngs [**RY68, Ri74**]. The Heawood conjecture reduces to determining the genus and the nonorientable genus of complete graphs. The solution by Ringel and Youngs involves constructions of minimum genus embeddings of these graphs and introduces techniques that can be also used for constructions of small genus embeddings of other graphs (see Chapter 4).

Finally, we outline coloring results and problems of a more specialized nature.

### 8.1. Planar graphs are 5-choosable

Euler's formula implies that every planar graph contains a vertex of degree five or less. Consequently, every planar graph is 6-colorable. Heawood [**He890**] proved that every planar graph is 5-colorable. In this section we present its generalization to list colorings due to Thomassen [**Th94c**].

---

[1] This notation should not be confused with $\chi(\Pi)$, the Euler characteristic of an embedding $\Pi$.

Let $G$ be a graph and suppose that, for each vertex $v \in V(G)$, $L(v) \subseteq \mathbb{N}$ is a *list* of colors. An *L-coloring* of $G$ is an assignment of a color to each vertex of $G$ such that adjacent vertices get distinct colors and such that each vertex $v$ receives a color from the list $L(v)$. Such colorings are also called *list colorings*. The graph $G$ is *k-choosable* if it always has a list coloring provided each list $L(v)$, $v \in V(G)$, contains (at least) $k$ colors.

In 1975, Vizing raised the question whether or not every planar graph is 5-choosable (see [**JT95**]). Erdős, Rubin, and Taylor [**ERT80**] conjectured that every planar graph is 5-choosable, but not necessarily 4-choosable. In 1993, Voigt [**Vo93**] proved the last part of this conjecture by exhibiting examples of planar graphs that are not 4-choosable. (Subsequently, even 3-colorable non-4-choosable planar graphs have been found, see Voigt and Wirth [**VW97**] and Mirzakhani [**Mi96**].) Thomassen verified the former part of the conjecture:

THEOREM 8.1.1 (Thomassen [**Th94c**]). *Every planar graph is 5-choosable.*

Theorem 8.1.1 follows immediately from Theorem 8.1.2 below. Let us recall that a plane graph $G$ is a *near-triangulation* if it is 2-connected and all faces, except possibly the outer face, are triangles.

THEOREM 8.1.2 (Thomassen [**Th94c**]). *Let $G$ be a near-triangulation with the outer cycle $C = v_1 v_2 \ldots v_p v_1$. Assume that $L(v_1) = \{1\}$ and $L(v_2) = \{2\}$ and that, for any other vertex $v$, $L(v)$ is a list of at least 3 colors if $v \in V(C) \backslash \{v_1, v_2\}$ and at least 5 colors if $v \in V(G - C)$. Then $G$ has an L-coloring.*

PROOF (by induction on the number of vertices of $G$). If $p = 3$ and $G = C$ there is nothing to prove. So we proceed to the induction step.

If $C$ has a chord $v_i v_j$ where $2 \leq i \leq j - 2 \leq p - 1$ ($v_{p+1} = v_1$), then we apply the induction hypothesis to the cycle $C' = v_1 v_2 \ldots v_i v_j v_{j+1} \ldots v_1$ and its interior. We then use induction also to $C'' = v_j v_i v_{i+1} \ldots v_{j-1} v_j$ and its interior where each of the vertices $v_i$ and $v_j$ has the list consisting of the single color that was obtained by using induction to $C'$ and its interior.

Suppose now that $C$ is a chordless cycle. Let $v_1, u_1, u_2, \ldots, u_m, v_{p-1}$ be the neighbors of $v_p$ in that clockwise order around $v_p$. As the interior of $C$ is triangulated, $G$ contains the path $P = v_1 u_1 u_2 \ldots u_m v_{p-1}$. As $C$ is chordless, $C' = P \cup (C - v_p)$ is a cycle. Let $x, y$ be two distinct colors in $L(v_p) \backslash \{1\}$. Now define $L'(u_i) = L(u_i) \backslash \{x, y\}$, $1 \leq i \leq m$, and $L'(v) = L(v)$ if $v \in V(G) \backslash \{u_1, \ldots, u_m\}$. Then we apply the induction hypothesis to $C'$ and its interior and the new lists $L'$. We complete the obtained $L'$-coloring to an $L$-coloring of $G$ by assigning $x$ or $y$ to $v_p$ such that $v_p$ and $v_{p-1}$ get distinct colors. $\quad \square$

Theorem 8.1.1 implies, in particular, that every planar graph is 5-colorable.

## 8.2. The Four Color Theorem

In 1852 F. Guthrie asked if the regions of every map on the sphere can be colored with four colors in such a way that no two regions of the map with common boundary receive the same color. By duality, Guthrie asked if every planar graph is 4-colorable. This easily stated problem became known as the Four Color Problem[2] and it remained unsolved for more than a century. It was finally answered in positive by Appel and Haken [**AH76a, AH76b, AH77, AHK77**]. An updated version of their proof is collected in [**AH89**]. A simpler proof was recently presented by Robertson, Sanders, Seymour, and Thomas [**RSST96a, RSST97**].

**The Four Color Theorem.** *Every planar graph is 4-colorable.*

We shall explain the main ideas behind the proof(s) of the Four Color Theorem. Unfortunately, the complete proof cannot be checked without a computer. There are several hundreds of cases to be considered, and many of the cases need a computer checking going through thousands (sometimes even hundreds of thousands) of possibilities.

We shall establish some basic connectivity results for planar graphs.

LEMMA 8.2.1. *Let $S$ be a minimal separating set of vertices in a graph which contains no subdivision of $K_{3,3}$. If $|S| \geq 3$, then $G - S$ has precisely two components.*

PROOF. Let $v_1, v_2, v_3$ be distinct vertices in $S$ and let T be a spanning tree in a component of $G - S$. Since $S$ is a minimal separating set, there are edges $v_i u_i$ where $u_i \in V(T)$, $i = 1, 2, 3$. Clearly, $T \cup \{v_1 u_1, v_2 u_2, v_3 u_3\}$ contains a vertex and three internally disjoint paths joining this vertex to $v_1, v_2, v_3$. If $G - S$ has at least three components, this gives a subdivision of $K_{3,3}$ in $G$. □

COROLLARY 8.2.2. *If $S$ is a minimal separating set of vertices in a 3-connected planar graph $G$, then $G - S$ has precisely two components.*

PROPOSITION 8.2.3. *If $S$ is a minimal separating set of vertices in a plane triangulation $G$, then $G(S)$ is an induced, nonfacial cycle. Conversely, if $S$ is the vertex set of an induced nonfacial cycle, then $S$ is a minimal separating set, and $G - S$ has precisely two connected components.*

PROOF. Assume first that $S$ is a minimal separating set of vertices and let $v \in S$. Since $G - (S \setminus \{v\})$ is connected, $G - S$ has two vertices $u_1, u_2$ in distinct components of $G - S$ joined to $v$. The union of the facial triangles containing $v$ contains a cycle $C'$ whose vertices are precisely all neighbors

---

[2]Attempts to solve the famous Four Color Problem resulted in a number of important graph theory results. For an extensive discussion on the history and the impact of the Four Color Problem on graph theory we refer to Biggs, Lloyd, and Wilson [**BLW86**]; cf. also Ore [**Ore67**], Saaty and Kainen [**SK77**], and Woodall and Wilson [**WW78**].

of $v$. Clearly, both components of $C' - \{u_1, u_2\}$ intersect $S$. Hence $v$ has degree at least 2 in $G(S)$. The same argument shows the following. If $v_1 v_2$ is any edge of $G(S)$, then $v_2$ has a neighbor $v_3$ in $S \setminus \{v_1\}$ such that $G$ has edges $e_1, e_2$ joining $v_2$ with distinct components of $G - S$ such that $e_1$ and $e_2$ leave $v_2$ on different sides of the path $v_1 v_2 v_3$. The edge $v_2 v_3$ can be extended to a path $v_2 v_3 v_4$ with the same property, etc. We stop the first time there is a repetition in the sequence $v_1, v_2, v_3, \ldots$. This results in a separating cycle $C$ such that $V(C) \subseteq S$. The minimality of $S$ implies that $V(C) = S$ and that $C$ is chordless.

Suppose, conversely, that $C$ is an induced nonfacial cycle. Clearly, both $\text{int}(C)$ and $\text{ext}(C)$ contain vertices of $G$. Hence $V(C)$ is a separating set. As any minimal separating subset of $V(C)$ induces a cycle, $V(C)$ is a minimal separating set. By Corollary 8.2.2, $G - V(C)$ has precisely two components.     □

A planar graph $G$ is called a *minimal counterexample* if it is not 4-colorable and every planar graph $G'$ with $|V(G')| + |E(G')| < |V(G)| + |E(G)|$ is 4-colorable. In proving that a minimal counterexample does not exist, the first step is to establish some connectivity properties of such a graph. We say that $G$ is *internally 6-connected* if it is 5-connected and for every set $U \subseteq V(G)$ of size 5, $G - U$ is either connected or consists of two connected components, one of which is just a vertex. Birkhoff proved:

LEMMA 8.2.4 (Birkhoff [**Bi13**]). *Every minimal counterexample is an internally 6-connected triangulation of the sphere.*

PROOF. Let $G$ be a minimal counterexample. It is easy to see that $G$ is 2-connected. If $G$ has a facial cycle $v_1 v_2 \ldots v_k$ where $k \geq 4$, then either $v_1 v_3 \notin E(G)$ or $v_2 v_4 \notin E(G)$. Assuming the former, let $G'$ be the graph obtained from $G$ by identifying $v_1$ and $v_3$. Then $G'$ is planar, and its 4-coloring determines a 4-coloring of $G$ where $v_1$ and $v_3$ have the color of the new vertex. Since $G$ is not 4-colorable, this proves that $G$ is a triangulation.

By Proposition 8.2.3, every minimal separating vertex set induces a cycle. If $C = v_1 v_2 \ldots v_k$ is such a cycle, let $G_1 = \text{Int}(C)$ and $G_2 = \text{Ext}(C)$. If $k = 3$, 4-colorings of $G_1$ and $G_2$ give rise to a 4-coloring of $G$ (after permuting the colors of $G_2$).

If $k = 4$, we argue similarly. We first identify $v_1$ and $v_3$ in $G_1$ and in $G_2$, respectively, and 4-color the resulting graphs. This gives 4-colorings $c_1$ of $G_1$ and $c_2$ of $G_2$, respectively, where $c_i(v_1) = c_i(v_3)$, $i = 1, 2$. If $c_1(v_2) \neq c_1(v_4)$ and $c_2(v_2) \neq c_2(v_4)$, then we get a 4-coloring of $G$ by combining $c_1$ and $c_2$ (after permuting the colors of $c_2$). We may therefore assume that in every 4-coloring of $G_1$ in which $v_1$ and $v_3$ have the same color, also $v_2$ and $v_4$ have the same color. The same arguments used after identifying $v_2$ and $v_4$ in $G_2$ show that in every 4-coloring of $G_1$ in which $v_2$ and $v_4$ have the same color, also $v_1$ and $v_3$ have the same color. Let $H(a, b)$

be the subgraph of $G_1$ induced by the vertices whose color is either the color $a = c_1(v_1)$ or the color $b$ distinct from $a$ and $c_1(v_2)$. Let $H'$ be the connected component of $H(a, b)$ containing $v_1$. By switching the colors $a$ and $b$ in $H'$ we get another 4-coloring $c_1'$ of $G_1$. (This operation is called the *Kempe change* at $v_1$ of colors $a$ and $b$.) By the above, $b = c_1'(v_1) = c_1'(v_3)$, and hence there is a path $P$ in $H'$ from $v_1$ to $v_3$. Now we can do the Kempe change at $v_2$ of $c_1(v_2)$ and the color distinct from $a, b, c_1(v_2)$. The corresponding subgraph $H'$ is disjoint from $P$. Therefore it cannot contain $v_4$. This contradiction proves that $G$ is 5-connected.

Suppose now that $k = 5$ and that each of $\text{int}(C)$ and $\text{ext}(C)$ contains at least two vertices. Let $\mathcal{C}_i$ be the set of 4-colorings of $C$ which can be extended to a 4-coloring of $G_i$ ($i = 1, 2$). For $1 \le t \le 5$, let $A_t$ (respectively $B_t$) be the set of 4-colorings of $C$ such that $v_{t-1}$ and $v_{t+1}$ (all indices are taken modulo 5) have the same color, and $v_t$ and $v_{t+2}$ have the same color (respectively $v_t, v_{t-2}$, and $v_{t+2}$ have distinct colors). By adding a vertex joined to $v_1, \ldots, v_5$ in $G_i$ and 4-coloring the resulting graph we conclude that $A_t \subseteq \mathcal{C}_i$ for some $t \in \{1, \ldots, 5\}$. By identifying $v_{t-1}$ and $v_{t+1}$ in $G_i$, we see as above that for $t = 1, \ldots, 5$,

$$\mathcal{C}_i \cap (A_t \cup A_{t-1} \cup B_t) \ne \emptyset. \tag{8.1}$$

Since $G$ is not 4-colorable, $\mathcal{C}_1 \cap \mathcal{C}_2 = \emptyset$. Hence we may assume that $A_2 \subseteq \mathcal{C}_1$ and that $A_1 \cap \mathcal{C}_1 = \emptyset$. Extend a coloring of $A_2$ to $G_1$. Let $a$ and $b$ be the colors of $v_2$ and $v_5$, respectively. Let $H'$ be the component of $H(a, b)$ containing $v_2$. After the Kempe change at $v_2$ of colors $a$ and $b$, we get a coloring of $G_1$ whose restriction to $C$ is not in $A_1$. Therefore, $v_5$ is a vertex of $H'$ and there is a path $P$ in $H'$ from $v_2$ to $v_5$. Now we can do the Kempe change at $v_1$ of the color of $v_1$ and the color which does not occur on $C$. The corresponding subgraph $H'$ is disjoint from $P$. Therefore it cannot contain $v_3$ and so the Kempe change does not change the coloring of $C - v_1$. This shows that $B_3 \subseteq \mathcal{C}_1$ and hence $B_3 \cap \mathcal{C}_2 = \emptyset$. By (8.1) and since $A_2 \cap \mathcal{C}_2 = \emptyset$ we conclude that $A_3 \subseteq \mathcal{C}_2$.

We have shown that the assumptions $A_2 \subseteq \mathcal{C}_1$ and $A_1 \cap \mathcal{C}_1 = \emptyset$ imply that $A_3 \subseteq \mathcal{C}_2$ and $A_2 \cap \mathcal{C}_2 = \emptyset$. Repeating this argument implies that $A_4 \subseteq \mathcal{C}_1$, $A_5 \subseteq \mathcal{C}_2$, $A_1 \subseteq \mathcal{C}_1$. This contradiction proves that $G$ is internally 6-connected. $\qquad\square$

Let $G$ be a 4-connected triangulation of the sphere and let $H$ be a near-triangulation with the outer cycle $C$. Suppose that there is a mapping $p$ which maps vertices of $H$ to vertices of $G$ and edges of $H$ to edges of $G$ such that the incidence is preserved, i.e., for each edge $uv \in E(H)$, the ends of $p(uv)$ are the vertices $p(u)$ and $p(v)$.[3] Suppose, furthermore, that every facial triangle of $H$ is mapped to a facial triangle of $G$ and that $p$ is 1-1 on the edges $E(H) \backslash E(C)$. Then we say that $G$ *contains* $H$. Denote by

---

[3]Such maps are called *graph homomorphisms*.

$\mathrm{Int}_p(C)$ the subgraph of $G$ with vertex set $p(V(H))$ and edges $p(E(H))$, and let $\mathrm{Ext}_p(C)$ be the subgraph of $G$ such that $\mathrm{Ext}_p(C) \cup \mathrm{Int}_p(C) = G$ and $\mathrm{Ext}_p(C) \cap \mathrm{Int}_p(C) = p(C)$.

Lemma 8.2.4 and the following two results (proved in [**RSST97**]) imply the Four Color Theorem.

THEOREM 8.2.5 (Robertson, Sanders, Seymour, and Thomas [**RSST97**]). *If $G$ is a minimal counterexample, then $G$ contains none of the 633 near-triangulations shown in Appendix B.*

THEOREM 8.2.6 (Robertson, Sanders, Seymour, and Thomas [**RSST97**]). *Every internally 6-connected triangulation of the sphere contains at least one of the 633 near-triangulations shown in Appendix B.*

The original proof of the Four Color Theorem by Appel and Haken [**AH77, AHK77**] is analogous except that their set of 1834 near-triangulations is different.

The basic method in the proof of Theorem 8.2.5 is the Kempe change argument used first by Kempe [**Ke879**] and Heawood [**He890**] and also used in the proof of Lemma 8.2.4. A near-triangulation $H$ is said to be *reducible* if $H$ cannot be contained in any minimal counterexample to the Four Color Conjecture. Let $C$ be the outer cycle of $H$. Suppose that $c$ is a 4-coloring of $C$. We say that $c$ is *Kempe extendible* if the following holds: If $H$ is contained in a triangulation $G$ (where $p$ is the corresponding mapping), then for every 4-coloring of $\mathrm{Ext}_p(C)$ whose restriction to $p(C)$ is $c$, there is a sequence of Kempe changes in $\mathrm{Ext}_p(C)$ that results in a coloring of $\mathrm{Ext}_p(C)$ whose restriction to $p(C)$ can be extended to $\mathrm{Int}_p(C)$. If every 4-coloring of $C$ is Kempe extendible, then the near-triangulation $H$ is said to be *D-reducible*. Clearly, this implies that $H$ is reducible.

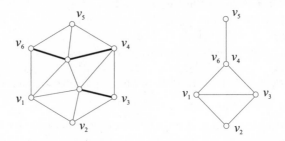

FIGURE 8.1. Contraction of edges for C-reducibility

More generally, we may consider only certain 4-colorings of $C$. For example, in the case of the near-triangulation $H$ of Figure 8.1, after contracting the thick edges and removing the resulting parallel edges, we transform $H$ into the graph $H'$ shown on the right. We may restrict our attention only to 4-colorings of $C$ which are obtained from 4-colorings of $H'$. Such 4-colorings have $v_4$ and $v_6$ colored the same and $v_1, v_3$ colored differently. Again, if every such coloring is Kempe extendible, then $H$ is reducible. The near-triangulation $H$ is called *C-reducible* in this case. The C-reducible near-triangulations shown in Appendix B are distinguished from the D-reducible near-triangulations by a set of thick edges which have to be contracted.

These types of reducibility were used in the computer proof of Theorem 8.2.5 and in the corresponding result of Appel and Haken [**AH89**].

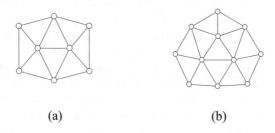

(a)                              (b)

FIGURE 8.2. An unavoidable set of near-triangulations

We now turn to Theorem 8.2.6. Heesch [**He69**] introduced a procedure called *discharging* for generating sets of near-triangulations one of which must occur in every internally 6-connected triangulation $G$ of the sphere. Such a set of near-triangulations is said to be *unavoidable*. We assign to each vertex $v$ of $G$ the number $c(v) = 6 - \deg(v)$, called the *charge* of $v$. Euler's formula implies that the total charge assigned to $G$ is positive (namely, 12). *Discharging* is a rule that redistributes the charge of the vertices of $G$ so that the total charge remains the same. Therefore, there is a vertex with positive charge after the redistribution process. This can be used to prove that $G$ contains certain near-triangulations. To illustrate the technique with a simple and well known example, let us prove that the set of near-triangulations in Figure 8.2 is unavoidable. Assume that $G$ is internally 6-connected. We claim that $G$ has either two adjacent vertices of degree 5 or a triangle with degrees 5, 6, and 6, respectively. Define discharging on $G$ as follows. Send $\frac{1}{3}$ of a unit of charge from each vertex of degree 5 to each of its neighbors of degree 7 or more. Assuming that $G$ contains neither of the near-triangulations of Figure 8.2, each vertex of degree 5 has at least 3 neighbors of degree $\geq 7$. Thus, its charge becomes nonpositive. Vertices of degree 6 retain charge 0. Every vertex of degree $d \geq 7$ has at most $\lfloor d/2 \rfloor$ neighbors of degree 5 because the near-triangulation of Figure 8.2(a) is not present. Therefore its charge remains nonpositive. Hence the total charge is nonpositive. This contradiction proves that one of the configurations in Figure 7.2 must be present. Further information about these techniques is given by Woodall and Wilson [**WW78**].

Table 8.1 shows the 32 discharging rules used by Robertson et al. [**RSST97**]. They should be interpreted as follows. If $G$ is an internally 6-connected triangulation, we say that a rule $T$ of Table 8.1 can be *placed on* $G$ if there is a subgraph of $G$ isomorphic to $T$ such that the degrees in $G$ satisfy the restrictions on the degrees shown by the numbers at the vertices of $T$ (where $d^+$ and $d^-$ means that the degree is at least $d$ and at most $d$, respectively, and the black vertices have degree 5). For each rule $T$ and for each possibility to place $T$ on $G$, we send the charge $\frac{1}{10}$ (or $\frac{2}{10}$ in the case of the first rule) along the edge of $G$ corresponding to the edge of $T$ marked with an arrow. Having the initial charge $c(v) = 6 - \deg(v)$, $v \in V(G)$, this process defines a new set of charges with the same total sum equal to 12. If $v$ is a vertex with positive charge, then the computer-assisted proof of [**RSST97**] shows that $v$ and its first and second neighborhood contains one of the near-triangulations in Appendix B. The authors of [**RSST97**] estimate that the discharging part of the proof could also be checked without a computer in a few months.

The complete proof of the Four Color Theorem is available from ftp://math.gatech.edu/pub/users/thomas/four. It includes the encoding of the unavoidable set of the 633 near-triangulations, the discharging rules, the computer programs, and other helpful information needed for the proof of Theorem 8.2.6.

An important side result of the proof of the Four Color Theorem is a polynomially bounded algorithm to 4-color planar graphs.

THEOREM 8.2.7 (Robertson, Sanders, Seymour, Thomas [**RSST96b**]). *For an arbitrary planar graph of order $n$, a 4-coloring can be found in $O(n^2)$ steps.*

Recall that, in contrast to Theorem 8.2.7, deciding if a planar graph is 3-colorable is **NP**-complete [**GJ79**].

There are many other results and many open problems about colorings of planar graphs. We refer the reader to [**JT95**].

### 8.3. Color critical graphs and the Heawood formula

Heawood [**He890**] proved the following analogue of the Four Color Theorem for general surfaces.

THEOREM 8.3.1 (Heawood [**He890**]). *Let $S$ be a surface with Euler genus $g > 0$ and let $G$ be a graph embedded in $S$. Then*

$$\chi(G) \leq \left\lfloor \frac{7 + \sqrt{1 + 24g}}{2} \right\rfloor.$$    (8.2)

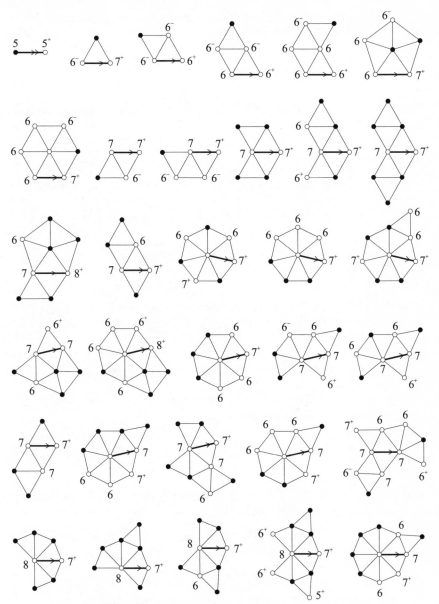

TABLE 8.1. The discharging rules of Robertson, Sanders, Seymour, and Thomas

PROOF. Let $c = \chi(G)$. We may assume that each vertex of $G$ has degree at least $c - 1$. (If $\deg(v) < c - 1$ then $\chi(G - v) = \chi(G)$, so we may reduce the problem to $G-v$.) Proposition 4.4.4 implies that $q \le 3n-6+3g$, where $n = |V(G)|$ and $q = |E(G)|$. Since $2q \ge (c - 1)n$,

$$(c - 7)n + 12 - 6g \le 0. \tag{8.3}$$

Since $g > 0$, the right hand side of (8.2) is at least 6, and so we may assume that $c \ge 7$. Since $n \ge c$, (8.3) implies that $c^2 - 7c + 12 - 6g \le 0$. This proves (8.2). $\qquad\square$

The bound (8.2) holds also for the sphere by the Four Color Theorem. The *list chromatic number* $\chi_l(G)$ of the graph $G$ is the minimum integer $k$ such that $G$ is $k$-choosable. The proof of Theorem 8.3.1 shows that every graph on $S$ has a vertex of degree less or equal to $\frac{1}{2}(5 + \sqrt{1 + 24g})$. Therefore

$$\chi_l(G) \le \left\lfloor \frac{7 + \sqrt{1 + 24g}}{2} \right\rfloor. \tag{8.4}$$

Heawood conjectured that (8.2) is best possible. Theorem 4.4.5 shows that for every surface, except the Klein bottle, there exists a complete graph $G$ for which equality holds in (8.2). Below we discuss this result in more detail in terms of color critical graphs.

Dirac introduced the concept of a $k$-*color-critical* or just a $k$-*critical* graph as follows: The graph $G$ is $k$-*critical* if $\chi(G) = k$ but $\chi(G') < k$ for every proper subgraph of $G$.

If $v$ is a vertex of a $k$-critical graph $G$, then $G-v$ has a $(k-1)$-coloring. Since this coloring cannot be extended to a $(k - 1)$-coloring of $G$, all $k - 1$ colors must be present at the neighbors of $v$. In particular, $v$ has degree at least $k - 1$. If $G$ has $n$ vertices and $q$ edges, then this implies

$$2q \ge (k - 1)n. \tag{8.5}$$

Equality holds for $K_k$. One of Dirac's results on $k$-critical graphs is the following improvement of (8.5).

THEOREM 8.3.2 (Dirac [**Di57b**]). *If $G$ is a $k$-critical graph with $n$ vertices and $q$ edges, and $G \ne K_k$, then*

$$2q \ge (k - 1)n + k - 3. \qquad\square$$

For a short proof see, e.g., Kronk and Mitchem [**KM72**]. We include a proof of the following powerful result of Gallai.

THEOREM 8.3.3 (Gallai [**Ga63a, Ga63b**]). *Let $G$ be a $k$-critical graph and let $H$ be the subgraph of $G$ induced by the vertices of degree $k-1$. Then every block of $H$ is either a complete graph or an odd cycle.*

PROOF. We prove that $H$ has the following property:

(P) If $C$ is an even cycle in $H$ and $v \in V(C)$, then $C$ has a chord incident with $v$.

Suppose that $C = v_1 v_2 \ldots v_p v_1$. Let $c$ be a $(k-1)$-coloring of $G - v_1$. Then the $k - 1$ colors are all present at the neighbors of $v_1$. So, if we move the color of $v_2$ to $v_1$, we obtain a $(k-1)$-coloring of $G - v_2$. Then we move the color of $v_3$ to $v_2$, the color of $v_4$ to $v_3, \ldots$, the color of $v_1$ to $v_p$. As this results in a new $(k-1)$-coloring of $G - v_1$, we conclude that all $k - 1$ colors are present at the neighbors of $v_1$. If there is no chord of $C$ incident with $v_1$, this implies that $\{c(v_p), c(v_2)\} = \{c(v_2), c(v_3)\}$. Repeating this argument, we conclude that $\{c(v_2), c(v_3)\} = \{c(v_3), c(v_4)\}$, etc. Hence $\{c(v_2), c(v_3), \ldots, c(v_p)\}$ has cardinality 2. Since $c(v_2) \neq c(v_p)$, this is impossible if $p$ is even. So $H$ has property (P).

We next prove, by induction, that every block $H$ with property (P) is a complete graph or an odd cycle. This is trivial if $|V(H)| \leq 4$ or if $H$ is a cycle. So assume that $|V(H)| \geq 5$ and that $H$ is not a cycle. Hence $H$ has two vertices joined by three internally disjoint paths. Two of these form an even cycle. That even cycle has at least two chords. Since $H \neq K_4$, we can then find an even cycle $C$ in $H$ such that $V(C) \neq V(H)$. By the induction hypothesis $C$ induces a complete graph in $H$. Let $K$ be a maximal complete graph in $H$. We claim that $K = H$. For otherwise we pick a vertex $v \in V(H) \backslash V(K)$ and let $P_1, P_2$ be two paths from $v$ to $K$ such that $P_1 \cap P_2 = \{v\}$ and $|V(P_1)| + |V(P_2)|$ is smallest possible. We now extend $P_1 \cup P_2$ to an even cycle $C'$ by adding either an edge in $K$ or a vertex in $K$ that is not a neighbor of $v$. Then $C'$ has a chord incident with $v$ and we obtain a contradiction to the minimality of $P_1 \cup P_2$ unless each of $P_1$ and $P_2$ has length 1. But then $v$ is joined to all vertices of $K$, a contradiction to the maximality of $K$.  $\square$

Let $L$ be an assignment of lists $L(v)$ to the vertices of $G$. We say that $G$ is $L$-critical if $G$ is not $L$-colorable but every proper subgraph of $G$ is $L$-colorable. Thomassen [**Th97b**] observed that the proof of Theorem 8.3.3 extends to $L$-critical graphs if each list has cardinality $k - 1$.

COROLLARY 8.3.4 (Brooks [**Br41**]). *Let $G$ be a connected graph and $k \geq 2$ an integer. If all vertices of $G$ have degree at most $k - 1$, then $G$ can be $(k-1)$-colored unless $G$ is the complete graph $K_k$ or an odd cycle (when $k = 3$).*

PROOF. The case $k = 2$ is easy. If $\chi(G) \geq k$, then $G$ contains a $k$-critical subgraph $G'$. As every $k$-critical graph is 2-connected (when $k \geq 3$), $G'$ is 2-connected. By Theorem 8.3.3, $G'$ is an odd cycle (if $k = 3$) or a $K_k$. As $G$ has maximum degree $k - 1$ and $G'$ has minimum degree $k - 1$, $G'$ is a component of $G$. Hence $G$ is an odd cycle or a complete graph.  $\square$

While Theorem 8.3.2 is useful when $n$ is small, Theorem 8.3.6 below is useful for $n$ large. We need the following lemma which is easy to prove by induction on the number of blocks.

LEMMA 8.3.5. *Let $k \geq 4$ be an integer and let $G$ be a graph with $n$ vertices and of maximum degree less than $k$. If each block of $G$ is an odd cycle or a complete graph of order at most $k - 1$, then*

$$|E(G)| \leq \left( \frac{k-2}{2} + \frac{1}{k-1} \right) n - 1 \,. \qquad \Box$$

THEOREM 8.3.6 (Gallai [**Ga63a, Ga63b**]). *Let $k \geq 4$ be an integer. If $G$ is a $k$-critical graph with $n$ vertices and $q$ edges, and $G \neq K_k$, then*

$$2q \geq (k-1)n + \frac{k-3}{k^2-3}n + \frac{2(k-1)}{k^2-3} \,.$$

PROOF (DUE TO G. A. DIRAC). Let $H$ be the subgraph of $G$ induced by the vertices of degree $k - 1$. Put $n' = |V(H)|$. Counting the degrees in $G$ of the vertices in $H$ we get

$$q \geq (k-1)n' - |E(H)| \geq (k-1)n' - \left( \frac{k-2}{2} + \frac{1}{k-1} \right) n' + 1 \qquad (8.6)$$

where the second inequality follows from Lemma 8.3.5.

Counting all the degrees in $G$ yields

$$2q \geq (k-1)n' + k(n-n') = kn - n' \,. \qquad (8.7)$$

Multiplying both sides of (8.7) by $k/2 - 1/(k-1)$ and adding the resulting inequality to (8.6) gives

$$\left( k - \frac{2}{k-1} + 1 \right) q \geq \left( \frac{k}{2} - \frac{1}{k-1} \right) kn + 1$$

which is equivalent to the inequality of the theorem. $\qquad \Box$

Put $H(g) = \lfloor \frac{1}{2}(7 + \sqrt{1+24g}) \rfloor$. Heawood's theorem asserts that every graph $G$ on a surface of Euler genus $g$ has chromatic number at most $H(g)$. Dirac proved the following extension for all values of $g$ except 1 and 3 which were done later by Albertson and Hutchinson.

THEOREM 8.3.7 (Dirac [**Di52, Di57a**], Albertson, Hutchinson [**AH79**]). *If $G$ is a graph embedded in the surface of Euler genus $g \geq 1$, then $\chi(G) < H(g)$ unless $G$ contains the complete graph of order $H(g)$ as a subgraph.*

PROOF. We present Dirac's proof which works for $g \neq 1, 3$. Assume that $h := \chi(G) = H(g)$, and let $H$ be an $h$-critical graph in $G$. Let $n = |V(H)|$ and $q = |E(H)|$. Then

$$q \leq 3n - 6 + 3g \qquad (8.8)$$

by Proposition 4.4.4. Assume that $H \neq K_h$. By Theorem 8.3.2,

$$2q \geq (h-1)n + h - 3.$$

Hence

$$(h-1)n + h - 3 \leq 6n - 12 + 6g.$$

Since $g \geq 2$, we have $h \geq 7$. Since $n \geq h + 1$,

$$(h-7)(h+1) \leq 6g - 9 - h. \tag{8.9}$$

It can be verified that (8.9) leads to a contradiction for $g = 2$ and $4 \leq g \leq 10$. (We leave the calculation to the reader.) If $g \geq 11$, then $h \geq 11$ and, therefore,

$$\frac{(h-3)(h+4)}{h^2 - 3} > 1.$$

Consequently, because of (8.9), $h^2 - 3h - 15 - 6g < 0$ implying that

$$h < \frac{3 + \sqrt{24g + 69}}{2}.$$

On the other hand, $h = H(g) \geq \frac{1}{2}(5 + \sqrt{24g+1})$. Hence $2 + \sqrt{24g+1} < \sqrt{24g+69}$. This implies that $4\sqrt{24g+1} < 64$ and, therefore, $g < 255/24 < 11$, a contradiction. $\square$

Since the complete graph $K_7$ cannot be embedded in the Klein bottle (cf. Theorem 4.4.6), Theorem 8.3.7 implies in particular that the chromatic number of graphs which can be embedded in the Klein bottle is at most 6.

Škrekovski [**Šk99p**] considered $(H(g)-1)$-critical graphs on a surface of Euler genus $g$ and proved that for $g \geq 10$ the only such graphs are $K_{H(g)-1}$ and, in some exceptional cases, the join $K_{H(g)-4} + C_5$. Theorem 8.3.7 was extended to list colorings by Böhme, Mohar, and Stiebitz [**BMS99**] (except for the case of Euler genus $g = 3$).

## 8.4. Coloring in few colors

Dirac [**Di57c**] observed the last assertion of Theorem 8.4.1 below which implies that for each fixed surface and each natural number $k \geq 8$, there are only finitely many $k$-critical graphs on the surface.

THEOREM 8.4.1. *Let $S$ be a surface of Euler genus $g \geq 3$. Every 7-critical graph on $S$ has less than $69(g-2)$ vertices, and for $k \geq 8$, every $k$-critical graph on $S$ has at most $6(g-2)/(k-7)$ vertices.*

PROOF. Let $G$ be a $k$-critical graph on $S$, where $k \geq 7$. By (8.8), $6n - 12 + 6g \geq 2|E(G)| \geq (k-1)n$ which implies the last statement of Theorem 8.4.1. If $k = 7$, we apply Theorem 8.3.6. $\square$

COROLLARY 8.4.2. *Let $S$ be a surface and $k \geq 7$ an integer. Then there are only finitely many $k$-critical graphs on $S$. There exists a polynomially bounded algorithm for testing if an arbitrary graph on $S$ is $(k-1)$-colorable.*

PROOF. The first assertion follows from Theorem 8.4.1. The last assertion follows from the first because a graph is $(k-1)$-colorable if and only if it does not contain a $k$-critical subgraph. □

The algorithm in Corollary 8.4.2 for testing $(k-1)$-colorability can be extended to a linear time algorithm which also returns a $(k-1)$-coloring if one exists. Given the graph $G$ on $S$, remove successively its vertices of degree less than $k-1$ until no such vertex exists. If $k=7$, then also remove a vertex of degree 6 if all neighbors have degree 6 and induce a cycle. Then the number of vertices in the remaining graph $H$ is bounded above by a constant depending on $S$ only, and every $(k-1)$-coloring of $H$ can be extended (in linear time) to $G$. Hence $G$ can be $(k-1)$-colored if and only $H$ has a $(k-1)$-coloring (which can be discovered in constant time by checking all possibilities). A similar linear time algorithm was proposed by Edwards [**Ed92**] who also discussed in detail how to perform the removal of vertices in linear time.

Theorem 8.4.1 says that a graph on a fixed surface is 6-colorable unless the graph contains a small obstruction. We now describe a 6-color theorem by Fisk and Mohar [**FM94b**] in terms of the width. Let $\mathbf{ew}_1(G)$ denote the length of a shortest surface nonseparating cycle of the embedded graph $G$.

LEMMA 8.4.3. *Let $G$ be a connected graph embedded in a surface $S$ of Euler genus $g$. Denote by $\delta$ the minimum vertex degree of $G$. Suppose that one of the following conditions is satisfied:*

(a) *$\delta \geq 6$ and each vertex of degree 6 is either contained in a nontriangular face, or has a neighbor of degree 7 or more.*

(b) *$\delta \geq 4$, the girth[4] of $G$ is at least 4, and each vertex $v$ of degree 4 is either contained in a nonquadrangular face, or there is a vertex at distance at most 2 from $v$ whose degree is at least 5.*

(c) *$\delta \geq 3$, the girth of $G$ is at least 6, and each vertex $v$ of degree 3 is either contained in a nonhexagonal face, or there is a vertex at distance at most 3 from $v$ whose degree is at least 4.*

*Then $\mathbf{ew}_1(G) \leq c \log g$, where $c$ is a universal constant.*

PROOF. It is easy to see that a graph $G$ satisfying (a), (b), or (c) is nonplanar and is not projective planar. Therefore $g \geq 2$ and $G$ contains a surface nonseparating cycle by the remark after Lemma 4.2.4.

Fix a vertex $v \in V(G)$. Denote by $B_i$ the set of all vertices of $G$ which have distance at most $i$ from $v$ ($i = 0, 1, 2, \ldots$). Let $G_i$ be the subgraph of $G$ induced by $B_i$. Suppose that $i$ is an integer such that $G_i$ contains a surface nonseparating cycle. Let $C$ be a shortest one. If two vertices on $C$, say $x, y$, are joined by a path $P$ in $G_i$ whose length is smaller than the distance between $x$ and $y$ on $C$, then we may assume

---

[4]The *girth* of a graph is the length of a shortest cycle in the graph.

that $E(P) \cap E(C) = \emptyset$, and hence each of the two cycles $C_1$, $C_2$ of $C \cup P$ different from $C$ is shorter than $C$, and at least one of them is surface nonseparating by the 3-path-property. Therefore $C$ is *isometric* in $G_i$, i.e. the distance in $G_i$ between any two vertices of $C$ is equal to their distance on $C$. Let $x, y$ be (almost) diametrically opposite vertices on $C$. Since their distance from $v$ is at most $i$, their distance in $G_i$ is at most $2i$. Since $C$ is isometric in $G_i$, the length of $C$ is at most $4i + 1$. In particular, $4i + 1 \geq \mathbf{ew}_1(G)$.

The absence of surface nonseparating cycles in $G_i$ for $i < (\mathbf{ew}_1(G) - 1)/4$ implies that the induced embedding of $G_i$ has Euler genus zero. Denote by $v_i$, $q_i$, and $f_i$ the number of vertices, edges, and faces, respectively, of $G_i$ of this induced embedding. Moreover, let $f_{i,j}$ ($j \geq 3$) be the number of facial walks of $G_i$ of length $j$, and let $v_{i,j}$ ($j \geq 1$) be the number of vertices of degree $j$ in $G_i$. Observe that the degrees in $G_i$ of vertices of $B_{i-1}$ are equal to the degrees in $G$. Clearly,

$$v_i = \sum_{j \geq 1} v_{i,j} , \qquad f_i = \sum_{j \geq 3} f_{i,j} \tag{8.10}$$

and

$$2q_i = \sum_{j \geq 1} j v_{i,j} = \sum_{j \geq 3} j f_{i,j} . \tag{8.11}$$

Let us now prove the sufficiency of (a). Using Euler's formula $v_i - q_i + f_i = 2$ and applying (8.10) and (8.11) we get:

$$12 = (6v_i - 2q_i) - (4q_i - 6f_i) \tag{8.12}$$

$$= -\sum_{j \geq 1}(j - 6)v_{i,j} - 2\sum_{j \geq 4}(j - 3)f_{i,j} \tag{8.13}$$

$$\leq 5\sum_{j=1}^{5} v_{i,j} - \sum_{j \geq 7}(j - 6)v_{i,j} - 2\sum_{j \geq 4}(j - 3)f_{i,j} \tag{8.14}$$

$$\leq 5\sum_{j=1}^{5} v_{i,j} - \frac{1}{43}\sum_{j \geq 7}(6j + 1)v_{i,j} - \frac{1}{2}\sum_{j \geq 4} j f_{i,j} . \tag{8.15}$$

To establish (8.15) we use the inequalities $43(j - 6) \geq 6j + 1$ (if $j \geq 7$) and $4(j - 3) \geq j$ (if $j \geq 4$). It follows that the number in (8.15) is positive. Let $t$ denote the number of vertices of degree 6 in $B_{i-1}$. Then (8.12)–(8.15) imply

$$v_i - v_{i-1} \geq \sum_{j=1}^{5} v_{i,j} \tag{8.16}$$

$$> \frac{1}{215} \sum_{j \geq 7} v_{i,j} + \frac{1}{215} \sum_{j \geq 7} 6j v_{i,j} + \frac{1}{10} \sum_{j \geq 4} j f_{i,j} \tag{8.17}$$

$$\geq \frac{1}{215} v_{i-1} - \frac{1}{215} t + \frac{1}{215} \sum_{j \geq 7} 6j v_{i,j} + \frac{1}{10} \sum_{j \geq 4} j f_{i,j} \tag{8.18}$$

If $u \in B_{i-1}$ is a vertex of degree 6 (in $G_i$), let $u'$ be an arbitrary neighbor of $u$ in $B_{i-2}$ (we assume that $i \geq 3$). Then one of the following holds:

(1) $\deg(u') \geq 7$.

(2) $\deg(u') = 6$, all faces containing $u'$ are triangles, but $u'$ has a neighbor of degree 7 or more.

(3) $\deg(u') = 6$ and $u'$ is on a facial walk of $G$ of length 4 or more. Since $u' \in B_{i-2}$, $u'$ is on a facial walk of $G_i$ of length $> 3$.

We say that $u$ is of type $s$ in Case $(s)$, $s = 1, 2, 3$. Let $t_s$ denote the number of vertices in $B_{i-1}$ of degree 6, which are of type $s$. Now clearly

$$\sum_{j \geq 7} 6j v_{i,j} \geq t_1 + t_2 \quad \text{and} \quad \sum_{j \geq 4} 6j f_{i,j} \geq t_3 . \tag{8.19}$$

It follows that the last three terms in (8.18) add up to a nonnegative number. Therefore

$$v_i \geq \frac{216}{215} v_{i-1} . \tag{8.20}$$

The above proof applies only for $3 \leq i < \frac{1}{4}(\mathbf{ew}_1(G) - 1)$ but (8.20) also holds for $i = 2$. So

$$v_i \geq \left( \frac{216}{215} \right)^i \tag{8.21}$$

for $i < \frac{1}{4}(\mathbf{ew}_1(G) - 1)$.

Euler's formula for the surface $S$, combined with the condition (a) imply

$$|V(G)| = O(g). \tag{8.22}$$

Now (8.21) and (8.22) imply that $i = O(\log g)$.

Conditions (b) and (c) are disposed of in a similar way. The details are left to the reader. $\quad\square$

THEOREM 8.4.4 (Fisk and Mohar [**FM94b**]). *There is a universal constant $c$ such that every graph $G$ embedded in a surface of Euler genus $g > 0$ with $\mathbf{ew}_1(G) > c \log g$ is 6-colorable.*

PROOF. Since all projective graphs are 6-colorable, we may assume that $g \geq 2$. Let $c$ be a constant of Lemma 8.4.3 such that $c \log 2 > 3$. If $G$ is not 6-colorable, it contains a 7-critical subgraph $G'$. The minimum vertex degree of $G'$ is at least 6. By Lemma 8.4.3 we may assume that $G'$ contains a vertex $v$ of degree 6 whose neighbors all have degree 6 and the faces containing $v$ are triangles. By Theorem 8.3.3, the blocks of the subgraph of $G'$ induced by the vertices of degree 6 in $G'$ are complete graphs or odd cycles. It follows that the block containing $v$ is $K_7$. Therefore $\mathbf{ew}_1(G) \leq \mathbf{ew}_1(G') = 3 < c \log 2 \leq c \log g$.                                    □

We now turn to 5-colorings.

FIGURE 8.3. The 4-critical graph $H_7$

If $G$ is $k$-critical and $H$ is $l$-critical, then their join[5] $G + H$ is $(k + l)$-critical. Therefore the graphs $C_3 + C_5$, $K_2 + H_7$ (where $H_7$ is the graph of Figure 8.3) are 6-critical. Figure 8.4 shows that they can be embedded in the torus.[6] Thomassen [**Th94d**] characterized those toroidal graphs which are not 5-colorable, a problem raised by Albertson and Hutchinson [**AH80**].

THEOREM 8.4.5 (Thomassen [**Th94d**]). *There are precisely four 6-critical toroidal graphs, namely* $K_6$, $C_3 + C_5$, $K_2 + H_7$ *(where $H_7$ is the graph of Figure 8.3), and the graph $T_{11}$ which is obtained from the 11-cycle* $x_0 x_1 \ldots x_{10} x_0$ *by adding all chords $x_i x_{i+j}$ $(i = 0, \ldots, 10;\ j = 2, 3)$.*

Albertson and Stromquist [**AS82**] proved that every graph $G$ which is embedded in the torus such that $\mathbf{ew}(G) \geq 8$ is 5-colorable. Since all 6-critical graphs in Theorem 8.4.5 contain noncontractible triangles in all their embeddings in the torus, as can easily be checked, we get a stronger result:

COROLLARY 8.4.6. *If a graph $G$ is embedded in the torus such that* $\mathbf{ew}(G) \geq 4$, *then $G$ is 5-colorable.*

Hutchinson [**Hu84b**] had earlier established an analogue of Corollary 8.4.6 for general orientable surfaces. She proved that if $G$ has an embedding in $\mathbb{S}_g$ $(g \geq 1)$ in the standard $4g$-gonal representation of $\mathbb{S}_g$ (see

---

[5]If $G$ and $H$ are graphs, the *join* $G + H$ is the graph obtained from the disjoint union of $G$ and $H$ by adding all edges $uv$ $(u \in V(G),\ v \in V(H))$ between $G$ and $H$.

[6]The black vertices of the second graph correspond to $K_2$.

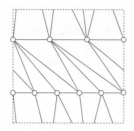
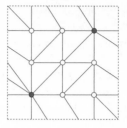

FIGURE 8.4. The graphs $C_3 + C_5$ and $K_2 + H_7$ on the torus

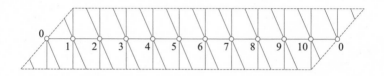

FIGURE 8.5. The graph $T_{11}$ on the torus

Figure 3.4(a)), where the regular $4g$-sided fundamental polygon of $\mathbb{S}_g$ has sides of length 1, such that each edge of $G$ has length at most $\frac{1}{5}$, then $G$ is 5-colorable.

Thomassen [**Th93b**] extended this result (except for the constant $\frac{1}{5}$) by replacing Hutchinson's geometrical condition by a purely combinatorial condition.

THEOREM 8.4.7 (Thomassen [**Th93b**]). *Let $G$ be a graph that is embedded in an orientable surface of genus $g$ such that the edge-width is at least $2^{14g+6}$. Then $G$ is 5-colorable.*

The proof of Theorem 8.4.7 uses Theorem 5.11.1 and an extension of the 5-color theorem for planar graphs. The reader is referred to [**Th93b**] for details.

Theorem 8.4.7 has been extended as follows.

THEOREM 8.4.8 (Thomassen [**Th97b**]). *For each surface $S$, there are only finitely many 6-critical graphs that can be embedded in $S$.*

The complete list of 6-critical graphs in Theorem 8.4.8 is known only for the sphere where there are none, the projective plane, where there is only one, namely $K_6$ (by Theorem 8.3.7), and for the torus (by Theorem 8.4.5).

COROLLARY 8.4.9 (Thomassen [**Th97b**]). *For each fixed surface $S$ there is a polynomially bounded algorithm for deciding if a graph on $S$ can be 5-colored.*

As mentioned earlier, the problem of 3-coloring graphs on the sphere (and hence any other surface) is **NP**-complete, see [**GJ79**]. So, Corollary 8.4.9 suggests the following problem.

PROBLEM 8.4.10. *Let S be a fixed surface. Does there exist a polynomially bounded algorithm for deciding if a graph on S can be 4-colored.*

The answer to Problem 8.4.10 is known only for the sphere, see Theorem 8.2.7. An affirmative answer cannot be obtained in the same way as Corollary 8.4.9 is derived from Theorem 8.4.8. The results below show that there is no 4-color analogue of Theorem 8.4.8.

THEOREM 8.4.11 (Fisk [**Fi78**]). *Let G be a triangulation of some surface. If G has exactly two vertices of odd degree, and these two vertices are adjacent, then G is not 4-colorable.*

The proof of this result follows immediately from the following more general lemma found independently by Ballantine [**Ba30**] and Fisk [**Fi78**]. Ballantine formulated Lemma 8.4.12 for planar triangulations only.

LEMMA 8.4.12 (Ballantine [**Ba30**], Fisk [**Fi78**]). *Let G be a 4-colored triangulation of some surface. Then the number of vertices of odd degree colored 1 has the same parity as the number of odd degree vertices of any other color.*

PROOF. Pick a pair of colors, say 1 and 2, and consider all facial triangles of $G$ with a pair of vertices colored 1 and 2. These triangles are divided into pairs having an edge in common, and the two triangles in every such pair either have the same third color, or one of them is colored 1, 2, 3 the other 1, 2, 4. It follows that the number of facial triangles colored 1, 2, 3 has the same parity as the number of facial triangles colored 1, 2, 4. Since our choice of 1, 2 was arbitrary, we see that also the parity of the number of facial triangles colored with any other triple of colors is the same.

Let $t_i$ $(i = 1, 2, 3, 4)$ be the number of facial triangles containing a vertex colored $i$. By the observation above, the numbers $t_1, t_2, t_3, t_4$ have the same parity. On the other hand, $t_i$ is equal to the sum of degrees of vertices colored $i$. This implies the lemma.  □

Note that the proof of Lemma 8.4.12 also works for every 4-colored graph in which each edge is in precisely two triangles.

COROLLARY 8.4.13. *Let S be a surface distinct from the sphere, and let $k \in \{3, 4, 5\}$. Then there are infinitely many k-critical graphs on S.*

PROOF. The odd cycles are 3-critical and the wheels of even order are 4-critical. Fisk's examples in Theorem 8.4.11 give rise to triangulations of $S$ which are not 4-colorable and are of arbitrarily large edge-width. Each such triangulation $G$ contains a 5-critical subgraph $G'$. As $G'$ is nonplanar (by the Four Color Theorem) we have

$$|V(G')| \geq \mathbf{ew}(G') \geq \mathbf{ew}(G)$$

and hence there are infinitely many 5-critical graphs on $S$.          □

Stromquist [**St79**] asked if the 5-critical graphs of large edge-width on orientable surfaces are precisely the examples by Fisk. Thomassen [**Th94d**] found other examples. More generally, Mohar proved that Fisk's examples are never 5-critical when the edge-width and connectivity are at least 5. To see this, let $M$ be a graph on $S$ such that all vertices have even degree and all faces except one are triangles and every 3-cycle in $M$ is facial. The exceptional face of $M$ is bounded by a cycle of length 5. Then $M$ is not 4-colorable and no 5-critical subgraph of $M$ is a triangulation. Such graphs $M$ can be constructed by taking a 4-connected triangulation as in Theorem 8.4.11 and then deleting a path of length 2 from $x$ to $y$, where $x$ and $y$ are the adjacent vertices of odd degree. To see that $M$ is not 4-colorable, assume it has a 4-coloring. Then some vertex, say $u$, of the facial cycle of length 5 has a color which does not appear on other vertices of the 5-cycle. By adding two edges incident with $u$, we obtain a triangulation with a 4-coloring and with precisely two adjacent vertices of odd degree, a contradiction to Theorem 8.4.11. So none of Fisk's examples of edge-width and connectivity at least 5 are 5-critical, and Stromquist's question may be replaced by the following:

PROBLEM 8.4.14 (Thomassen [**Th97b**]). *Does some surface have 5-critical triangulations of arbitrarily large edge-width?*

PROBLEM 8.4.15 (Albertson [**Al81**]). *Let $S$ be any surface. Does there exist a natural number $q = q(S)$ such that any graph $G$ on $S$ contains a set $A$ of at most $q$ vertices such that $G - A$ is 4-colorable?*

Problem 8.4.15 is open even for the torus where possibly $q = 3$ will do as conjectured by Albertson [**Al81**].

## 8.5. Graphs without short cycles

In this section we discuss the chromatic number of graphs on a fixed surface when conditions on the girth are imposed on the graphs.

THEOREM 8.5.1 (Grötzsch [**Gr59**]). *Every planar graph of girth at least 4 is 3-colorable.*

Thomassen [**Th94b**] gave a short proof of Grötzsch's theorem. The difficult part of the proof, which is conducted by induction, is when the graph has girth 5. Therefore Theorem 8.5.1 can also be derived quickly from the following list color theorem.

THEOREM 8.5.2 (Thomassen [**Th95**]). *Every planar graph of girth 5 is 3-choosable.*

Voigt [**Vo95**] gave an example of a planar graph of girth 4 which is not 3-choosable.

Inspired by Theorems 8.5.1 and 8.4.7, Hutchinson proved the following.

THEOREM 8.5.3 (Hutchinson [**Hu95**]). *For every positive integer $g$ there exists a number $f(g)$ such that the following holds: If $G$ is embedded in the orientable surface $\mathbb{S}_g$ of genus $g$ such that $\mathbf{ew}(G) > f(g)$ and all facial walks have even length, then $G$ is 3-colorable.*

While the 5-color result in Theorem 8.4.7 holds both for orientable and nonorientable surfaces, the 3-color result of Hutchinson does not extend to nonorientable surfaces because of the following counter-intuitive result of Youngs.

THEOREM 8.5.4 (Youngs [**Yo96**]). *If $G$ is a connected loopless multi-graph embedded in the projective plane such that every facial walk is of length 4, then $G$ has chromatic number 2 or 4. In other words, if $G$ is not bipartite, then $G$ is not 3-colorable.*

PROOF. Suppose (reductio ad absurdum) that $G$ is not bipartite, that it is 3-colored, and that $|V(G)|$ is minimum subject to these conditions. Suppose first that $G$ has a facial walk $xyzwx$ which is a 4-cycle. Assume without loss of generality that $x$ and $z$ have the same color. If $x \neq z$, then we delete the edges $xy$ and $xw$, and identify $x$ and $z$. The resulting multigraph is a loopless nonbipartite 3-colored quadrangulation of the projective plane, a contradiction to the minimality of $G$.

So we may assume that every facial walk has only three (or two) distinct vertices. Again, let $F = xyzwx$ be a facial walk and assume that $x = z$. Then there is a simple closed curve $C$ in $F$ which has precisely $x$ in common with $G$ and which has $y$ and $w$ on distinct sides. If $C$ is contractible, then $x$ is a cutvertex of $G$. We choose the notation such that $y$ is in the interior of $C$. The subgraph of $G$ in the interior of $C$ is bipartite. Now we delete that part of the graph and also remove one of the edges between $x$ and $w$. The resulting nonbipartite graph contradicts the minimality of $G$. So, we may assume that $C$ is noncontractible. As no facial walk is a cycle, such a curve $C$ can be chosen in any other face as well. Since the projective plane has no two disjoint noncontractible curves, it follows that any such curve contains $x$. Consequently, any edge of $G$ is incident with $x$, contradicting the assumption that $G$ is nonbipartite. $\square$

Theorem 8.5.4 does not extend to graphs on the torus. This follows from Theorem 8.5.3 or just by taking the Cartesian product of two cycles. An easy modification of this example also shows that Theorem 8.5.4 does not extend to graphs on the Klein bottle. However, Klavžar and Mohar [**KM95**, Theorem 2.4] proved that "twisted Cartesian products" of two cycles of arbitrarily large edge-width on the Klein bottle sometimes need four colors. Recently, Mohar and Seymour [**MS99p**] obtained a counterpart of Theorem 8.5.3 for nonorientable surfaces by proving that every non-3-colorable graph of large edge-width with all facial walks even contains a quadrangulation which is not 3-colorable, and then characterizing completely the non-3-colorable quadrangulations.

Using the proof of Theorem 8.5.1 in [**Th94b**], Gimbel and Thomassen extended Theorem 8.5.4 to obtain a complete characterization of those graphs in the projective plane which have girth at least 4 and are not 3-colorable.

THEOREM 8.5.5 (Gimbel and Thomassen [**GT97**]). *Let $G$ be a graph of girth at least 4 in the projective plane. Then $G$ is 3-colorable if and only if $G$ does not contain a nonbipartite quadrangulation of the projective plane.*

Horst Sachs has informed us that Theorem 8.5.5 was conjectured by Rademacher [**Ra74**].

Gimbel and Thomassen [**GT97**] also gave an upper bound for the chromatic number of triangle-free graphs on a fixed surface, analogous to Theorem 8.3.1 except that the bound is not sharp. For simplicity we formulate it for orientable surfaces only.

THEOREM 8.5.6 (Gimbel and Thomassen [**GT97**]). *There exist positive constants $c_1, c_2$ such that the following statements hold:*

(1) *Every triangle-free graph of genus $g$ has chromatic number at most $c_2 \sqrt[3]{g/\log g}$.*

(2) *For each $g \geq 1$, there exists a triangle-free graph of genus $g$ and chromatic number at least $c_1 \sqrt[3]{g}/\log g$.*

PROOF. Erdős [**Er61**] proved the existence of a triangle-free graph of order $\lfloor \sqrt[3]{g^2} \rfloor$ having at most $g$ edges and the independence number[7] less than $c\sqrt[3]{g}\log g$. Clearly, such a graph has genus less than $g$. Since the vertices of each color form an independent vertex set, dividing the order of $G$ by the independence number establishes the lower bound.

In [**AKS80, AKS81**] it is proved that there exists a constant $c_1$ such that any triangle-free graph of order $n$ contains an independent set of order at least $c_1 \sqrt{n \log n}$. Suppose that $G$ is a triangle-free graph of genus $g$. Let us set $s = \sqrt[3]{g/\log g}$. We wish to show the existence of a constant $c_2$ such that $\chi(G) \leq c_2 s$. Now, successively remove from $G$ vertices of degree less than $s$. Let $H$ denote the graph that remains. Then $|E(H)| \geq$

---
[7]A vertex set $U \subseteq V(G)$ is called *independent* if no two vertices in $U$ are adjacent. The *independence number* of the graph $G$ is the maximum cardinality of an independent set in $G$.

$s|V(H)|/2$. Note that $G - H$ can be colored with $s$ colors. Assuming $H$ is nonempty, we will color $H$ with at most $c_3 s$ colors. By Euler's formula, $|E(H)| \leq 3|V(H)| - 6 + 6g$. Hence, $|V(H)| < c_4 \sqrt[3]{g^2 \log g} =: w$. Suppose that some triangle-free graph has order $t$, where $w/4^{m+1} \leq t < w/4^m$. Then it must contain an independent set of order at least

$$c_1 \frac{\sqrt{w}}{2^{m+1}} \sqrt{\log w - (m+1) \log 4}$$

by [**AKS80, AKS81**]. After assigning this set a color, let us remove it and repeat the process until at most $w/4^{m+1}$ vertices remain. In doing so, we have used at most

$$\frac{\sqrt{w}}{c_1 2^{m-1} \sqrt{\log w - (m+1) \log 4}}$$

colors. Let us apply this process to $H$, increasing $m$ until $w/4^m \leq s$. Such a process generates at most

$$\sum_{m=0}^{M} \frac{\sqrt{w}}{c_1 2^{m-1} \sqrt{\log w - (m+1) \log 4}}$$

color classes, where $M = \lceil \log_4(w/s) \rceil + 1$. But this sum is bounded above by $\frac{8}{c_1} \sqrt{w / \log w}$. Furthermore, this expression is less than some multiple of $s$.

At this point, at most $s$ vertices are uncolored. We may assign each a distinct color, and in doing so, we color $G$ with fewer than some multiple of $s$ colors. $\square$

Let $G$ be a 2-connected graph of girth $m$ with $n$ vertices and $q$ edges embedded on a surface of Euler genus $g$. If $f$ denotes the number of facial walks, then

$$2q \geq mf.$$

Combined with Euler's formula this implies that

$$q(1 - \tfrac{2}{m}) \leq n - 2 + g.$$

This inequality combined with Theorem 8.3.6 implies

THEOREM 8.5.7 (Fisk, Mohar [**FM94b**]; Gimbel, Thomassen [**GT97**]).

(1) *If $S$ is a surface and $k \geq 5$, then there are only finitely many $k$-critical, triangle-free graphs on $S$.*

(2) *If $S$ is a surface and $k \geq 4$, then there are only finitely many $k$-critical graphs of girth at least 6 on $S$.*

Theorem 8.5.7(1) implies that, for each positive integer $g$, there exists a (smallest) integer $w(g)$ such that any triangle-free graph of edge-width

$\geq w(g)$ on a surface of Euler genus $g$ can be 4-colored. Using Lemma 8.4.3(b), Fisk and Mohar [**FM94b**] proved that

$$w(g) < c \log g$$

where $c$ is defined in Lemma 8.4.3. An analogous result holds for 3-colorings of graphs of girth at least 6.

Theorem 8.5.7 also shows that, for each fixed surface $S$, the chromatic number of a graph of girth 6 on $S$ can be found in polynomial time. Also, it can be decided in polynomial time if a triangle-free graph on $S$ can be 4-colored. This raises the following questions.

PROBLEM 8.5.8 (Gimbel and Thomassen [**GT97**]). *Does there exist a surface $S$ and an infinite family of 4-critical graphs of girth 5 that can be embedded in $S$?*

Thomassen [**Th94b**] proved that there are no such graphs on the torus or the projective plane.

PROBLEM 8.5.9 (Gimbel and Thomassen [**GT97**]). *Let $S$ be any fixed surface. Can the chromatic number of a triangle-free graph on $S$ be found in polynomial time?*

# The minimal forbidden subgraphs for the projective plane

This appendix contains drawings of the 103 minimal forbidden subgraphs for the embeddability in the projective plane found by Glover, Huneke, and Wang [**GHW79**]. Following [**GHW79**], the graphs are listed according to their *cyclomatic number* $\alpha(G) = |E(G)| - |V(G)| + \omega(G)$ where $\omega(G)$ denotes the number of components of $G$. The graphs denoted by $A_i$ ($1 \leq i \leq 5$) have $\alpha(G) = 12$, $B_i$ ($1 \leq i \leq 11$) have $\alpha(G) = 11$, etc. The graphs with the same cyclomatic number are ordered lexicographically according to their degree sequence (the nonincreasing sequence of their vertex-degrees). For each of the graphs, the numbers of its vertices and edges, respectively, are given in parenthesis.

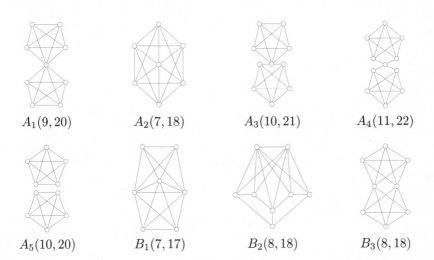

| $A_1(9,20)$ | $A_2(7,18)$ | $A_3(10,21)$ | $A_4(11,22)$ |

| $A_5(10,20)$ | $B_1(7,17)$ | $B_2(8,18)$ | $B_3(8,18)$ |

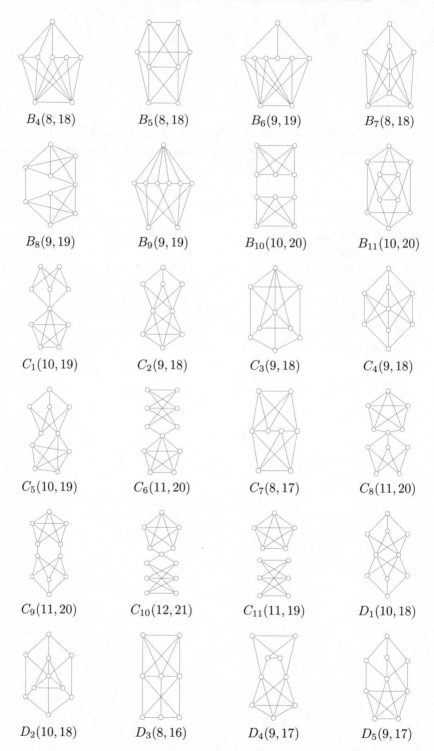

$B_4(8,18)$    $B_5(8,18)$    $B_6(9,19)$    $B_7(8,18)$

$B_8(9,19)$    $B_9(9,19)$    $B_{10}(10,20)$    $B_{11}(10,20)$

$C_1(10,19)$    $C_2(9,18)$    $C_3(9,18)$    $C_4(9,18)$

$C_5(10,19)$    $C_6(11,20)$    $C_7(8,17)$    $C_8(11,20)$

$C_9(11,20)$    $C_{10}(12,21)$    $C_{11}(11,19)$    $D_1(10,18)$

$D_2(10,18)$    $D_3(8,16)$    $D_4(9,17)$    $D_5(9,17)$

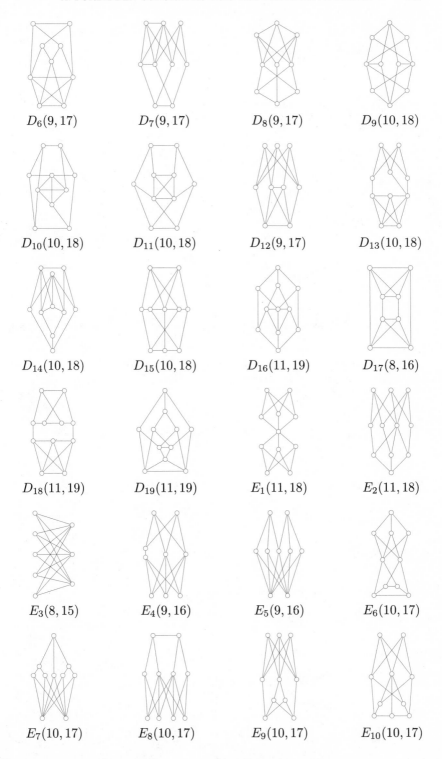

$D_6(9,17)$      $D_7(9,17)$      $D_8(9,17)$      $D_9(10,18)$

$D_{10}(10,18)$      $D_{11}(10,18)$      $D_{12}(9,17)$      $D_{13}(10,18)$

$D_{14}(10,18)$      $D_{15}(10,18)$      $D_{16}(11,19)$      $D_{17}(8,16)$

$D_{18}(11,19)$      $D_{19}(11,19)$      $E_1(11,18)$      $E_2(11,18)$

$E_3(8,15)$      $E_4(9,16)$      $E_5(9,16)$      $E_6(10,17)$

$E_7(10,17)$      $E_8(10,17)$      $E_9(10,17)$      $E_{10}(10,17)$

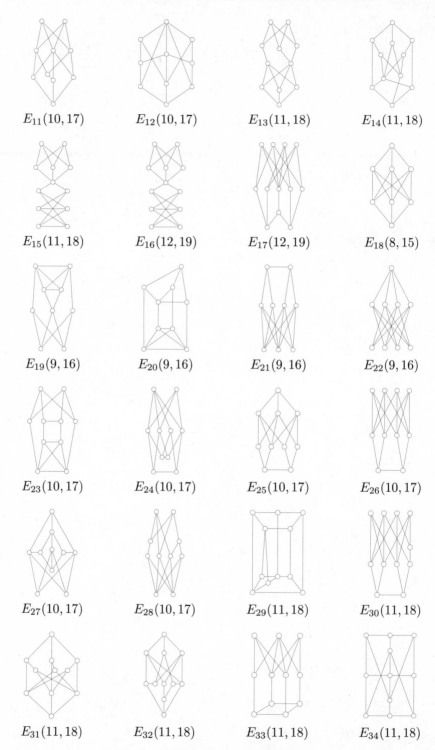

$E_{11}(10,17)$      $E_{12}(10,17)$      $E_{13}(11,18)$      $E_{14}(11,18)$

$E_{15}(11,18)$      $E_{16}(12,19)$      $E_{17}(12,19)$      $E_{18}(8,15)$

$E_{19}(9,16)$      $E_{20}(9,16)$      $E_{21}(9,16)$      $E_{22}(9,16)$

$E_{23}(10,17)$      $E_{24}(10,17)$      $E_{25}(10,17)$      $E_{26}(10,17)$

$E_{27}(10,17)$      $E_{28}(10,17)$      $E_{29}(11,18)$      $E_{30}(11,18)$

$E_{31}(11,18)$      $E_{32}(11,18)$      $E_{33}(11,18)$      $E_{34}(11,18)$

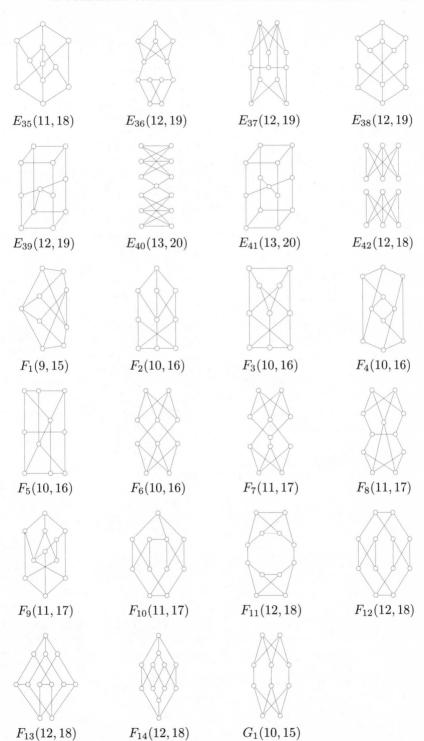

$E_{35}(11,18)$ $E_{36}(12,19)$ $E_{37}(12,19)$ $E_{38}(12,19)$

$E_{39}(12,19)$ $E_{40}(13,20)$ $E_{41}(13,20)$ $E_{42}(12,18)$

$F_1(9,15)$ $F_2(10,16)$ $F_3(10,16)$ $F_4(10,16)$

$F_5(10,16)$ $F_6(10,16)$ $F_7(11,17)$ $F_8(11,17)$

$F_9(11,17)$ $F_{10}(11,17)$ $F_{11}(12,18)$ $F_{12}(12,18)$

$F_{13}(12,18)$ $F_{14}(12,18)$ $G_1(10,15)$

APPENDIX B

# The unavoidable configurations in planar triangulations

The 633 unavoidable near-triangulations used by Robertson, Sanders, Seymour, and Thomas [**RSST97**] in their (new) proof of The Four Color Theorem (cf. Theorem 8.2.6) are presented below. Instead of drawing an entire near-triangulation, the vertices of its outer cycle are removed, and only the resulting subgraph is shown. The vertices are marked as follows:

- • denotes a vertex of degree 5,
- · denotes a vertex of degree 6,
- ○ denotes a vertex of degree 7,
- □ denotes a vertex of degree 8,
- ▽ denotes a vertex of degree 9,
- ⌂ denotes a vertex of degree 10.

Only the very last near-triangulation contains a vertex of degree 11 for which no special symbol has been introduced. Some of the drawings contain a cutvertex. In all such cases, precisely two edges incident with the cutvertex $w$ are not shown. Each of these edges joins $w$ to the outer cycle and, in the clockwise ordering around $w$, they are nonconsecutive and occur between the edges of the two blocks. These conventions determine uniquely the near-triangulation corresponding to each of the drawings. For example, the drawing

determines the following near-triangulation:

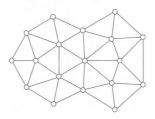

253

In the case of a C-reducible near-triangulation (cf. p. 229), the contracted edges used in proving C-reducibility are shown by thick lines. In the case when such an edge $uv$ has one of its ends (say $v$) on the outer cycle, it is represented as a halfedge at $u$. The other edges joining $u$ with the outer cycle are also displayed as halfedges so that it is possible to determine the position of the edge $uv$ in the clockwise ordering around $u$.

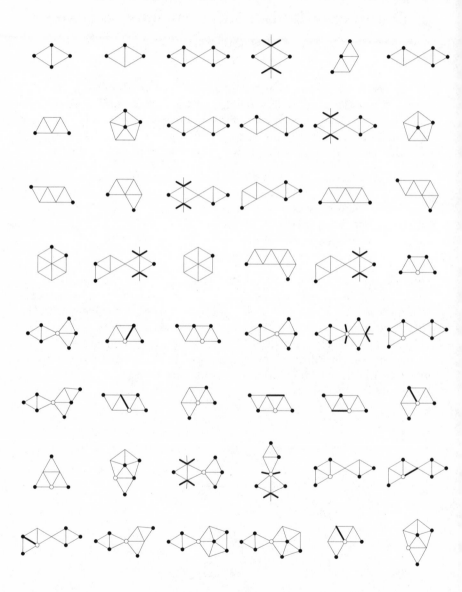

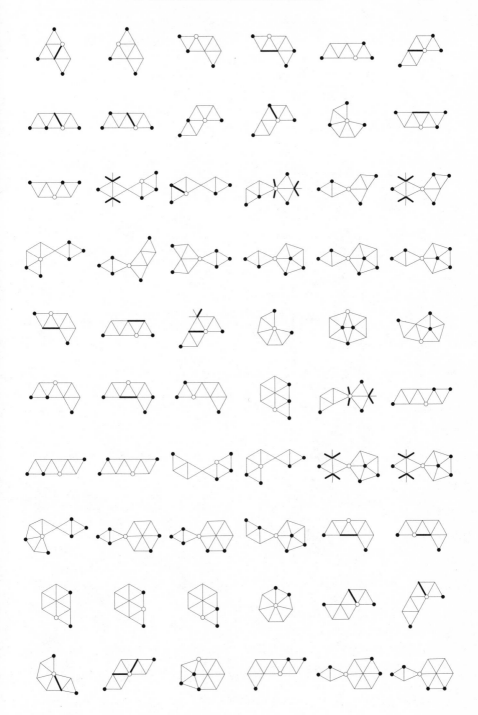

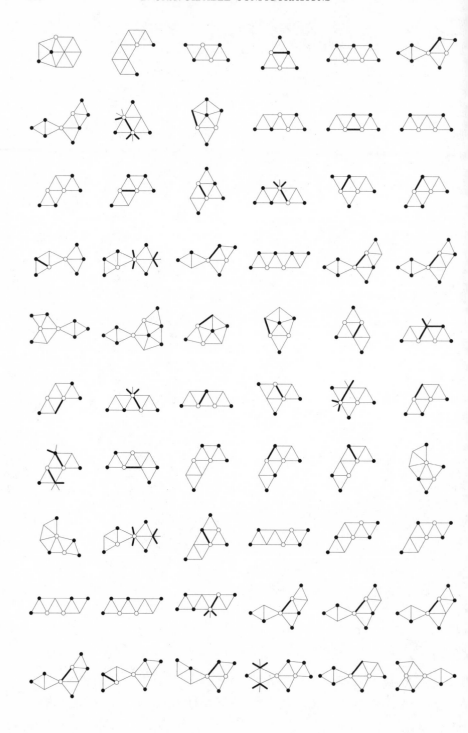

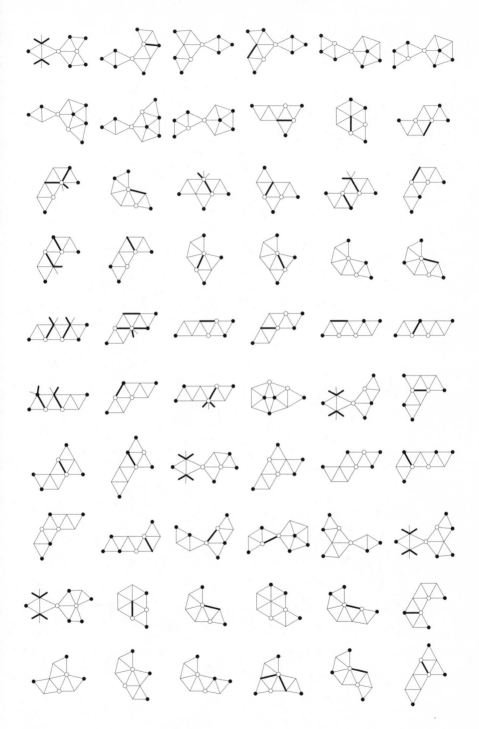

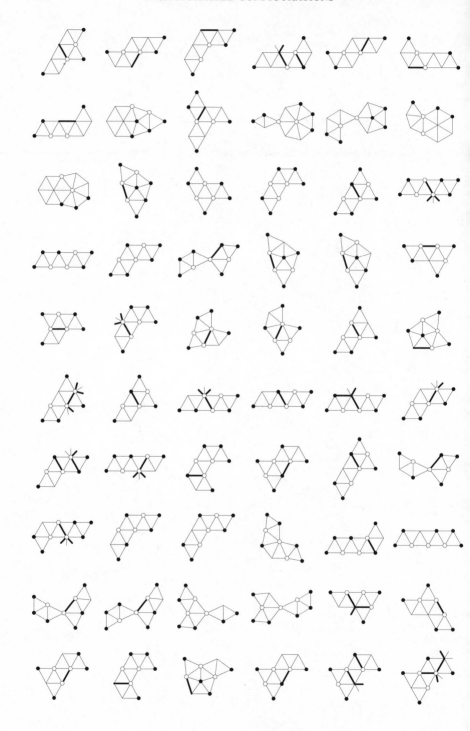

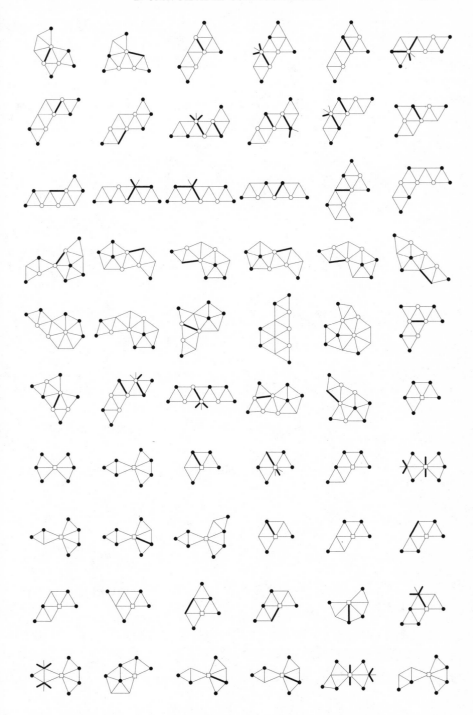

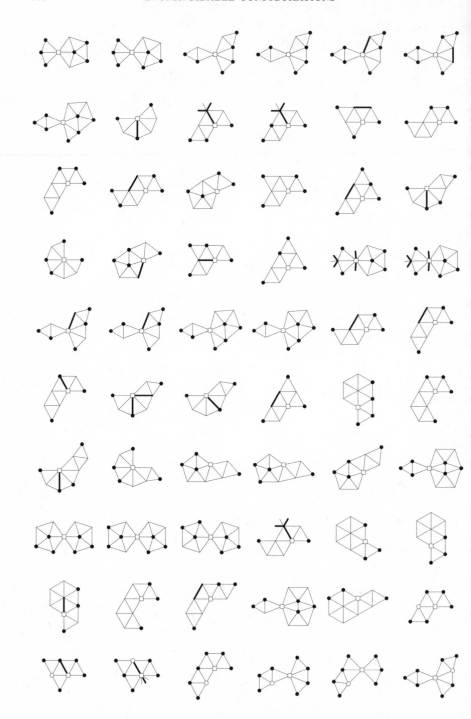

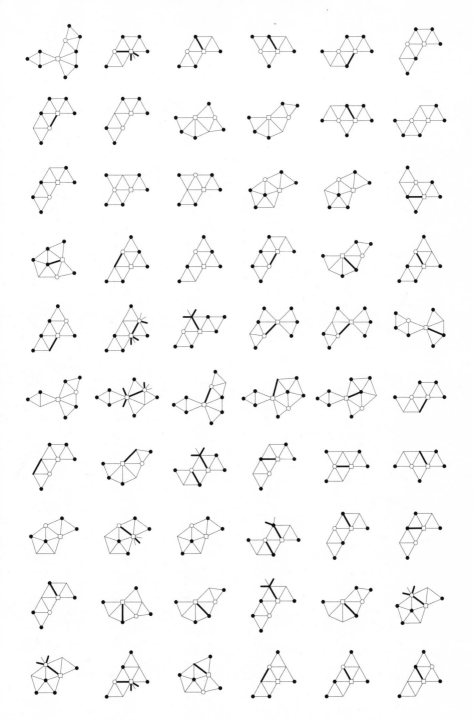

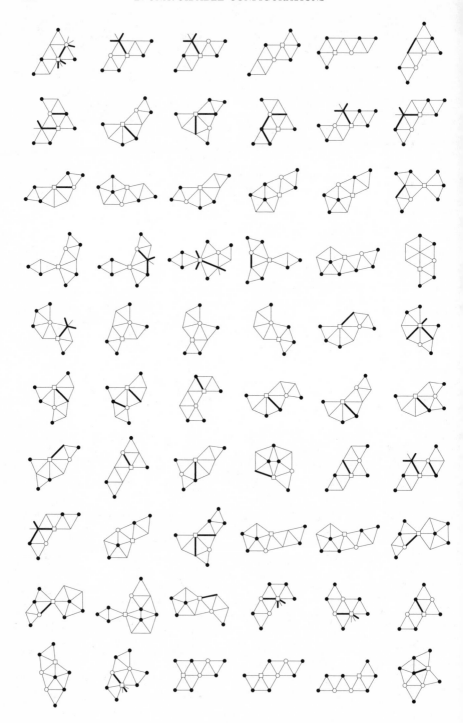

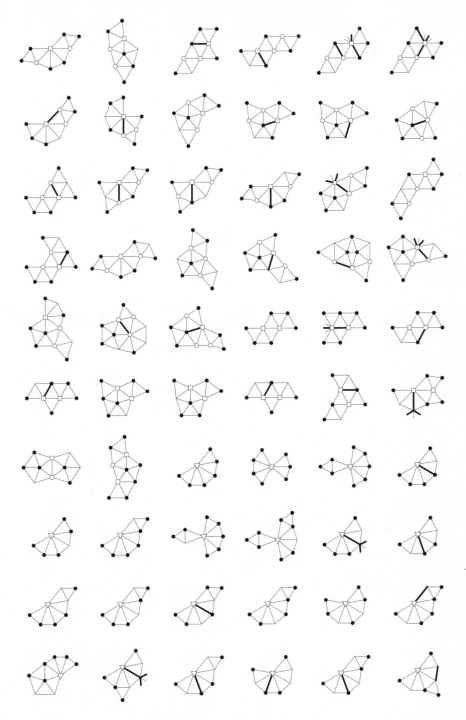

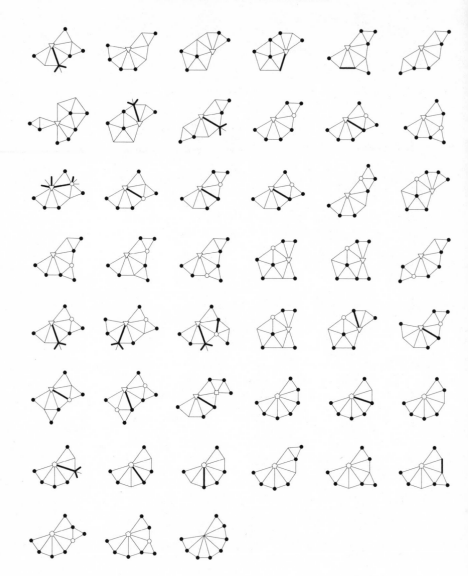

# Bibliography

[Ah90]   D. Aharonov, The hexagonal packing lemma and discrete potential theory, Canad. Math. Bull. 33 (1990) 247–252.

[AHU74]  A. V. Aho, J. E. Hopcroft, J. D. Ullman, The Design and Analysis of Computer Algorithms, Addison-Wesley, Reading, Mass., 1974.

[AKS80]  M. Ajtai, J. Komlós, E. Szemerédi, A note on Ramsey numbers, J. Combin. Theory Ser. A 29 (1980) 354–360.

[AKS81]  M. Ajtai, J. Komlós, E. Szemerédi, A dense infinite Sidon sequence, Europ. J. Combinatorics 2 (1981) 1–11.

[Al81]   M. O. Albertson, Open problem 2, in "The Theory and Applications of Graphs," Ed. G. Chartrand et al., Wiley, 1981, p. 609.

[AH79]   M. O. Albertson, J. P. Hutchinson, The three excluded cases of Dirac's mapcolor theorem, Ann. N. Y. Acad. Sci. 319 (1979) 7–17.

[AH80]   M. O. Albertson, J. P. Hutchinson, On 6-chromatic triangulations, Proc. London Math. Soc. 41 (1980) 533–556.

[AS82]   M. O. Albertson, W. R. Stromquist, Locally planar toroidal graphs are 5-colorable, Proc. Amer. Math. Soc. 84 (1982) 449–457.

[AGZ94]  B. Alspach, L. A. Goddyn, C.-Q. Zhang, Graphs with the circuit cover property, Trans. Amer. Math. Soc. 344 (1994) 131–154.

[AZ93]   B. Alspach, C.-Q. Zhang, Cycle covers of cubic multigraphs, Discrete Math. 111 (1993) 11–17.

[Al72]   A. Altshuler, Hamiltonian circuits in some maps on the torus, Discrete Math. 1 (1972) 299–314.

[An70a]  E. M. Andreev, On convex polyhedra in Lobačevskiĭ spaces, Mat. Sb. (N. S.) 81 (1970) 445–478; Engl. transl. in Math. USSR Sb. 10 (1970) 413–440.

[An70b]  E. M. Andreev, On convex polyhedra of finite volume in Lobačevskiĭ space, Mat. Sb. (N. S.) 83 (1970) 256–260; Engl. transl. in Math. USSR Sb. 12 (1970) 255–259.

[AH76a]  K. Appel, W. Haken, Every planar map is four colorable, Bull. Amer. Math. Soc. 82 (1976) 711–712.

[AH76b]  K. Appel, W. Haken, The existence of unavoidable sets of geographically good configurations, Ill. J. Math. 20 (1976) 218–297.

[AH77]   K. Appel, W. Haken, Every planar map is four colorable. Part I: Discharging, Ill. J. Math. 21 (1977) 429–490.

[AH89]   K. Appel, W. Haken, Every Planar Map is Four Colorable, Contemp. Math. 98, AMS, Providence, R.I., 1989.

[AHK77]  K. Appel, W. Haken, J. Koch, Every planar map is four colorable. Part II: Reducibility, Ill. J. Math. 21 (1977) 491–567.

[Ar80]   D. Archdeacon, A Kuratowski theorem for the projective plane, Ph. D. Thesis, The Ohio State University, Ohio, 1980.

[Ar81]   D. Archdeacon, A Kuratowski theorem for the projective plane, J. Graph Theory 5 (1981) 243–246.

266                                    BIBLIOGRAPHY

[Ar84]    D. Archdeacon, Face colorings of embedded graphs, J. Graph Theory 8 (1984)
          387–398.
[Ar86a]   D. Archdeacon, The nonorientable genus is additive, J. Graph Theory 10
          (1986) 363–383.
[Ar86b]   D. Archdeacon, The orientable genus is nonadditive, J. Graph Theory 10
          (1986) 385–401.
[Ar90]    D. Archdeacon, The complexity of the graph embedding problem, in "Topics
          in Combinatorics and Graph Theory," R. Bodendiek and R. Henn (Eds.),
          Physica-Verlag, Heidelberg, 1990, pp. 59–64.
[Ar92]    D. Archdeacon, Densely embedded graphs, J. Combin. Theory Ser. B 54
          (1992) 13–36.
[ABL93]   D. Archdeacon, C. P. Bonnington, C. H. C. Little, Cycles, cocycles and
          diagonals: A characterization of planar graphs, in "Planar Graphs", Ed.
          W. T. Trotter, DIMACS Series Vol. 9, Amer. Math. Soc., Providence, R. I.,
          1993, pp. 1–3.
[AHL96]   D. Archdeacon, N. Hartsfield, C. H. C. Little, Nonhamiltonian triangulations
          with large connectivity and representativity, J. Combin. Theory Ser. B 68
          (1996) 45–55.
[ArH89]   D. Archdeacon, P. Huneke, A Kuratowski theorem for nonorientable surfaces,
          J. Combin. Theory Ser. B 46 (1989) 173–231.
[AR92]    D. Archdeacon, R. B. Richter, Construction and classification of self-dual
          spherical polyhedra, J. Combin. Theory Ser. B 54 (1992) 37–63.
[AS98]    D. Archdeacon, J. Širáň, Characterizing planarity using theta graphs,
          J. Graph Theory 27 (1998) 17–20.
[ABY63]   L. Auslander, I. A. Brown, J. W. T. Youngs, The imbedding of graphs in
          manifolds, J. Math. Mech. 12 (1963) 629–634.
[Ba30]    J. P. Ballantine, A postulational introduction to the four color problem,
          University of Washington Publications in Mathematics, Vol. 2, University of
          Washingthon Press, Seattle, WA, 1930, pp. 1–15.
[Ba66]    D. Barnette, Trees in polyhedral graphs, Canad. J. Math. 18 (1966) 731–736.
[Ba82a]   D. Barnette, Generating the triangulations of the projective plane, J. Com-
          bin. Theory Ser. B 33 (1982) 222–230.
[Ba82b]   D. Barnette, Polyhedral maps on 2-manifolds, in "Convexity and Related
          Combinatorial Geometry (Norman, Okla., 1980)," Dekker, New York, 1982,
          pp. 7–19.
[Ba87]    D. W. Barnette, Generating closed 2-cell embeddings in the torus and the
          projective plane, Discr. Comput. Geom. 2 (1987) 233–247.
[Ba88]    D. W. Barnette, Decomposition theorems for the torus, projective plane and
          Klein bottle, Discrete Math. 70 (1988) 1–16.
[Ba89]    D. W. Barnette, Unique embeddings of simple projective plane polyhedral
          maps, Israel J. Math. 67 (1989) 251–256.
[Ba91a]   D. W. Barnette, Generating projective plane polyhedral maps, J. Combin.
          Theory Ser. B 51 (1991) 277–291.
[Ba91b]   D. W. Barnette, The minimal projective plane polyhedral maps, in "Applied
          Geometry and Discrete Mathematics," Eds. P. Gritzmann and B. Sturmfels,
          Amer. Math. Soc., Providence, R.I., 1991, pp. 63–70.
[Ba92]    D. W. Barnette, 3-trees in polyhedral maps, Israel J. Math. 79 (1992) 251–
          256.
[Ba94]    D. W. Barnette, 2-connected spanning subgraphs of planar 3-connected
          graphs, J. Combin. Theory Ser. B 61 (1994) 210–216.
[BE88]    D. W. Barnette, A. Edelson, All orientable 2-manifolds have finitely many
          minimal triangulations, Israel J. Math. 62 (1988) 90–98.

[BE89]    D. W. Barnette, A. Edelson, All 2-manifolds have finitely many minimal triangulations, Israel J. Math. 67 (1989) 123–128.

[BETT94]  G. Di Battista, P. Eades, R. Tamassia, I. G. Tollis, Algorithms for drawing graphs: An annotated bibliography, Comput. Geom. 4 (1994) 235–282.

[BHKY62]  J. Battle, F. Harary, Y. Kodama, J. W. T. Youngs, Additivity of the genus of a graph, Bull. Amer. Math. Soc. 68 (1962) 565–568.

[BS91]    A. F. Beardon, K. Stephenson, The Schwartz–Pick lemma for circle packings, Ill. J. Math. 35 (1991) 577–606.

[BH65]    L. Beineke, F. Harary, The genus of the $n$-cube, Canad. J. Math. 17 (1965) 494–496.

[BGR94]   E. A. Bender, Z. Gao, L. B. Richmond, Almost all rooted maps have large representativity, J. Graph Theory 18 (1994) 545–555.

[BLW86]   N. L. Biggs, E. K. Lloyd, R. J. Wilson, Graph Theory. 1736–1936, 2nd Edition, Oxford Univ. Press, New York, 1986.

[Bi13]    G. D. Birkhoff, The reducibility of maps, Amer. J. Math. 35 (1913) 114–128.

[BW86]    R. Bodendiek, K. Wagner, A characterization of the minimal basis of the torus, Combinatorica 6 (1986) 245–260.

[BMM99p]  T. Böhme, J. Maharry, and B. Mohar, $K_{a,k}$ minors in graphs of bounded tree-width, preprint, 1999.

[BMS99]   T. Böhme, B. Mohar, and M. Stiebitz, Dirac's map-color theorem for choosability, J. Graph Theory 32 (1999) 327–339.

[BMT98p]  T. Böhme, B. Mohar, and C. Thomassen, On longest cycles in 4-connected embedded graphs, preprint, 1998.

[BM81]    J. A. Bondy, U. S. R. Murty, Graph Theory with Applications, North-Holland, New York, 1981.

[BL95]    C. P. Bonnington, C. H. C. Little, The Foundations of Topological Graph Theory, Springer, New York, 1995.

[BL76]    K. S. Booth, G. S. Lueker, Testing for the consecutive ones property, interval graphs, and graph planarity using $PQ$-tree algorithms, J. Comput. System Sci. 13 (1976) 335–379.

[Bo78a]   A. Bouchet, Orientable and nonorientable genus of the complete bipartite graph, J. Combin. Theory Ser. B 24 (1978) 24–33.

[Bo78b]   A. Bouchet, Triangular imbeddings into surfaces of a join of equicardinal independent sets following an eulerian graph, in "Theory and Applications of Graphs," Y. Alavi and D. R. Lick, Eds., Springer-Verlag, New York, 1978, pp. 86–115.

[Bo82]    A. Bouchet, Constructions of covering triangulations with folds, J. Graph Theory 6 (1982) 57–75.

[BM90]    A. Bouchet, B. Mohar, Triangular embeddings of tensor products of graphs, in "Topics in Combinatorics and Graph Theory", Eds. R. Bodendiek, R. Henn, Physica-Verlag, Heidelberg, 1990, pp. 129–135.

[Br21]    H. R. Brahana, Systems of circuits on two-dimensional manifolds, Ann. Math. 23 (1921) 144–168.

[BS93]    G. R. Brightwell, E. R. Scheinerman, Representations of planar graphs, SIAM J. Disc. Math. 6 (1993) 214–229.

[Br41]    R. L. Brooks, On colouring the nodes of a network, Proc. Cambr. Phil. Soc. 37 (1941) 194–197.

[BEGMR95]  R. Brunet, M. N. Ellingham, Z. Gao, A. Metzlar, R. B. Richter, Spanning planar subgraphs of graphs in the torus and Klein bottle, J. Combin. Theory Ser. B 65 (1995) 7–22.

[BMR96]   R. Brunet, B. Mohar, R. B. Richter, Separating and nonseparating disjoint homotopic cycles in graph embeddings, J. Combin. Theory Ser. B 66 (1996) 201–231.

[BR95]   R. Brunet, R. B. Richter, Hamiltonicity of 5-connected toroidal triangula-
         tions, J. Graph Theory 20 (1995) 267–286.
[BRS96]  R. Brunet, R. B. Richter, J. Širáň, Covering genus-reducing edges by Kura-
         towski subgraphs, J. Graph Theory 22 (1996) 39–45.
[Br89]   V. W. Bryant, Straight line representations of planar graphs, El. Math. 44
         (1989) 64–65.
[Br83]   A. Brøndsted, An Introduction to Convex Polytopes, Springer-Verlag, 1983.
[Bu78]   M. Burstein, Kuratowski-Pontrjagin theorem on planar graphs, J. Combin.
         Theory Ser. B 14 (1978) 228–231.
[CJPX83] O. Chevalier, F. Jaeger, C. Payan, N. H. Xuong, Odd rooted orientations and
         upper-embeddable graphs, in "Combinatorial Mathematics", North-Holland,
         Amsterdam, 1983, pp. 177–181.
[CNAO85] N. Chiba, T. Nishizeki, S. Abe, T. Ozawa, A linear algorithm for embedding
         planar graphs using $PQ$-trees, J. Comput. System Sci. 30 (1985) 54–76.
[CYN84]  N. Chiba, T. Yamanouchi, T. Nishizeki, Linear algorithms for convex draw-
         ings of planar graphs, in "Progress in Graph theory", Eds. J. A. Bondy and
         U. S. R. Murty, Academic Press, Toronto, 1984, pp. 153–173.
[CEMS91] L. H. Clark, R. C. Entringer, J. E. McCanna, L. A. Székely, Extremal
         problems for local properties of graphs, Australasian J. Combin. 4 (1991)
         25–31.
[CL72]   M. Cohn, A. Lempel, Cycle decomposition by disjoint transpositions, J. Com-
         bin. Theory Ser. A 13 (1972) 83–89.
[CV87]   Y. Colin de Verdière, Constructions de laplaciens dont une partie finie du
         spectre est donnée, Ann. Sci. École Norm. Sup. 20 (1987) 599–615.
[CV89]   Y. Colin de Verdière, Empilements de cercles: Convergence d'une méthode
         de point fixe, Forum Math. 1 (1989) 395–402.
[CV90]   Y. Colin de Verdière, Sur un nouvel invariant des graphes et un critère de
         planarité, J. Combin. Theory Ser B 50 (1990) 11–21.
[CV91]   Y. Colin de Verdière, Un principe variationnel pour les empilements des
         cercles, Invent. Math. 104 (1991) 655–669.
[CV93]   Y. Colin de Verdière, On a new graph invariant and a criterion for planarity,
         in "Graph Structure Theory", Ed. N. Robertson and P. Seymour, Contemp.
         Math. 147, Amer. Math. Soc., Providence, R. I., pp. 137–147.
[CS98p]  C. R. Collins, K. Stephenson, A circle packing algorithm, preprint.
[De90]   N. Dean, Distribution of contractible edges in $k$-connected graphs, J. Combin.
         Theory Ser. B 48 (1990) 1–5.
[De78]   R. Decker, On the orientable genus of a graph, Ph. D. Thesis, Ohio State
         University, 1978.
[DGH81]  R. W. Decker, H. H. Glover, J. P. Huneke, The genus of 2-amalgamations of
         graphs, J. Graph Theory 5 (1981) 95–102.
[Di97]   R. Diestel, Graph Theory, Springer-Verlag, New York, 1997.
[DJGT99] R. Diestel, T. R. Jensen, K. Y. Gorbunov, and C. Thomassen, Highly con-
         nected sets and the excluded grid theorem, J. Combin. Theory, Ser. B 75
         (1999) 61–73.
[Di52]   G. A. Dirac, Map colour theorems, Canad. J. Math. 4 (1952) 480–490.
[Di57a]  G. A. Dirac, Short proof of the map colour theorem, Canad. J. Math. 9
         (1957) 225–226.
[Di57b]  G. A. Dirac, A theorem of R. L. Brooks and a conjecture of H. Hadwiger,
         Proc. London Math. Soc. (3) 7 (1957) 161–195.
[Di57c]  G. A. Dirac, Map colour theorems related to the Heawood colour formula,
         J. London Math. Soc. 32 (1957) 436–455.
[Di60]   G. A. Dirac, Generalisations du théorème du Menger, Comptes Rendus Paris
         250, no. 26 (1960) 4252–4253.

[Di64]  G. A. Dirac, Homomorphism theorems for graphs, Math. Ann. 153 (1964) 69–80.

[Di67]  G. A. Dirac, Minimally 2-connected graphs, J. reine angew. Math. 228 (1967) 204–216.

[DSch54]  G. A. Dirac, S. Schuster, A theorem of Kuratowski, Indag. Math. 16 (1954) 343–348.

[DS95a]  T. Dubejko, K. Stephenson, The branched Schwarz lemma: a classical result via circle packing, Michigan Math. J. 42 (1995) 211–234.

[DS95b]  T. Dubejko, K. Stephenson, Circle packing: experiments in discrete analytic function theory, Experiment. Math. 4 (1995) 307–348.

[Du66]  R. A. Duke, The genus, regional number, and Betti number of a graph, Canad. J. Math. 18 (1966) 817–822.

[DH72]  R. A. Duke, G. Haggard, The genus of subgraphs of $K_8$, Israel J. Math. 11 (1972) 452–455.

[DuMi41]  B. Dushnik, E. W. Miller, Partially ordered sets, Amer. J. Math 63 (1941) 600–610.

[Ed60]  J. R. Edmonds, A combinatorial representation for polyhedral surfaces, Notices Amer. Math. Soc. 7 (1960) 646.

[Ed65]  J. Edmonds, Minimum partition of a matroid into independent subsets, J. Res. Nat. Bur. Standards Sect. B 69 (1965) 67–72.

[Ed92]  K. Edwards, The complexity of some graph colouring problems, Discrete Appl. Math. 36 (1992) 131–140.

[EOSY95]  Y. Egawa, K. Ota, A. Saito, X. Yu, Non-contractible edges in a 3-connected graph, Combinatorica 15 (1995) 357–364.

[ES91]  Y. Egawa, A. Saito, Contractible edges in nonseparating cycles, Combinatorica 11 (1991) 389–392.

[EG94]  M. N. Ellingham, Z. Gao, Spanning trees in locally planar triangulations, J. Combin. Theory Ser. B 61 (1994) 178–198.

[Er61]  P. Erdős, Graph theory and probability II, Canad. J. Math. 13 (1961) 346–352.

[ERT80]  P. Erdős, A. L. Rubin, H. Taylor, Choosability in graphs, Proceedings of the West Coast Conference on Combinatorics, Graph Theory and Computing, Arcata, California, 5.–7. 9. 1979, Congr. Numer. 26 (1980) 125–157.

[Fa48]  I. Fáry, On straight representations of planar graphs, Acta Sci. Math. Szeged 11 (1948) 229–233.

[FL89]  M. R. Fellows, M. A. Langston, An analogue of the MyHill-Nerode theorem and its use in computing finite-basis characterizations, in "30th Ann. Symp. on Foundations of Comp. Sci.", pp. 520–525, 1989.

[FHRR95]  J. R. Fiedler, J. P. Huneke, R. B. Richter, N. Robertson, Computing the orientable genus of projective graphs, J. Graph Theory 20 (1995) 297–308.

[FMR79]  I. S. Filotti, G. L. Miller, J. Reif, On determining the genus of a graph in $O(v^{O(g)})$ steps, in "Proc. 11th Ann. ACM STOC," Atlanta, Georgia (1979) pp. 27–37.

[Fi78]  S. Fisk, The non-existence of colorings, J. Combin. Theory Ser. B 24 (1978) 247–248.

[FM94a]  S. Fisk, B. Mohar, Surface triangulations with isometric boundary, Discrete Math. 134 (1994) 49–62.

[FM94b]  S. Fisk, B. Mohar, Coloring graphs without short non-bounding cycles, J. Combin. Theory Ser. B 60 (1994) 268–276.

[FMN94]  S. Fisk, B. Mohar, R. Nedela, Minimal locally cyclic triangulations of the projective plane, J. Graph Theory 18 (1994) 25–35.

[Fo74]  J. C. Fournier, Une relation de separation entre cocircuits d'un matroide, J. Combin. Theory Ser. B 16 (1974) 181–190.

[FR82]   H. de Fraysseix, P. Rosenstiehl, A depth-first-search characterization of planarity, Ann. Discrete Math. 13 (1982) 75–80.

[FR85]   H. de Fraysseix, P. Rosenstiehl, A characterization of planar graphs by Trémaux orders, Combinatorica 5 (1985) 127–135.

[FS30]   O. Frink, P. A. Smith, Abstract 179, Bull. Amer. Math. Soc. 36 (1930) 214.

[FGM88]  M. L. Furst, J. L. Gross, L. A. McGeoch, Finding a maximum-genus graph imbedding, J. Assoc. Comput. Mach. 35 (1988) 523–534.

[Ga63a]  T. Gallai, Kritische Graphen. I., Magyar Tud. Akad. Mat. Kutató Int. Közl. 8 (1963) 165–192.

[Ga63b]  T. Gallai, Kritische Graphen. II., Magyar Tud. Akad. Mat. Kutató Int. Közl. 8 (1963) 373–395.

[Ga95]   Z. Gao, 2-connected coverings of bounded degree in 3-connected graphs, J. Graph Theory 20 (1995) 327–338.

[GRT91]  Z. Gao, L. B. Richmond, C. Thomassen, Irreducible triangulations and triangular embeddings on a surface, Research Report CORR 91–07, University of Waterloo, 1991.

[GR94]   Z. Gao, R. B. Richter, 2-walks in circuit graphs, J. Combin. Theory Ser. B 62 (1994) 259–267.

[GRS96]  Z. Gao, R. B. Richter, P. D. Seymour, Irreducible triangulations of surfaces, J. Combin. Theory Ser. B 68 (1996) 206–217.

[GRY95]  Z. Gao, R. B. Richter, X. Yu, 2-walks in 3-connected planar graphs, Australas. J. Combin. 11 (1995) 117–122.

[GW94]   Z. Gao, N. C. Wormald, Spanning Eulerian subgraphs of bounded degree in triangulations, Graphs Combin. 10 (1994) 123–131.

[GJ79]   M. R. Garey, D. S. Johnson, Computers and Intractability: A Guide to the Theory of the NP-Completeness, W. H. Freeman, San Francisco, 1979.

[Ga78]   B. L. Garman, Voltage graph embeddings and the associated block designs, J. Graph Theory 2 (1978) 181–187.

[GGW00p] J. F. Geelen, A. M. H. Gerards, and G. Whittle, Branch width and well-quasi-ordering in matroids, manuscript, 2000.

[Gi82]   R. Giles, Optimum matching forests I: Special weights, Math. Programming 22 (1982) 1–11.

[GT97]   J. Gimbel, C. Thomassen, Coloring graphs with fixed genus and girth, Trans. Amer. Math. Soc. 349 (1997) 4555–4564.

[GH75]   H. H. Glover, J. P. Huneke, Cubic irreducible graphs for the projective plane, Discrete Math 13 (1975) 341–355.

[GH77]   H. Glover, J. P. Huneke, Graphs with bounded valence that do not embed in projective plane, Discrete Math. 18 (1977) 155–165.

[GH78]   H. Glover, J. P. Huneke, The set of irreducible graphs for the projective plane is finite, Discrete Math. 22 (1978) 243–256.

[GHW79]  H. H. Glover, J. P. Huneke, C.-S. Wang, 103 graphs that are irreducible for the projective plane, J. Combin. Theory Ser. B 27 (1979) 332–370.

[GT79]   R. Z. Goldstein, E. C. Turner, Applications of topological graph theory to group theory, Math. Z. 165 (1979) 1–10.

[Gr94]   M. de Graaf, Graphs and curves on surfaces, Ph. D. Thesis, University of Amsterdam, Amsterdam, 1994.

[GS94]   M. de Graaf, A. Schrijver, Grid minors of graphs on the torus, J. Combin. Theory Ser. B 61 (1994) 57–62.

[GA73]   J. L. Gross, S. R. Alpert, Branched coverings of graph imbeddings, Bull. Amer. Math. Soc. 79 (1973) 942–945.

[GT87]   J. L. Gross, T. W. Tucker, Topological Graph Theory, Wiley – Interscience, New York, 1987.

[Gr59]   H. Grötzsch, Ein Dreifarbensatz für dreikreisfreie Netze auf der Kugel, Wiss. Z. Martin Luther-Univ. Halle Wittenberg, Math.-Nat. Reihe 8 (1959) 109–120.

[Gr67]   B. Grünbaum, Convex Polytopes, Interscience, 1967.

[Gr69]   B. Grünbaum, Conjecture 6, in "Recent Progress in Combinatorics", Ed. W. T. Tutte, Academic Press, New York, 1969, p. 343.

[Gr70]   B. Grünbaum, Polytopes, graphs and complexes, Bull. Amer. Math. Soc. 76 (1970) 1131–1201.

[GS87]   B. Grünbaum, G. C. Shephard, Some problems on polyhedra, J. Geometry 29 (1987) 182–190.

[Ha77]   G. Haggard, Edmonds' characterization of disc embeddings, Congr. Numer. 19 (1977) 291–302.

[Ha69]   R. Halin, Zur Theorie der n-fach zusammenhängenden Graphen, Abh. Math. Sem. Hamburg Univ. 33 (1969) 133–164.

[Ha43]   D. W. Hall, A note on primitive skew curves, Bull. Amer. Math. Soc. 49 (1943) 935–937.

[HS93]   Zh.-X. He, O. Schramm, Fixed points, Koebe uniformization, and circle packings, Ann. Math. 137 (1993) 369–406.

[HS95]   Zh.-X. He, O. Schramm, Koebe uniformization for "almost circle domains", Amer. J. Math. 117 (1995) 653–667.

[He890]  P. J. Heawood, Map-colour theorem, Quart. J. Pure Appl. Math. 24 (1890) 332–338.

[He69]   H. Heesch, Untersuchungen zum Vierfarbenproblem, B. I. Hochschulscripten 810/810a/810b, Bibliographisches Institut, Mannheim, Wien, Zürich, 1969.

[He891]  L. Heffter, Über das Problem der Nachbargebiete, Math. Ann. 38 (1891) 477–508.

[Hl97]   A. L. Hlavacek, 9-Vertex Irreducible Graphs on the Torus, Ph. D. Thesis, Ohio State University, 1997.

[Hl98]   P. Hliněný, $K_{4,4} - e$ has no finite planar cover, J. Graph Theory 27 (1998) 51–60.

[HR84]   P. Hoffman, B. Richter, Embedding graphs in surfaces, J. Combin. Theory Ser. B 36 (1984) 65–84.

[Ho95]   H. van der Holst, A short proof of the planarity characterization of Colin de Verdière, J. Combin. Theory, Ser. B 65 (1995) 269–272.

[HLS99]  H. van der Holst, L. Lovász, A. Schrijver, The Colin de Verdière graph parameter, in "Graph theory and combinatorial biology" J. Bolyai Math. Soc., Budapest, 1999, pp. 29–85.

[HL77]   D. A. Holton, C. H. C. Little, A new characterization of planar graphs, Bull. Amer. Math. Soc. 83 (1977) 137–138.

[HT73]   J. E. Hopcroft, R. E. Tarjan, Dividing a graph into triconnected components, SIAM J. Comput. 2 (1973) 135–158.

[HT74]   J. E. Hopcroft, R. E. Tarjan, Efficient planarity testing, J. Assoc. Comput. Mach. 21 (1974) 549–568.

[HW41]   W. Hurewicz, H. Wallman, Dimension Theory, Princeton Univ. Press, Princeton, N.J., 1941.

[Hu84a]  J. P. Hutchinson, Automorphism properties of embedded graphs, J. Graph Theory 8 (1984) 35–49.

[Hu84b]  J. P. Hutchinson, A five-color theorem for graphs on surfaces, Proc. Amer. Math. Soc. 90 (1984) 497–504.

[Hu88]   J. P. Hutchinson, On short noncontractible cycles in embedded graphs, SIAM J. Discrete Math. 1 (1988) 185–192.

[Hu89]    J. P. Hutchinson, On genus-reducing and planarizing algorithms for embedded graphs, in "Graphs and Algorithms", Contemp. Math. 89, AMS, Providence, 1989, pp. 19–26.

[Hu95]    J. P. Hutchinson, Three-coloring graphs embedded on surfaces with all faces even-sided, J. Combin. Theory Ser. B 65 (1995) 139–155.

[Ja80]    B. Jackson, Triangular embeddings of $K((i-2)n, n, \ldots, n)$, J. Graph Theory 4 (1980) 21–30.

[JR84]    B. Jackson, G. Ringel, Colorings of circles, Amer. Math. Monthly 91 (1984) 42–49.

[Ja79]    F. Jaeger, Interval matroids and graphs, Discrete Math. 27 (1979) 331–336.

[Ja85]    F. Jaeger, A survey of the cycle double cover conjecture, in "Cycles in Graphs" (B. Alspach and C. Godsil, Eds.), Ann. Discrete Math. 27 (1985) 1–12.

[JT95]    T. Jensen, B. Toft, Graph Coloring Problems, John Wiley, New York, 1995.

[Ju70]    H. A. Jung, Eine Verallgemeinerung des $n$-fachen Zusammenhangs für Graphen, Math. Ann. 187 (1970) 95–103.

[Ju75]    M. Jungerman, The genus of the symmetric quadripartite graph, J. Combin. Theory Ser. B 19 (1975) 181–187.

[JR78]    M. Jungerman, G. Ringel, The genus of the $n$-octahedron: regular cases, J. Graph Theory 2 (1978) 69–75.

[Ju95]    M. Juvan, Algorithms and obstructions for embedding graphs in the torus (in Slovene), Ph. D. Thesis, University of Ljubljana, 1995.

[JMM96]    M. Juvan, A. Malnič, B. Mohar, Systems of curves on surfaces, J. Combin. Theory Ser. B 68 (1996) 7–22.

[JMM95]    M. Juvan, J. Marinček, B. Mohar, Embedding graphs in the torus in linear time, in "Integer programming and combinatorial optimization," E. Balas, J. Clausen (Eds.), Lect. Notes in Computer Science, Vol. 920, Springer, Berlin, 1995, pp. 360–363.

[JMM94]    M. Juvan, J. Marinček, B. Mohar, Obstructions for simple embeddings, preprint, 1994.

[JM95p]    M. Juvan, B. Mohar, Extending 2-restricted partial embeddings of graphs, preprint, 1995.

[Ka90]    A. Karabeg, Classification and detection of obstructions to planarity, Linear and Multilinear Algebra 26 (1990) 15–38.

[KR96]    J. Keir, R. B. Richter, Walks through every edge twice II, J. Graph Theory 21 (1996) 301–310.

[Ke80]    A. K. Kelmans, Concept of a vertex in a matroid and 3-connected graphs, J. Graph Theory 4 (1980) 13–19.

[Ke84a]    A. K. Kelmans, A problem on a strengthening of Kuratowski's planarity criterion, in "Finite and Infinite Sets", Colloq. Math. Soc. J. Bolyai 37, North-Holland, 1984, pp. 881–882.

[Ke84b]    A. K. Kelmans, A strengthening of the Kuratowski planarity criterion for 3-connected graphs, Discrete Math. 51 (1984) 215–220.

[Ke93]    A. K. Kelmans, Graph planarity and related topics, in "Graph Structure Theory", Ed. N. Robertson and P. Seymour, Contemp. Mathematics 147, Amer. Math. Soc., Providence, R. I., 1993, pp. 635–667.

[Ke97]    A. K. Kelmans, Some strengthenings of the Kuratowski planarity criterion, 28th Southeastern International Conference on Combinatorics, Graph Theory and Computing, Boca Raton, FL, 1997.

[Ke879]    A. B. Kempe, On the geographical problem of the four-colors, Amer. J. Math. 2 (1879) 193–200.

[Ke23]    B. Kerékjártó, Vorlesungen über Topologie, Springer, Berlin, 1923.

[KM95]   S. Klavžar, B. Mohar, The chromatic numbers of graph bundles over cycles, Discrete Math. 138 (1995) 301–314.

[Ko36]   P. Koebe, Kontaktprobleme der konformen Abbildung, Ber. Verh. Sächs. Akad. Wiss. Leipzig, Math.–Phys. Kl. 88 (1936) 141–164.

[Kö36]   D. König, Theorie der endlichen und unendlichen Graphen, Akademische Verlagsgesellschaft, 1936; reprinted by Chelsea, New York, 1950.

[KM72]   H. V. Kronk, J. Mitchem, On Dirac's generalization of Brooks' theorem, Canad. J. Math. 24 (1972) 805–807.

[Ku30]   K. Kuratowski, Sur le problème des courbes gauches en topologie, Fund. Math. 15 (1930) 271–283.

[La87a]  S. Lavrenchenko, An infinite set of torus triangulations of connectivity 5 whose graphs are not uniquely embeddable in the torus, Discrete Math. 66 (1987) 299–301.

[La87b]  S. Lawrencenko, The irreducible triangulations of the torus (in Russian), Ukrain. Geom. Sb. 30 (1978) 52–62.

[La92]   S. Lawrencenko, The variety of triangular embeddings of a graph in the projective plane, J. Combin. Theory Ser. B 54 (1992) 196–208.

[LN97]   S. Lawrencenko, S. Negami, Irreducible triangulations of the Klein bottle, J. Combin. Theory Ser. B 70 (1997) 265–291.

[LN99]   S. Lawrencenko, S. Negami, Constructing the graphs that triangulate both the torus and the Klein bottle, J. Combin. Theory Ser. B 77 (1999) 211–218.

[LS90]   J. Lehel, H. Sachs, Problem 16, in: Unsolved problems presented in connection with the Julius Petersen Graph Theory Conference, Eds. B. Toft and J. Bang-Jensen, Odense, 1990, pp. 5–6.

[LEC67]  A. Lempel, S. Even, I. Cederbaum, An algorithm for planarity testing of graphs, in "Theory of Graphs, Int. Symp. Rome, July, 1966," Ed. P. Rosenstiehl, Gordon and Breach, New York, 1967, pp. 215–232.

[Li81]   S. Lins, A minimax theorem on circuits in projective graphs, J. Combin. Theory Ser. B 30 (1981) 253–262.

[Li88]   C. Little, Cubic combinatorial maps, J. Combin. Theory Ser. B 44 (1988) 44–63.

[Lo81]   L. Lovász, The matroid matching problem, in "Algebraic Methods in Graph Theory", North-Holland, Amsterdam, 1981, pp. 495–517.

[ML37]   S. MacLane, A combinatorial condition for planar graphs, Fund. Math. 28 (1937) 22–32.

[Ma99]   W. Mader, An extremal problem for subdivisions of $K_5^-$, J. Graph Theory 30 (1999) 261–276.

[Ma98]   W. Mader, $3n - 5$ edges do force a subdivision of $K_5$, Combinatorica 18 (1998) 569–595.

[Ma97]   Y. Makarychev, A short proof of Kuratowski's graph planarity criterion, J. Graph Theory 25 (1997) 129–131.

[MM92]   A. Malnič, B. Mohar, Generating locally cyclic triangulations of surfaces, J. Combin. Theory Ser. B 56 (1992) 147–164.

[MN95]   A. Malnič, R. Nedela, $k$-Minimal triangulations of surfaces, Acta Math. Univ. Comenian. 64 (1995) 57–76.

[Ma71]   P. Mani, Automorphismen von polyedrischen Graphen, Math. Ann. 192 (1971) 279–303.

[Ma81]   M. L. Marx, Graph imbeddings and branched extensions of curves in orientable surfaces, Congr. Numer. 33 (1981) 205–212.

[Ma67]   W. S. Massey, Algebraic Topology: An Introduction, Harcourt, Brace and World, 1967.

[MC84]   W. McCuaig, A simple proof of Menger's theorem, J. Graph Theory 8 (1984) 427–429.

[Me27]    K. Menger, Zur allgemeinen Kurventheorie, Fund. Math. 10 (1927) 96–115.

[Mi72]    M. Milgram, Irreducible graphs, J. Combin. Theory Ser. B 12 (1972) 6–31.

[Mi73]    M. Milgram, Irreducible graphs – Part 2, J. Combin. Theory Ser. B 14 (1973)
          7–45.

[Mi87]    G. L. Miller, An additivity theorem for the genus of a graph, J. Combin.
          Theory Ser. B 43 (1987) 25–47.

[Mi96]    M. Mirzakhani, A small non-4-choosable planar graph, Bull. ICA 17 (1996)
          15–18.

[Mo88a]   B. Mohar, Nonorientable genus of nearly complete bipartite graphs, Discr.
          Comput. Geom. 3 (1988) 137–146.

[Mo88b]   B. Mohar, Embeddings of infinite graphs, J. Combin. Theory Ser. B 44 (1988)
          29–43.

[Mo89]    B. Mohar, An obstruction to embedding graphs in surfaces, Discrete Math. 78
          (1989) 135–142.

[Mo92]    B. Mohar, Combinatorial local planarity and the width of graph embeddings,
          Canad. J. Math. 44 (1992) 1272–1288.

[Mo93]    B. Mohar, Projective planarity in linear time, J. Algorithms 15 (1993) 482–
          502.

[Mo94]    B. Mohar, Obstructions for the disk and the cylinder embedding extension
          problems, Combin. Probab. Comput. 3 (1994) 375–406.

[Mo94p]   B. Mohar, Universal obstructions for embedding extension problems,
          preprint, 1994.

[Mo95]    B. Mohar, Uniqueness and minimality of large face-width embeddings of
          graphs, Combinatorica 15 (1995) 541–556.

[Mo96]    B. Mohar, Embedding graphs in an arbitrary surface in linear time, Proc.
          28th Ann. ACM STOC, Philadelphia, ACM Press, 1996, pp. 392–397.

[Mo97a]   B. Mohar, Circle packings of maps in polynomial time, Europ. J. Combin.
          18 (1997) 785–805.

[Mo97b]   B. Mohar, Apex graphs with embeddings of face-width three, Discrete Math.
          176 (1997) 203–210.

[Mo97c]   B. Mohar, Face-width of embedded graphs, Math. Slovaca 47 (1997) 35–63.

[Mo98]    B. Mohar, On the orientable genus of graphs with bounded nonorientable
          genus, Discrete Math. 182 (1998) 245–253.

[Mo99]    B. Mohar, A linear time algorithm for embedding graphs in an arbitrary
          surface, SIAM J. Discrete Math. 12 (1999) 6–26.

[Mo98p]   B. Mohar, Face covers and the genus of apex graphs, preprint, 1998.

[Mo00]    B. Mohar, Existence of polyhedral embeddings of graphs, Combinatorica
          (2000).

[Mo01]    B. Mohar, Graph minors and graphs on surfaces, in "Surveys in Combina-
          torics 2001," Ed. J. Hirschfeld, Cambridge Univ. Press, Cambridge, 2001.

[MPP85]   B. Mohar, T. D. Parsons, T. Pisanski, The genus of nearly complete bipartite
          graphs, Ars Combin. 20-B (1985) 173–183.

[MPŠ88]   B. Mohar, T. Pisanski, M. Škoviera, The maximum genus of graph bundles,
          Europ. J. Combin. 9 (1988) 215–224.

[MR93]    B. Mohar, N. Robertson, Disjoint essential circuits in toroidal maps, in "Pla-
          nar Graphs", Ed. W. T. Trotter, Dimacs Series in Discrete Math. and Theor.
          Comp. Sci. 9, Amer. Math. Soc., Providence, R. I., 1993, pp. 109–130.

[MR96a]   B. Mohar, N. Robertson, Planar graphs on nonplanar surfaces, J. Combin.
          Theory Ser. B 68 (1996) 87–111.

[MR96b]   B. Mohar, N. Robertson, Disjoint essential cycles, J. Combin. Theory Ser. B
          68 (1996) 324–349.

[MR98p]   B. Mohar, N. Robertson, Flexibility of polyhedral embeddings of graphs in
          surfaces, preprint, 1998.

[MS99p]   B. Mohar, P. D. Seymour, Coloring locally bipartite graphs on nonorientable surfaces, preprint, 1999.

[MRV96]   B. Mohar, N. Robertson, R. P. Vitray, Planar graphs on the projective plane, Discrete Math. 149 (1996) 141–157.

[Mo77]   E. E. Moise, Geometric Topology in Dimensions 2 and 3, Springer-Verlag, New York, 1977.

[NO95]   A. Nakamoto, K. Ota, Note on irreducible triangulations of surfaces, J. Graph Theory 20 (1995) 227–233.

[NW61]   C. St. J. A. Nash-Williams, Edge-disjoint spanning trees of finite graphs, J. London Math. Soc. 36 (1961) 445–450.

[NW73]   C. St. J. A. Nash-Williams, Unexplored and semi-explored territories in graph theory, in "New Directions in Graph Theory," Ed. F. Harary, Academic Press, New York, 1973, pp. 149–186.

[Ne81a]   L. Nebeský, Every connected, locally connected graph is upper embeddable, J. Graph Theory 5 (1981) 205–207.

[Ne81b]   L. Nebeský, A new characterization of the maximum genus of a graph, Czech. Math. J. 31 (106) (1981) 604–613.

[Ne93]   L. Nebeský, Characterizing the maximum genus of a connected graph, Czech. Math. J. 43 (118) (1993) 177–185.

[Ne83]   S. Negami, Uniqueness and faithfulness of embedding of toroidal graphs, Discrete Math. 44 (1983) 161–180.

[Ne85]   S. Negami, Unique and faithful embeddings of projective-planar graphs, J. Graph Theory 9 (1985) 235–243.

[Ne88a]   S. Negami, Re-embedding of projective-planar graphs, J. Combin. Theory Ser. B 44 (1988) 276–299.

[Ne88b]   S. Negami, The spherical genus and virtually planar graphs, Discrete Math. 70 (1988) 159–168.

[NC88]   T. Nishizeki, N. Chiba, Planar Graphs: Theory and Algorithms, Ann. Discr. Math. 32, North-Holland, 1988.

[Ore51]   O. Ore, A problem regarding the tracing of graphs, Elem. Math. 6 (1951) 49–53.

[Ore67]   O. Ore, The Four Color Problem, Academic Press, New York, 1967.

[Pa71]   T. D. Parsons, On planar graphs, Amer. Math. Monthly 78 (1971) 176–178.

[PPPV87]   T. D. Parsons, G. Pica, T. Pisanski, A. G. S. Ventre, Orientably simple graphs, Math. Slovaca 37 (1987) 391–394.

[Pi80]   T. Pisanski, Genus of Cartesian products of regular bipartite graphs, J. Graph Theory 4 (1980) 31–42.

[Pi82]   T. Pisanski, Nonorientable genus of Cartesian products of regular graphs, J. Graph Theory 6 (1982) 391–402.

[PP97]   T. M. Przytycka, J. H. Przytycki, A simple construction of high representativity triangulations, Discrete Math. 173 (1997) 209–228.

[Ra74]   E. Rademacher, Zu Färbungsproblemen von Graphen auf der projektiven Ebene, dem Torus und dem Kleinschen Schlauch, Dissertation, Technische Hochschule Ilmenau, 1974.

[Ra25]   T. Radó, Über den Begriff der Riemannschen Fläche, Acta Litt. Sci. Szeged 2 (1925) 101–121.

[Ra97]   S. P. Randby, Minimal embeddings in the projective plane, J. Graph Theory 25 (1997) 153–163.

[Ri63]   I. Richards, On the classification of noncompact surfaces, Trans. Amer. Math. Soc. 106 (1963) 259–269.

[Ri87a]   R. B. Richter, On the non-orientable genus of a 2-connected graph, J. Combin. Theory Ser. B 43 (1987) 48–59.

[Ri87b] R. B. Richter, On the Euler genus of a 2-connected graph, J. Combin. Theory Ser. B 43 (1987) 60–69.

[RS84] R. B. Richter, H. Shank, The cycle space of an embedded graph, J. Graph Theory 8 (1984) 365–369.

[Ri79] R. D. Ringeisen, Survey of results on the maximum genus of a graph, J. Graph Theory 3 (1979) 1–13.

[Ri65a] G. Ringel, Das Geschlecht des vollständigen paaren Graphen, Abh. Math. Sem. Univ. Hamburg 28 (1965) 139–150.

[Ri65b] G. Ringel, Der vollständige paare Graph auf nichtorientierbaren Flächen, J. reine angew. Math. 220 (1965) 88–93.

[Ri65c] G. Ringel, Über drei kombinatorische Probleme am $n$-dimensionalen Würfel und Wurfelgitter, Abh. Math. Sem. Univ. Hamburg 20 (1965) 10–19.

[Ri74] G. Ringel, Map Color Theorem, Springer-Verlag, Berlin, 1974.

[Ri77a] G. Ringel, The combinatorial map color theorem, J. Graph Theory 1 (1977) 141–155.

[Ri77b] G. Ringel, On the genus of $K_n \times K_2$, or the $n$-prism, Discrete Math. 20 (1977) 287–294.

[Ri78] G. Ringel, Non-existence of graph embeddings, in "Theory and Applications of Graphs", Lecture Notes Math. 642, Springer-Verlag, Berlin, 1978, pp. 465–476.

[Ri85] G. Ringel, 250 Jahre Graphentheorie, in "Graphen in Forschung und Unterricht", Ed. R. Bodendiek et al., Verlag Barbara Franzbecker, Bad Salzdetfurth, 1985, pp. 136–152.

[RY68] G. Ringel, J. W. T. Youngs, Solution of the Heawood map-coloring problem, Proc. Nat. Acad. Sci. U.S.A. 60 (1968) 438–445.

[Ri94p] A. Riskin, On the nonembeddability and crossing numbers of some toroidal graphs on the Klein bottle, preprint, 1994.

[RSST96a] N. Robertson, D. P. Sanders, P. Seymour, R. Thomas, A new proof of the four-colour theorem, Electron. Res. Announc. Amer. Math. Soc. 2 (1996) 17–25.

[RSST96b] N. Robertson, D. P. Sanders, P. Seymour, R. Thomas, Efficiently four-coloring planar graphs, Proc. ACM Symp. Theory Comput. 28 (1996) 571–575.

[RSST97] N. Robertson, D. Sanders, P. Seymour, R. Thomas, The four-colour theorem, J. Combin. Theory Ser. B 70 (1997) 2–44.

[RS84a] N. Robertson, P. D. Seymour, Generalizing Kuratowski's theorem, Congr. Numer. 45 (1984) 129–138.

[RS85] N. Robertson, P. D. Seymour, Graph minors — a survey, in "Surveys in Combinatorics 1985," Cambridge Univ. Press, Cambridge, 1985, pp. 153–171.

[RS83] N. Robertson, P. D. Seymour, Graph minors. I. Excluding a forest, J. Combin. Theory Ser. B 35 (1983) 39–61.

[RS86a] N. Robertson, P. D. Seymour, Graph minors. II. Algorithmic aspects of tree-width, J. Algorithms 7 (1986) 309–322.

[RS84b] N. Robertson, P. D. Seymour, Graph minors. III. Planar tree-width, J. Combin. Theory Ser. B 36 (1984) 49–64.

[RS90a] N. Robertson, P. D. Seymour, Graph minors. IV. Tree-width and well-quasi-ordering, J. Combin. Theory Ser. B 48 (1990) 227–254.

[RS86b] N. Robertson, P. D. Seymour, Graph minors. V. Excluding a planar graph, J. Combin. Theory Ser. B 41 (1986) 92–114.

[RS86c] N. Robertson, P. D. Seymour, Graph minors. VI. Disjoint paths across a disc, J. Combin. Theory Ser. B 41 (1986) 115–138.

[RS88]      N. Robertson, P. D. Seymour, Graph minors. VII. Disjoint paths on a surface,
            J. Combin. Theory Ser. B 45 (1988) 212–254.
[RS90b]     N. Robertson, P. D. Seymour, Graph minors. VIII. A Kuratowski theorem
            for general surfaces, J. Combin. Theory Ser. B 48 (1990) 255–288.
[RS90c]     N. Robertson, P. D. Seymour, Graph minors. IX. Disjoint crossed paths,
            J. Combin. Theory Ser. B 49 (1990) 40–77.
[RS91]      N. Robertson, P. D. Seymour, Graph minors. X. Obstructions to tree-
            decomposition, J. Combin. Theory Ser. B 52 (1991) 153–190.
[RS94]      N. Robertson, P. D. Seymour, Graph minors. XI. Circuits on a surface,
            J. Combin. Theory Ser. B 60 (1994) 72–106.
[RS95a]     N. Robertson, P. D. Seymour, Graph minors. XII. Distance on a surface,
            J. Combin. Theory Ser. B 64 (1995) 240–272.
[RS95b]     N. Robertson, P. D. Seymour, Graph minors. XIII. The disjoint paths prob-
            lem, J. Combin. Theory Ser. B 63 (1995) 65–110.
[RS95c]     N. Robertson, P. D. Seymour, Graph minors. XIV. Extending an embedding,
            J. Combin. Theory Ser. B 65 (1995) 23–50.
[RS96]      N. Robertson, P. D. Seymour, Graph minors. XV. Giant steps, J. Combin.
            Theory Ser. B 68 (1996) 112–148.
[RS86p]     N. Robertson, P. D. Seymour, Graph minors. XVI. Excluding a non-planar
            graph, preprint, 1986.
[RS99]      N. Robertson, P. D. Seymour, Graph minors. XVII. Taming a vortex, J. Com-
            bin. Theory Ser. B 77 (1999) 162–210.
[RS88p]     N. Robertson, P. D. Seymour, Graph minors. XVIII. Tree-decompositions
            and well-quasi-ordering, preprint, 1988.
[RS89p]     N. Robertson, P. D. Seymour, Graph minors. XIX. Well-quasi-ordering on a
            surface, preprint, 1989.
[RS88q]     N. Robertson, P. D. Seymour, Graph minors. XX. Wagner's conjecture,
            preprint,1988.
[RS92p]     N. Robertson, P. D. Seymour, Graph minors. XXI. Graphs with unique link-
            ages, preprint, 1992.
[RS92q]     N. Robertson, P. D. Seymour, Graph minors. XXII. Irrelevant vertices in
            linkage problems, preprint, 1992.
[RS90d]     N. Robertson, P. D. Seymour, An outline of a disjoint paths algorithm, in:
            "Paths, Flows, and VLSI-Layout," B. Korte, L. Lovász, H. J. Prömel, and
            A. Schrijver Eds., Springer-Verlag, Berlin, 1990, pp. 267–292.
[RST93]     N. Robertson, P. D. Seymour, R. Thomas, A survey of linkless embeddings,
            in "Graph Structure Theory (Seattle, WA, 1991)", Contemp. Math. 147,
            Amer. Math. Soc., Providence, RI, 1993, pp. 125–136.
[RT91]      N. Robertson, R. Thomas, On the orientable genus of graphs embedded in
            the Klein bottle, J. Graph Theory 15 (1991) 407–419.
[RV90]      N. Robertson, R. P. Vitray, Representativity of surface embeddings, in:
            "Paths, Flows, and VLSI-Layout," B. Korte, L. Lovász, H. J. Prömel, and
            A. Schrijver Eds., Springer-Verlag, Berlin, 1990, pp. 293–328.
[RZZ95p]    N. Robertson, X. Zha, Y. Zhao, On the uniqueness of embeddings of graphs
            in the torus, preprint, 1995.
[Ro87]      B. Rodin, Schwarz's lemma for circle packings, Invent. Math. 89 (1987) 271–
            289.
[Ro89]      B. Rodin, Schwarz's lemma for circle packings. II, J. Diff. Geom. 30 (1989)
            539–554.
[RS87]      B. Rodin, D. Sullivan, The convergence of circle packings to the Riemann
            mapping, J. Diff. Geom. 26 (1987) 349–360.
[Ro76]      P. Rosenstiehl, Caractérisation des graphes planaires par une diagonale
            algébrique, Comp. Rend. Acad. Sci. Paris Sér. A 283 (1976) 417–419.

[RR78]     P. Rosenstiehl, R. C. Read, On the principal edge tripartition of a graph, Ann. Discrete Math. 3 (1978) 195–226.

[SK77]     T. L. Saaty, P. C. Kainen, The Four-Color Problem, Assaults and Conquest, Dover Publ., New York, 1977.

[Sa94]     H. Sachs, Coin graphs, polyhedra, and conformal mapping, Discrete Math. 134 (1994) 133–138.

[Sa90]     A. Saito, Covering contractible edges in 3-connected graphs. I: Covers of size three are cutsets, J. Graph Theory 14 (1990) 635–643.

[Sch89]    W. Schnyder, Planar graphs and poset dimension, Order 5 (1989) 323–343.

[Sc92]     O. Schramm, How to cage an egg, Invent. Math. 107 (1992) 543–560.

[Sc96]     O. Schramm, Conformal uniformization and packings, Israel J. Math. 93 (1996) 399–428.

[Sch90]    A. Schrijver, Homotopic routing methods, in: "Paths, Flows, and VLSI-Layout," B. Korte, L. Lovász, H. J. Prömel, and A. Schrijver Eds., Springer-Verlag, Berlin, 1990, pp. 329–371.

[Sch91a]   A. Schrijver, Disjoint circuits of prescribed homotopies in a graph on a compact surface, J. Combin. Theory Ser. B 51 (1991) 127–159.

[Sch91b]   A. Schrijver, Decomposition of graphs on surfaces and a homotopic circulation theorem, J. Combin. Theory Ser. B 51 (1991) 161–210.

[Sch92a]   A. Schrijver, Circuits in graphs embedded on the torus, Discrete Math. 106/107 (1992) 415–433.

[Sch92b]   A. Schrijver, On the uniqueness of kernels, J. Combin. Theory Ser. B 55 (1992) 146–160.

[Sch93]    A. Schrijver, Graphs on the torus and geometry of numbers, J. Combin. Theory Ser. B 58 (1993) 147–158.

[Sch94]    A. Schrijver, Classification of minimal graphs of given face-width on the torus, J. Combin. Theory Ser. B 61 (1994) 217–236.

[Sch97]    A. Schrijver, Minor-monotone graph invariants, in "Surveys in Combinatorics, 1997," Cambridge Univ. Press, 1997, pp. 163–196.

[Sch87]    E. Schulte, Analogues of Steinitz's theorem about non-inscribable polytopes, in "Intuitive Geometry", Colloq. Math. Soc. J. Bolyai 48 (1987) pp. 503–516.

[SS95]     Á. Seress, T. Szabó, Dense graphs with cycle neighborhoods, J. Combin. Theory Ser. B 63 (1995) 281–293.

[Se79]     P. D. Seymour, Sums of circuits, in "Graph Theory and Related Topics" (J.A. Bondy and U.S.R. Murty, Eds.), Academic Press, New York, 1979, pp. 341–355.

[Se80]     P. D. Seymour, Disjoint paths in graphs, Discrete Math. 29 (1980) 293–309.

[Se86]     P. D. Seymour, Adjacency in binary matroids, European J. Combin. 7 (1986) 171–176.

[Se95p]    P. D. Seymour, A bound on the excluded minors for a surface, submitted.

[ST96]     P. D. Seymour, R. Thomas, Uniqueness of highly representative surface embeddings, J. Graph Theory 23 (1996) 337–349.

[Sh80]     Y. Shiloach, A polynomial solution to the undirected two paths problem, J. Assoc. Comput. Mach. 27 (1980) 445–456.

[Šk92]     M. Škoviera, The decay number and the maximum genus of a graph, Math. Slovaca 42 (1992) 391–406.

[Šk99p]    R. Škrekovski, A generalization of the Dirac Map-Color Theorem, preprint, 1999.

[St78]     S. Stahl, Generalized embedding schemes, J. Graph Theory 2 (1978) 41–52.

[St80]     S. Stahl, Permutation-partition pairs: a combinatorial generalization of graph embeddings, Trans. Amer. Math. Soc. 259 (1980) 129–145.

[St81]  S. Stahl, Permutation-partition pairs II: bounds on the genus of the amalga-
        mation of graphs, Trans. Amer. Math. Soc. 271 (1981) 175–182.
[St83]  S. Stahl, A combinatorial analog of the Jordan Curve Theorem, J. Combin.
        Theory Ser. B 35 (1983) 28–38.
[SB77]  S. Stahl, L. W. Beineke, Blocks and the nonorientable genus of graphs,
        J. Graph Theory 1 (1977) 75–78.
[SW76]  S. Stahl, A. T. White, Genus embeddings for some complete tripartite graphs,
        Discrete Math. 14 (1976) 279–296.
[St51]  S. K. Stein, Convex maps, Proc. Amer. Math. Soc. 2 (1951) 464–466.
[St22]  E. Steinitz, Polyeder und Raumeinteilungen, Enzykl. math. Wiss., Part
        3AB12 (1922) pp. 1–139.
[St93]  K. Stephenson, Cumulative bibliography on circle packings, BIBTEX data
        base on http://www.math.utk.edu/~kens/.
[St79]  W. R. Stromquist, The four color problem for locally planar graphs, in
        "Graph Theory and Related Topics" (J. A. Bondy and U. S. R. Murty, eds.),
        Academic Press, New York, 1979, pp. 369–370.
[Sz73]  G. Szekeres, Polyhedral decompositions of cubic graphs, Bull. Austral. Math.
        Soc. 8 (1973) 367–387.
[TY94]  R. Thomas, X. Yu, 4-connected projective planar graphs are Hamiltonian,
        J. Combin. Theory Ser. B 62 (1994) 114–132.
[TY97]  R. Thomas, X. Yu, Five-connected toroidal graphs are Hamiltonian, J. Com-
        bin. Theory Ser. B 69 (1997) 79–96.
[Th80a] C. Thomassen, Planarity and duality of finite and infinite graphs, J. Combin.
        Theory Ser. B 29 (1980) 244–271.
[Th80b] C. Thomassen, 2-linked graphs, European J. Combin. 1 (1980) 371–378.
[Th81]  C. Thomassen, Kuratowski's theorem, J. Graph Theory 5 (1981) 225–241.
[Th84a] C. Thomassen, Plane representations of graphs, in "Progress in Graph The-
        ory", Eds. J. A. Bondy and U. S. R. Murty, Academic Press, 1984, pp. 43–69.
[Th84b] C. Thomassen, A refinement of Kuratowski's theorem, J. Combin. Theory
        Ser. B 37 (1984) 245–253.
[Th88]  C. Thomassen, Rectilinear drawings of graphs, J. Graph Theory 12 (1988)
        335–341.
[Th89]  C. Thomassen, The graph genus problem is NP-complete, J. Algorithms 10
        (1989) 568–576.
[Th90a] C. Thomassen, A link between the Jordan Curve Theorem and the Kura-
        towski planarity criterion, Amer. Math. Monthly 97 (1990) 216–218.
[Th90b] C. Thomassen, Embeddings of graphs with no short noncontractible cycles,
        J. Combin. Theory Ser. B 48 (1990) 155–177.
[Th90c] C. Thomassen, Bidirectional retracting-free double tracings and upper em-
        beddability of graphs, J. Combin. Theory Ser. B 50 (1990) 198–207.
[Th90d] C. Thomassen, A characterization of the 2-sphere in terms of Jordan Curve
        Separation, Mat-report 1990–19, Technical University of Denmark, 1990.
[Th92a] C. Thomassen, The Jordan-Schönflies Theorem and the classification of sur-
        faces, Amer. Math. Monthly 99 (1992) 116–130.
[Th92b] C. Thomassen, A characterization of the 2-sphere in terms of Jordan curve
        separation, Proc. Amer. Math. Soc. 115 (1992) 863–864.
[Th93a] C. Thomassen, Triangulating a surface with a prescribed graph, J. Combin.
        Theory Ser. B 57 (1993) 196–206.
[Th93b] C. Thomassen, Five-coloring maps on surfaces, J. Combin. Theory Ser. B 59
        (1993) 89–105.
[Th94a] C. Thomassen, Trees in triangulations, J. Combin. Theory Ser. B 60 (1994)
        56–62.

[Th94b]   C. Thomassen, Grötzsch's 3-color theorem and its counterparts for the torus and the projective plane, J. Combin. Theory Ser. B 62 (1994) 268–279.

[Th94c]   C. Thomassen, Every planar graph is 5-choosable, J. Combin. Theory Ser. B 62 (1994) 180–181.

[Th94d]   C. Thomassen, Five-coloring graphs on the torus, J. Combin. Theory Ser. B 62 (1994) 11–33.

[Th95]    C. Thomassen, 3-list coloring planar graphs of girth 5, J. Combin. Theory Ser. B 64 (1995) 101–107.

[Th97a]   C. Thomassen, The genus problem for cubic graphs, J. Combin. Theory Ser. B 69 (1997) 52–58.

[Th97b]   C. Thomassen, Color-critical graphs on a fixed surface, J. Combin. Theory Ser. B 70 (1997) 67–100.

[Th97c]   C. Thomassen, A simpler proof of the excluded minor theorem for higher surfaces, J. Combin. Theory Ser. B 70 (1997) 306–311.

[Th78]    W. P. Thurston, The Geometry and Topology of 3-manifolds, Princeton Univ. Lect. Notes, Princeton, NJ, 1978.

[Tr92]    W. T. Trotter, Combinatorics and Partially Ordered Sets. Dimension Theory, Johns Hopkins Univ. Press, Baltimore, 1992.

[Tr66]    D. J. Troy, On traversing graphs, Amer. Math. Monthly 73 (1966) 497–499.

[Tr80]    K. Truemper, On Whitney's 2-isomorphism theorem for graphs, J. Graph Theory 4 (1980) 43–49.

[Tu56]    W. T. Tutte, A theorem on planar graphs, Trans. Amer. Math. Soc. 82 (1956) 99–116.

[Tu58]    W. T. Tutte, Matroids and graphs, Trans. Amer. Math. Soc. 88 (1958) 144–174.

[Tu60]    W. T. Tutte, Convex representations of graphs, Proc. London Math. Soc. 10 (1960) 304–320.

[Tu61a]   W. T. Tutte, A theory of 3-connected graphs, Nederl. Akad. Wetensch. Proc. 64 (1961) 441–455.

[Tu61b]   W. T. Tutte, On the problem of decomposing a graph into $n$ connected factors, J. London Math. Soc. 36 (1961) 221–230.

[Tu63]    W. T. Tutte, How to draw a graph, Proc. London Math. Soc. 13 (1963) 743–768.

[Tu66a]   W. T. Tutte, Connectivity of Graphs, Univ. Toronto Press, Toronto, Ontario; Oxford Univ. Press, London, 1966.

[Tu66b]   W. T. Tutte, On the algebraic theory of graph colourings, J. Combin. Theory 1 (1966) 15–50.

[Tu75]    W. T. Tutte, Separation of vertices by a circuit, Discrete Math. 12 (1975) 173–184.

[Tu84]    W. T. Tutte, Graph Theory, Encyclopedia of Mathematics and Its Applications 21, Addison-Wesley, Menlo Park, CA, 1984.

[Tv80]    H. Tverberg, A proof of the Jordan Curve Theorem, Bull. London Math. Soc. 12 (1980) 34–38.

[Ve05]    O. Veblen, Theory of plane curves in non-metrical analysis situs, Trans. Amer. Math. Soc. 6 (1905) 83–98.

[Vi89]    A. Vince, Discrete Jordan Curve Theorems, J. Combin. Theory Ser. B 47 (1989) 251–261.

[Vi92]    R. Vitray, The 2- and 3-representative projective planar embeddings, J. Combin. Theory Ser. B 54 (1992) 1–12.

[Vi93]    R. P. Vitray, Representativity and flexibility on the projective plane, in "Graph Structure Theory (Seattle, WA, 1991)", Contemp. Math. 147, Amer. Math. Soc., Providence, RI, 1993, pp. 341–347.

[Vo93]    M. Voigt, List colourings of planar graphs, Discrete Math. 120 (1993) 215–219.

[Vo95]    M. Voigt, A not 3-choosable planar graph without 3-cycles, Discrete Math. 146 (1995) 325–328.

[VW97]    M. Voigt, B. Wirth, On 3-colorable non-4-choosable planar graphs, J. Graph Theory (1997) 233–235.

[Vo91]    H.–J. Voss, Cycles and Bridges in Graphs, Kluwer Academic Publishers, Dordrecht, 1991.

[Wa36]    K. Wagner, Bemerkungen zum Vierfarbenproblem, Jber. Deutsch. math. Verein. 46 (1936) 26–32.

[Wa37a]   K. Wagner, Über eine Eigenschaft der ebenen Komplexe, Math. Ann. 114 (1937) 570–590.

[Wa37b]   K. Wagner, Über eine Erweiterung eines Satzes von Kuratowski, Deutsche Math. 2 (1937) 253–280.

[Wh69]    A. T. White, The genus of the complete tripartite graph $K_{mn,n,n}$, J. Combin. Theory 7 (1969) 283–285.

[Wh70]    A. T. White, The genus of repeated cartesian products of bipartite graphs, Trans. Amer. Math. Soc. 151 (1970) 393–404.

[Wh72]    A. T. White, On the genus of the composition of two graphs, Pacific J. Math. 41 (1972) 275–279.

[Wh73]    A. T. White, Graphs, Groups and Surfaces, North-Holland, 1973; Revised Edition: North-Holland, 1984.

[Wh33a]   H. Whitney, Planar graphs, Fund. Math. 21 (1933) 73–84.

[Wh33b]   H. Whitney, 2-isomorphic graphs, Amer. J. Math. 55 (1933) 245–254.

[Wi80]    S. G. Williamson, Embedding graphs in the plane — algorithmic aspects, Ann. Discrete Math. 6 (1980) 349–384.

[Wi84]    S. G. Williamson, Depth-first search and Kuratowski subgraphs, J. Assoc. Comput. Mach. 31 (1984) 681–693.

[Wi85]    S. G. Williamson, Combinatorics for Computer Science, Computer Science Press, Rockville, Md., 1985.

[Wi93]    S. G. Williamson, Canonical forms for cycles in bridge graphs, Linear Multilin. Algebra 34 (1993) 301–341.

[WW78]    D. R. Woodall, R. J. Wilson, The Appel-Haken proof of the four-color theorem, in "Selected Topics in Graph Theory" (Ed. L. W. Beineke and R. J. Wilson), Academic Press, London–New York, 1978, pp. 83–101.

[Xu79]    N. H. Xuong, How to determine the maximum genus of a graph, J. Combin. Theory Ser. B 26 (1979) 217–225.

[Ya95]    L. Yanpei, Embeddability in Graphs, Kluwer, Dordrecht, 1995.

[Yo96]    D. A. Youngs, 4-chromatic projective graphs, J. Graph Theory 21 (1996) 219–227.

[Yo63]    J. W. T. Youngs, Minimal imbeddings and the genus of a graph, J. Math. Mech. 12 (1963) 303–315.

[Yu97]    X. Yu, Disjoint paths, planarizing cycles, and spanning walks, Trans. Amer. Math. Soc. 349 (1997) 1333–1358.

[ZZ93]    X. Zha, Y. Zhao, On nonnull separating circuits in embedded graphs, in "Graph Structure Theory (Seattle, WA, 1991)", Contemp. Math. 147, Amer. Math. Soc., Providence, RI, 1993, pp. 349–362.

[Zh96]    C.-Q. Zhang, Integer Flows and Cycle Covers of Graphs, Marcel Dekker, New York, 1996.

# Index

Abe, 49
accessible, 24
adding a crosscap, 81
adding a handle, 80
adding a twisted handle, 80
adjacent, 3
Aharonov, 65
Albertson, 234, 239, 242
Alexander Horned Sphere, 18
Alpert, 119
Alspach, 153
Altschuler, 169
Andreev, 61
angle
  mixed, 113, 215
Π-angle, 100
angle graph, 147
apex graph, 160, 162
appearance, 187, 188
  mixed, 215
Appel, 225, 228, 229
arc, 18
  polygonal, 19
Archdeacon, 38, 114, 172, 180, 183, 197, 201, 203
arcwise connected, 18
attached, 187, 188
attachment, 7
Auslander, 113, 163
automorphism, 64
auxiliary cycle, 191
auxiliary vertex, 191

Böhme, 182, 235
Ballantine, 241
Barnette, 142, 143, 155, 166, 167, 169, 179–181
barycentric subdivision, 154
basic piece, 188
2-basis, 35
Battle, 113

Beardon, 65
Beineke, 113, 119
Bender, 134
bidirectional retracting-free double tracing, 127
Biggs, 225
bipartition, 9
Birkhoff, 226
block, 10, 147, 185
block-cutvertex tree, 11
Bonnington, 38
Booth, 49
Bouchet, 117, 119
boundary component, 95
boundary side, 83
Brahana, 107
branch, 185
branch size, 186, 188
branch vertex, 185
breadth-first-search, 111
bridge, 185
  local, 186
  nonplanar, 215
$H$-bridge, 7
Brightwell, 61, 62
Brooks, 233
Brown, 113, 163
Brunet, 166, 180, 203
Bryant, 33
$\mathbf{bs}_K(\Omega)$, 192
Burstein, 30

$C_n$, 4
can be contracted, 142
can be embedded, 19
Cederbaum, 49
center of the wheel, 4
centered at infinity, 52
charge, 229
Chevalier, 126
Chiba, 49

Whitney's 2-switching theorem, 49
Whitney's theorem, 136, 168, 169
Whittle, 184, 205
width, 206
Williamson, 34, 49
Wilson, 225, 230
Wirth, 224
Woodall, 225, 230
Wormald, 181

Xuong, 123, 126

Youngs, D., 243
Youngs, J.W.T., 95, 99, 113, 115, 163,
223
Yu, 179–182

Zha, 166, 169
Zhang, 153, 154
Zhao, 166, 169